Cladistics

A Guide to Biological Classification

Third Edition

This new edition of a foundational text presents a contemporary review of cladistics, as applied to biological classification. It provides a comprehensive account of the past 50 years of discussion on the relationship between classification, phylogeny and evolution. It covers cladistics in the era of molecular data, detailing new advances and ideas that have emerged over the last 25 years. Written in an accessible style by internationally renowned authors in the field, readers are straightforwardly guided through fundamental principles and terminology. Simple worked examples and easy-to-understand diagrams also help readers navigate complex problems that have perplexed scientists for centuries. This practical guide is an essential addition for advanced undergraduates, postgraduates and researchers in taxonomy, systematics, comparative biology, evolutionary biology and molecular biology.

DAVID M. WILLIAMS is a researcher at the Natural History Museum, London, UK, specialising in diatom (Bacillariophyta) taxonomy and biogeography. He is the current president of the Systematics Association, London. He has written more than 240 scientific papers and 10 books.

MALTE C. EBACH is Senior Lecturer in Biogeography at the University of New South Wales (UNSW), Sydney. He has published extensively on the history, theory and methodology of biological systematics, taxonomy and biogeography. He is Associate Editor for the *Journal of Biogeography*, *Australian Systematic Botany* and Editor of the CRC Biogeography Book Series.

The Systematics Association Special Volume Series

SERIES EDITOR

GAVIN BROAD

Department of Life Sciences, The Natural History Museum, London, UK

The Systematics Association promotes all aspects of systematic biology by organising conferences and workshops on key themes in systematics, running annual lecture series, publishing books and a newsletter, and awarding grants in support of systematics research. Membership of the association is open globally to professionals and amateurs with an interest in any branch of biology, including palaeobiology. Members are entitled to attend conferences at discounted rates, to apply for grants and to receive the newsletter and mailed information; they also receive a generous discount on the purchase of all volumes produced by the association.

The first of the Systematics Association's publications, *The New Systematics* (1940), was a classic work edited by its then-president Sir Julian Huxley. Since then, more than 70 volumes have been published, often in rapidly expanding areas of science where a modern synthesis is required.

The association encourages researchers to organise symposia that result in multi-authored volumes. In 1997 the association organised the first of its international biennial conferences. This and subsequent biennial conferences, which are designed to provide for systematists of all kinds, included themed symposia that resulted in further publications. The association also publishes volumes that are not specifically linked to meetings and encourages new publications (including textbooks) in a broad range of systematics topics.

More information about the Systematics Association and its publications can be found at our website: www.systass.org

Previous Systematics Association publications are listed after the index for this volume.

Systematics Association Special Volumes published by Cambridge University Press:

SYSTEMATICS ASSOCIATION SPECIAL VOLUME 88

Cladistics

A Guide to Biological Classification

Third Edition

DAVID M. WILLIAMS

Natural History Museum, London

MALTE C. EBACH

University of New South Wales

THE
Systematics
ASSOCIATION

CAMBRIDGE
UNIVERSITY PRESS

CAMBRIDGE
UNIVERSITY PRESS

University Printing House, Cambridge CB2 8BS, United Kingdom

One Liberty Plaza, 20th Floor, New York, NY 10006, USA

477 Williamstown Road, Port Melbourne, VIC 3207, Australia

314–321, 3rd Floor, Plot 3, Splendor Forum, Jasola District Centre, New Delhi – 110025, India

79 Anson Road, #06-04/06, Singapore 079906

Cambridge University Press is part of the University of Cambridge.

It furthers the University's mission by disseminating knowledge in the pursuit of education, learning, and research at the highest international levels of excellence.

www.cambridge.org
Information on this title: www.cambridge.org/9781107008106
DOI: 10.1017/9781139047678

First published 1992 by Oxford University Press for the Systematics Association
Second edition 1998 by Oxford University Press for the Systematics Association
Third edition 2020

Printed in the United Kingdom by TJ International Ltd, Padstow Cornwall

A catalogue record for this publication is available from the British Library.

Library of Congress Cataloging-in-Publication Data

Names: Williams, David M. (David Mervyn), 1954- author. | Ebach, Malte C., author.

Title: Cladistics : a guide to biological classification / David M. Williams, Natural History Museum, London, Malte C. Ebach, University of New South Wales.

Description: Third edition. | Cambridge, United Kingdom ; New York : Cambridge University Press, 2020. | Series: The Systematics Association special volume ; 88 | Includes index.

Identifiers: LCCN 2020012877 (print) | LCCN 2020012878 (ebook) | ISBN 9781107008106 (hardback) | ISBN 9781107400412 (paperback) | ISBN 9781139047678 (epub)

Subjects: LCSH: Cladistic analysis. | Biogeography.

Classification: LCC QH83 .W53 2020 (print) | LCC QH83 (ebook) | DDC 578.01/2-dc23

LC record available at https://lccn.loc.gov/2020012877

LC ebook record available at https://lccn.loc.gov/2020012878

ISBN 978-1-107-00810-6 Hardback
ISBN 978-1-107-40041-2 Paperback

The author and publisher have acknowledged the sources of copyright material
where possible and are grateful for the permissions granted. While every effort
has been made, it has not always been possible to identify the sources of all the
material used, or to trace all copyright holders. We would appreciate any
omissions being brought to our attention.

For C. Linneaus, A.P. de Candolle and N.I. Platnick

Contents

Preface

Inadequate theory permitted belief that systematics was well understood. Remarkable is the fact that despite inadequate theory, systematics worked as well as it did through the centuries of its modern (post-enlightenment) development. Yet what human endeavour has ever required perfect understanding for its profitable pursuit?

(Nelson 2011, p. 140)

Cladistics: A Guide to Biological Classification is the third edition of the cladistics primer first published under the auspices of the Systematics Association, a London-based organisation dedicated to the promotion and development of systematics and taxonomy (*Nature* 140, pp. 163–164, 1937). The first edition, *Cladistics: A Practical Course in Systematics*, based on a week-long workshop given at the Natural History Museum, London, was published over 25 years ago (Forey et al. 1992); the second edition, *Cladistics: The Theory and Practice of Parsimony Analysis*, with a heavily revised content, was published 22 years ago (Kitching et al. 1998). With over two decades having passed since the second edition, why, now, offer a new version?

The first edition covered a series of topics considered to encapsulate the subject matter of cladistics, such as it was then understood in the early 1990s. The idea was that the book functioned as a manual and would form a baseline for teaching the rudiments of cladistics to neophytes: character coding, character polarity, tree-building techniques, tree statistics, DNA analysis, fossils, biogeography and classification. It appeared to be a success.

The second edition was revised, modified and updated, with the chapters on DNA analysis, fossils, biogeography and classification being retired (simply because other books covered those topics in more detail, such as Hillis and Moritz 1990 and Hillis et al. 1996, for molecular systematics; Humphries and Parenti 1986 for biogeography; Schoch 1986 and Smith 1994, for palaeontology) and, in addition to the technique-based chapters (character coding, character polarity, tree-building techniques, tree statistics), a few (then) topical and specialised problems related to analytical methods were added: the effect of missing values, character weighting, consensus trees, the 'simultaneous or partitioned analysis' of datasets discussion and three-item analysis. The second edition focused more on method than theory – although the way in which each new topic addressed theory was of some significance, as a close reading will detect growing differences among the contributors. In short, the second edition was primarily a discourse on what, with hindsight, might have been better called *Quantitative Cladistics* or *Quantitative Phyletics* (the former now used as a title for a teaching course on *Quantitative Cladistics and the use of TNT*, the latter after Kluge and Farris 1969, p. 1), a method associated directly with the workings of (in theory and practice) the Wagner parsimony algorithm and its implementation in various computer programs (of which *TNT*, '*Tree analysis using New Technology*', is its most recent incarnation, Goloboff et al. 2008). It was as if the intervening years between the two versions of this book – 1992 to 1998 –forged a direct and (almost) absolute equation between cladistics and parsimony, as implemented by the Wagner algorithm, in spite of the fact that some of the contributors to the 2nd edition were becoming violently opposed to such an equation. Of significance, too, is that those intervening years saw the increase in and use of many statistically based programs designed specifically to analyse the abundance of DNA sequence data that was becoming available. Those methods of data analysis now dominate the systematics landscape.

With this third edition, and the benefit of 20 odd years of reflection, we thought it appropriate to present a more holistic view of cladistics – in terms of its relationship to general taxonomic practice: that is, the discovery of taxa.

To articulate those aims we can do no better than re-state Gareth Nelson's 'Two Questions' concerning the original goals and tasks of cladistics:

> Question 1: In the simplest terms of three organisms, species or taxa, what is the evidence that two are related more closely to each other than to the third? Never mind if they are known from fossil or living material, alone or in combination.

> Question 2: If two of them are related more closely, what does this mean about the organisms, species or taxa: about their evolution, classification, even their nomenclature and usefulness and interest to humans? (Nelson 2004, p. 128)

If cladistics became associated with a particular viewpoint, rather than a discipline that focused on the generalities stated in these two questions, it is the fault of those who have promoted it as such – including, we are forced to say, the authors of the second edition of this book. Still, hopefully we all learn and move on. And in any case, we find the two questions above to still be of interest and their exploration still of some significance:

> These are questions, not answers, because the spirit of cladistics is enquiry, not ideology, propaganda, sloganizing, or the husbandry of sacred cows and clams. At least, that is what the spirit was, once upon a time, when it achieved general relevance lasting to the present. (Nelson 2004, p. 128)

In this short book we hope to outline and re-state – some might say re-found ('to found or establish again') – the rudiments of cladistics, demonstrating its wide utility, and offer reasons why and how it is of value, especially for taxonomy, a subject often characterised as being devoid of any great scientific merit.

Along the way, we hope to pose a few questions of our own, in the spirit of enquiry that launched cladistics in the first place, mainly because '... you never know enough about anything, and if for a few months or years you should ever believe that you do, you are either past it or in for a surprise ... Yesterday's secure knowledge is tomorrow's laughing matter' (Patterson 2011).

References

Forey, PL., Humphries, CJ., Kitching, IJ., Scotland, RW., Siebert, DJ. & Williams, DM. 1992. *Cladistics: A Practical Course in Systematics*. Oxford University Press, Oxford.

Goloboff, PA., Farris, JS. & Nixon, KC. 2008. TNT, a free program for phylogenetic analysis. *Cladistics* 24: 1–13.

Hillis, DM. & Moritz, C. 1990. *Molecular Systematics*. Sinauer Associates Incorporated, Sunderland, MA.

Hillis, DM., Moritz, C. & Mable, BK. 1996. *Molecular Systematics*. 2nd ed. Sinauer Associates Incorporated, Sunderland, MA.

Humphries, CJ. & Parenti, LR. 1986. *Cladistic Biogeography*. Oxford University Press, Oxford.

Kitching, IJ., Forey, P., Humphries, CJ. & Williams, DM. 1998. *Cladistics: The Theory and Practice of Parsimony Analysis*. Oxford University Press, Oxford.

Kluge, AG. & Farris, JS. 1969. Quantitative phyletics and the evolution of Anurans. *Systematic Zoology* 18: 1–32.

Nelson, GJ. 2004. Cladistics: its arrested development. In: Williams, DM. & Forey, PL. (eds), *Milestones in Systematics*. CRC Press, Florida, pp. 127–147.

Nelson, G. 2011. Resemblance as evidence of ancestry. *Zootaxa* 2946: 137–141.

Patterson, C. 2011. Adventures in the fish trade (edited and with an introduction by David M. Williams & Anthony C. Gill). *Zootaxa* 2946: 118–136.

Schoch, R. 1986. *Phylogeny Reconstruction in Palaeontology*. Van Nostrand Reinhold, New York.

Smith, AB. 1994. *Systematics and the Fossil Record: Documenting Evolutionary Patterns*. Wiley-Blackwell, Oxford.

Acknowledgements

We thank the following for reading one or more chapters and offering suggestions, advice, etc.: Paul Barrett, Greg Edgecombe, Ronald Jenner, Evgeny Mavrodiev, Gareth Nelson, Aleta Quinn, Jon Todd and Visotheary Ung.

We thank the following for providing information: Gavin Broad, Max Barclay, Mark Carine, Anne Jungblut and Dorothy Fouracre (Linnean Society).

Introduction: Carving Nature at Its Joints, or Why Birds Are Not Dinosaurs and Men Are Not Apes

We have a tendency to look back on the feuds among writers and artists [and scientists] of the past with a fond chuckle – to find something cute in the sparring, or perhaps poignant in the testimony such quarreling proffers to the notion that at least the world of ideas *mattered* to those tempestuous souls of the days gone by.

(George Prochnik 2014, The Impossible Exile, *p. 181)*

Some persons see the value of these recent developments [cladistics] as revolutionary, liberating us from the burdens of the past, and particularly from the necessity to understand the work and thought of our predecessors. I see recent developments as embedded in, and shedding light on, an old tradition, dating from the days of Cuvier and his colleagues, if not from the time of Linnaeus, or for that matter from the time of Aristotle.

(Nelson 1989, p. 61)

The phrase in the title above – 'carving nature at its joints' – comes from Plato's *Phaedrus* asking how and why people 'carve-up' and partition the organic world in the way they do. In short: "How do we classify the world?" There are, of course, many ways to classify, but the central question for biology is why are some groups of organisms, such as birds, recognised as real groups, when others, such as invertebrates, are rejected as such? This, of course, begs an additional question as to what 'real' might mean in terms of classification.

Children soon learn that animals with 'beaks' often have 'wings' and because they have wings, they usually fly: they learn that these animals are called birds. But some of the words used (beaks, wings) are open to interpretation: turtles also have 'beaks', beetles also have 'wings'. There are plenty of definitions of the word 'wing', but in biological terms it normally refers to the parts of an animal that allow it to become airborne, to fly. Beetles and birds both have wings, but the wing of a bird is formed from a set of bones, muscles, blood vessels and feathers, whereas the wings of beetles are membranous, having two sets, the forewings (elytra) are hardened and not used for flying. Animals with elytra are beetles; animals with feathers are birds.

These distinctions – what makes a bird a bird, a beetle a beetle – have occupied minds since Aristotle and are the basis for classifying scientifically. One might see in this simple example all that one needs to know of taxonomy: the union of evidence from the parts of organisms (elytra, feathers) with conclusions (beetles and birds).

One might still persist in thinking of classification as a harmless pursuit, an 'art-form' even, a question of taste, rather than a scientific topic. But there have been more contentious interactions on the topic of biological classification than almost any other area of biology.

For example, much ink has been spilt describing the famous 'battle' between the French anatomists Georges Cuvier (Jean Léopold Nicolas Frédéric Cuvier, 1769–1832) and Étienne Geoffroy Saint-Hilaire (1772–1844) in the 1830s, often (but mistakenly) portrayed as a tussle concerning evolution, with the 'creationist' Cuvier winning and the 'proto-evolutionist' Saint-Hilaire losing (a useful account can be found in Appel 1987 and Le Guyader 2004). As usual Thomas Henry Huxley (1825–1895), Charles Darwin's (1809–1882) bulldog, captured the poignancy of that fight and its significance:

> The Lecturer [Huxley] commenced by referring to a short essay by Goethe–the last which proceeded from his pen–containing a critical account of a discussion bearing upon the doctrine of the Unity of Organization of Animals, which had then (1830) just taken place in the French Academy. Goethe said that, for him, this controversy was of more importance than the Revolution of July which immediately followed it – a declaration which might almost be regarded as a prophecy; for while the *Charte* and those who established it have vanished as though they had never been, the Doctrine of Unity of Organization retains a profound interest and importance for those who study the science of life. (Huxley 1854, p. 72)

That was 1854. Huxley came to review and modify his views on the 'Unity of Organisation' after 1859 inspired by the publication of Darwin's *On the Origin of Species*:

> [...] that community of descent is the hidden bond which naturalists have been seeking, and not some unknown plan of creation, or the enunciation of general propositions, and the mere putting together and separating objects more or less alike. (Darwin 1859, p. 404)

And so 'community of descent' – 'descent with modification' – became the suggested organising principle for creating and naming groups of organisms, ushering in an apparent revolution in biology, if not the world at large.

It was the German biologist Ernst Haeckel (1834–1919) who took up one of Darwin's challenges by creating the first genealogical charts depicting the relationships between organisms, encouraging the view that birds, for example, 'evolved' from reptiles (Figures I.1a and b[1]). It was these harmless but frequently used phrases – 'birds evolved from reptiles', or its modern version, 'birds evolved from dinosaurs' – that inspired the next revolution in classification, one every bit as hard fought as the Cuvier–Hilaire battle of the 1830s.

The clash was between those who wished to recognise groups such as 'reptiles' and those who did not, the former persons eventually being called 'evolutionary taxonomists', the latter cladists. Broadly speaking, however, the issue was once again, "how do we classify?":

> Like the reds and the greens of Byzantium, or the Guelfs and Ghibellines in Dante's Italy, the cladists and their opponents have on occasion turned departments of paleontology into fields of passionate but obscure dispute. (Wade 1981, p. 35)

Nowadays the cladistic revolution has seemingly waned, disappeared altogether, the fruits of its labours absorbed silently into the everyday life of the comparative biologist, if such people still exist. Today molecular biologists, mathematicians and bio-informaticians conjure up massive genealogical trees, documenting the entire tree of life in one go – a recent effort utilised more than 30 000 genomes from all three currently recognised domains of life (Bacteria, Archaea, Eukarya) yielding a tree with 3000 organisms (Figure I1c, reproduced from Hug et al. 2016; see also Puigbò et al. 2009; McTavish et al. 2017), an impressive contribution towards the completion of Haeckel's original project, placing all organisms, fossil and Recent, from 'Monad to Man' (Ruse 1997), on a single gigantic tree (Lecointre & Le Guyader 2007). For cladists, those purveyors of 'passionate but obscure dispute', and with a phrase perhaps more relevant to the earlier battle between Cuvier and Hilaire but since adopted for the modern age (Nelson 2004, p. 127), their time was over: "Citoyens, la Revolution est fixée aux principes qui l'ont commencée. Elle est finie" (Napoleon Bonaparte 15 Dec 1799, in Nelson 2004, p. 127).

Or has it...

"la Revolution est fixée aux principes qui l'ont commence"?

---

[1] Haeckel's diagrams are not always easy to interpret, many having more than one tree on them. The images reproduced here as Figures I1a and b are taken from Haeckel (1866, Taf. VII). This diagram has two trees, one depicting the relationships of the organisms (reproduced here as Figure I1a), the other depicting an assumed 'genealogy' (reproduced here as Figure I1b) (see Richards 2008, for more details).

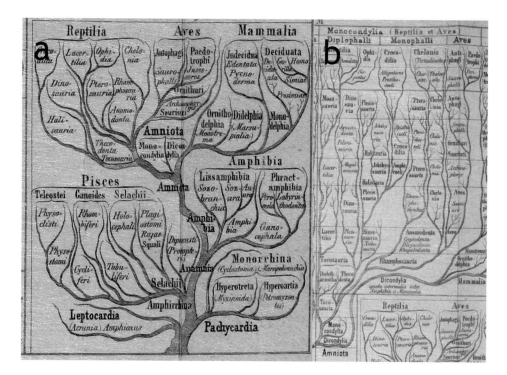

Figure I.1 This diagram of Haeckel's has two trees, one depicting the relationships of the organisms (a), the other depicting an assumed 'genealogy' (b). After Haeckel (1866, taf. VII). (c) "A current view of the tree of life, encompassing the total diversity represented by sequenced genomes", composed of "92 named bacterial phyla, 26 archaeal phyla and all five of the Eukaryotic supergroups". After Hug et al. (2016, fig. 1).

Received wisdom tells us that the cladistic revolution was born from the work of Willi Hennig (1913–1976), in particular the publication of his book *Phylogenetic Systematics* (Hennig 1966), an English translation of a revised version of the less well known and less easily obtained *Grundzüge einer Theorie der phylogenetischen Systematik* (Hennig 1950). The revolution's midwife (Wanntorp 1993), Lars Brundin (1907–1993), via his monograph *Transantarctic Relationships and their Significance, as Evidenced by Chironomid Midges* (Brundin 1966), was the conduit through which a group of comparative biologists – primarily ichthyologists and palaeichthyologists, Gareth Nelson, Colin Patterson, Niels Bonde, Philipe Janvier, Roger Miles, among others – were able to put Hennig's ideas to practical use (Greenwood et al. 1973).

Running parallel to these developments were the explorations of some more evolutionarily inclined zoologists who were attempting to utilise the many differ-ent aspects of the previous generation's numerical taxonomy and use them to

C

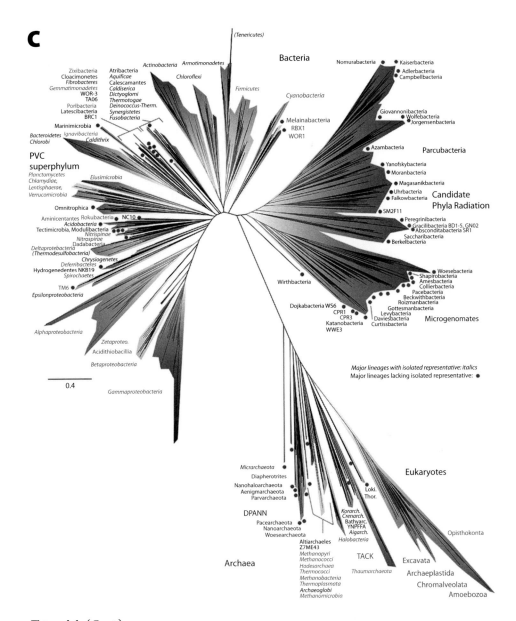

Figure I.1 (*Cont.*)

'reconstruct' phylogenies, which could then be used as the basis for classification. Of significance to these developments was something first called *Quantitative Phyletics*. The initial description of this approach was described so:

> Classical evolutionary taxonomy has been widely criticized for the lack of precision in its methods, while the far more precise numerical phenetic taxonomy has been even more widely censured for its failure to take into account the

> evolutionary basis of relationships among organisms. We believe it is worth while to develop still another taxonomic methodology, incorporating the precision of numerical techniques and the power of evolutionary inference. We refer to this hybrid methodology as *quantitative phyletic taxonomy* (Kluge & Farris 1969, p. 1)

Steve Farris went on to develop the Wagner parsimony program (Farris 1970), which searched for the shortest trees derived from a matrix of binary characters abstracted from the organisms under study (Farris 1988). Received wisdom tells us that over a few short decades Hennig's *Phylogenetic Systematics* became cladistics and cladistics became synonymous with Wagner parsimony:

> The 'grand alliance between the Wagner tradition and the Hennig tradition' (E. O. Wiley, pers. comm. 2001) formed the basis for the revolution in systematics during the last three decades of the 20th century (Schmitt 2003, p. 376)

With the profusion of numerous sophisticated statistically based methods to find – or reconstruct – phylogenetic trees, Wagner parsimony became (almost) redundant and hence cladistics too became (almost) redundant.

The trouble with this account is that while it contains elements of historical truth, it is, in its essence, false. True, Hennig's and Brundin's work did influence a generation of zoologists and palaeontologists; true, the method first called *quantitative phyletics* did evolve into Wagner parsimony; true, over time certain elements of the cladistic 'movement' did try to equate cladistics with Wagner parsimony (some parts of the first two editions of this book contributed to that movement); but significantly, the equation 'cladistics = Wagner parsimony' is patently false. To equate an entire world view with a single computer algorithm retrospectively justified with selected thoughts and ideas from Hennig's *Phylogenetic Systematics* is hopelessly narrow minded and would be the cause of its eventual demise, which does seem to be its destiny. Significant to the taxonomic enterprise, then, is the realisation that cladistics is a far more general field of endeavour than simply equating it with one particular approach to 'phylogeny reconstruction', or with one or another computer algorithm, or with one or another methodology – even with Hennig's 'phylogenetic systematics' itself (Wiley & Liebermann 2011).

Setting all that to one side for the moment, cladistics, in its most general form, has been enormously successful, if that success is judged not by methodology but by the parameters of a general approach to biological classification: monophyly based on evidence derived from homologues (homology), coupled with a precise understanding of relationship (more details of this in Chapter 3). Reflecting on an earlier era, historian Polly Winsor noted that

> [...] it may seem paradoxical that naturalists should use the word 'related' without agreeing on its meaning [...] (Winsor 2009, p. 1, we return to this in Chapter 2)

Paradoxical, too, that we now have a much better understanding of relationship but both monophyly and homology remain areas of contention and are still endlessly debated, even among those supposedly in the cladist camp (to sample these debates, see any issue of the journal *Cladistics* between the years 2011 and 2016). We attempt to tackle these topics afresh in later chapters of this book.

So cladistics might be seen differently. Not as revolution, but reform, specifically the reform of taxonomy, which we see as a modern version of comparative biology.

Revolution usually means the overthrow of something considered outdated, irrelevant even wrong in its basic tenets; reform means the improvement of something that might be considered to have become corrupted, or mutated in the wrong way, something that at its core is sound, save for faulty progress mistaken as profound. Although we are not certain, the idea of a 'Cladistic Revolution' was first given currency by Beverly Halstead (1933–1991), not known as a friend of cladistic taxonomy ('The cladistic revolution: Can it make the grade?', was the title of Beverly Halstead's first diatribe against cladistics, Halstead 1978). The phrase 'Cladistic Revolution' has since been used, and in many different contexts (e.g., Cartmill 2018). It is true that some have seen early developments in cladistics as the 'reform of palaeontology' (Nelson & Platnick 1981; Williams & Ebach 2004), but we have since come to understand cladistics as greater in reach than that: from the start it was, and remains, an attempt to reform taxonomy, an attempt to understand past efforts and eliminate contemporary confusion.

By way of a brief example, consider these words of William Whewell (1794–1866, which we return to later in Chapter 2):

> The basis of all Natural Systems of Classification is the Idea of Natural Affinity. The Principle which this Idea involves is this:–Natural arrangements, obtained from *different* sets of characters, must *coincide* with each other

If we interpret the words above such that 'Natural Systems of Classification' means roughly the same as 'taxonomy/systematics', and that 'Natural Affinity' refers to 'natural relationships', or more simply 'relationships', then the subject matter is identical to what became cladistics in the sense we wish to use it in the following chapters of this book:

> If cladistics is merely a restatement of the principles of natural classification, why has cladistics been the subject of argument? I suspect that the argument is largely misplaced, and that the misplacement stems, as [A.P.] de Candolle suggests, from confounding the goals of artificial and natural systems. (Nelson 1979, p. 20)

We will return to this subject – the 'goals of artificial and natural systems' of classification – in several chapters.

The generality of cladistics was recognised years, decades, even centuries ago (under different names, of course), and was most recently clearly articulated by a group of taxonomists, blessed or cursed, depending on your point of view, with the label 'Pattern Cladists'. What was not obvious, or at least not made abundantly clear at the time, was that the subject matter under discussion should have been *Taxonomy, the Scientific Study of Classification*, a subject we believe subsumes systematics, comparative biology, phylogeny reconstruction and so on. We are aware that each of these subject areas – systematics, comparative biology, phylogeny – are beset with varying definitions, but it is possible to see the commonality as expressed by the term 'relationship', discussed in more detail later.

Taxonomy has itself been the victim of much discussion and, in our view at least, harmed immensely by what can only be called historical inaccuracies. For example, discussions of the central role of essentialism and typology and their role in how taxonomists were supposed to have operated prior to any evolutionary considerations have been almost entirely debunked by some historians of science. These discussions have come largely from the work of Polly Winsor and, more recently, Joeri Witteveen; we lean heavily on their historical research when discussing matters relating to essentialism and typology.

Our task, then, is to describe cladistics as a manifestation of what was, and will probably always be, the effort to find a natural classification of organisms. With respect to the 'goals of artificial and natural systems', we hope to assist in disentangling the very many modern versions of artificial classifications from what we understand as *the* natural classification.

We are painfully aware of the many and varied critiques of 'Pattern Cladistics', most, in our view, mistaken in their understanding of the generalities as well as the particulars. We tackle some of these issues later, but there is no better way of capturing the tone of those critiques than words written some while ago by Richard Dawkins in his immensely popular and influential *The Blind Watchmaker*:

> My own interpretation is that they [pattern cladists] enjoy an exaggerated idea of the importance of taxonomy in biology [. . .] (Dawkins 1986, p. 286, 1996, p. 402)

With this book we hope, at the very least, to suggest why we believe taxonomy is not just exceedingly important, not just central to all biology, but has a vibrant research agenda of its own (rather than a subject that simply needs to embrace modern technology, e.g., compare Bik 2017 to Grimaldi & Engel 2007); we will attempt to document what taxonomy is, its basis, methods and problems via the lens of what became known as cladistics but in the broader interpretation we adopt here. This book is an attempt to articulate the approaches used, but left largely unexplained, in most of the contributions to contemporary taxonomic journals, a methodology we understand that extends in one form or another far back, its basis being found in the writings of Carl Linnaeus (1707–1778), Michel

Adanson (1727–1806), Antoine-Laurent de Jussieu (1748–1836), Augustin Pyramus de Candolle (1778–1841) and many others – perhaps extending even further back than that (Pavord 2005; Ogilvie 2006; Williams & Ebach 2017).

These topics will be the subjects of this book: taxonomy, systematics and natural affinity; in short, scientific classification, which we believe is synonymous with cladistics, providing answers as to why Birds are not Dinosaurs and Men are not Apes.

References

Appel, TA. 1987. *The Cuvier-Geoffrey Debate: French Biology in the Decades before Darwin.* Oxford University Press, Oxford.

Bik, HM. 2017. Let's rise up to unite taxonomy and technology. *PLoS Biology* 15(8): e2002231

Brundin, L. 1966. Transantarctic relationships and their significance as evidenced by chironomid midges. *Kungliga Svenska Vetenskapsakademiens Handlinger* 11 (Series 4): 1–472.

Cartmill, M. 2018. A sort of revolution: systematics and physical anthropology in the 20th century. *American Journal of Physical Anthropology* 165: 677–687.

Darwin, C. 1859. *On the Origin of Species by Means of Natural Selection, or, the Preservation of Favoured Races in the Struggle for Life.* John Murray, London.

Dawkins, R. 1986. *The Blind Watchmaker.* Longman Scientific and Technical, Harlow, Essex.

Dawkins, R. 1996. *The Blind Watchmaker.* W. W. Norton & Company, Inc., New York.

Farris, JS. 1970. Methods for computing Wagner trees. *Systematics Zoology* 19: 83–92.

Farris, JS. 1988. *HENNIG 86*, version 1.5.

Greenwood, PH., Miles, RS. & Patterson, C. (eds) 1973. *Interrelationships of Fishes.* Academic Press, London.

Grimaldi, DA. & Engel, MS. 2007. Why descriptive science still matters. *BioScience* 57(8): 646–647.

Halstead, LB. 1978. The cladistic revolution: can it make the grade? *Nature* 276: 759–760.

Hennig, W. 1950. *Grundzüge einer Theorie der phylogenetischen Systematik.* Deutscher Zentralverlag, Berlin.

Hennig, W. 1966. *Phylogenetic Systematics.* University of Illinois Press, Urbana.

Hug, LA., Baker, BJ., Anantharaman, K., Brown, CT., Probst, AJ., Castelle, CJ., Butterfield, CN., Hernsdorf, AW., Amano, Y., Ise, K., Suzuki, Y., Dudek, N., Relman, DA., Finstad, KM., Amundson, R., Thomas, BC. & Banfield, JF. 2016. A new view of the tree of life. *Nature Microbiology* 1: 16048. http://dx.doi.org/10.1038/nmicrobiol.2016.48

Huxley, TH. 1854. On the common plan of animal forms. *Annals and Magazine of Natural History* 14 (2nd ser.): 72–74.

Kluge, AG. & Farris, JS. 1969. Quantitative phyletics and the evolution of Anurans. *Systematic Zoology* 18: 1–32.

Lecointre, G. & Le Guyader, H. 2007. *The Tree of Life: A Phylogenetic Classification.* Belknap Press, Cambridge MA.

Le Guyader, H. 2004. *Geoffroy Saint-Hilaire: A Visionary Naturalist.* University of Chicago Press, Chicago.

McTavish, EJ., Drew, BT., Redelings, B. & Cranstom, KA. 2017. How and why to build a unified tree of life. *BioEssays* 2017: 1700114. https://doi.org/10.1002/bies.201700114

Nelson, G. 1979. Cladistic analysis and synthesis: principles and definitions, with a historical note on Adanson's *Familles des Plantes*. *Systematic Zoology* 28: 1–21.

Nelson, G. 1989. Species and taxa: systematics and evolution. In: Otte, D & Endler, J (eds), *Speciation and Its Consequences*. Sinauer, Sunderland, pp. 60–81.

Nelson, GJ. 2004. Cladistics: its arrested development. In: Williams, DM & Forey, PL (eds), *Milestones in Systematics*. CRC Press, Florida, pp. 127–147.

Nelson, G. & Platnick, NI. 1981. *Systematics and Biogeography: Cladistics and Vicariance*. Columbia University Press, New York.

Ogilvie, BW. 2006. *The Science of Describing: Natural History in Renaissance Europe*. University of Chicago Press, Chicago.

Pavord, A. 2005. *The Naming of Names: The Search for Order in the World of Plants*. Bloomsbury, London.

Puigbò, P., Wolf, YI. & Koonin, EV. 2009. Search for a 'Tree of Life' in the thicket of the phylogenetic forest. *Journal of Biology* 8: 59. https://doi:10.1186/jbiol159

Richards, R. 2008. *The Tragic Sense of Life: Ernst Haeckel and the Struggle over Evolutionary Thought*. University of Chicago Press, Chicago.

Ruse, M. 1997. *From Monad to Man: The Concept of Progress in Evolutionary Biology*. Harvard University Press, Cambridge, MA.

Schmitt, M. 2003. Willi Hennig and the rise of cladistics. In: Legakis, A., Stenthourakis, S., Polymeni, R. & Thessalou-Legaki, M. (eds), *The New Panorama of Animal Evolution*. Pensoft Publishers, Sofia, Moscow, pp. 369–379.

Wade, N. 1981. Dinosaur battle erupts in British Museum. *Science* 211 (4477): 35–36.

Wanntorp, H-E. 1993. Lars Brundin 30 May 1907 – 17 November 1993. *Cladistics* 9: 357–367.

Wiley, EO. & Liebermann, BS. 2011. *Phylogenetics: Theory and Practice of Phylogenetic Systematics*, 2nd ed. Wiley-Blackwell, Hoboken, NJ.

Williams, DM. & Ebach, MC. 2004. The reform of palaeontology and the rise of biogeography – 25 years after 'Ontogeny, Phylogeny, Paleontology and the Biogenetic law' (Nelson 1978). *Journal of Biogeography* 31: 685–712.

Williams, DM. & Ebach, MC. 2017. What is intuitive taxonomic practice? *Systematic Biology* 66: 637–643.

Winsor, MP. 2009. Taxonomy was the foundation of Darwin's evolution. *Taxon* 58: 43–49.

PART I

The Interrelationships of Organisms

1

What This Book Is About

... qu'il faloit chercher dans la nature elle-même son Systême, s'il étoit vrai qu'ele en eût un ... [...search in nature herself for her system, if it is true she really has one ...]

(Michel Adanson 1763, p. clvij; translation in Nelson 1979, p. 21)

Is it not extraordinary that young taxonomists are trained like performing monkeys, almost wholly by imitation, and that in only the rarest cases are they given any instruction in taxonomic theory?

(Cain 1959, p. 243, quoted in Simpson 1961, p. vii and Felsenstein 1982, p. 379)

None whatever ... the training I received in taxonomy was simply the attitude that this is purely incidental, everybody knows how to do it, there is no necessary theory, you need to know the rules of nomenclature, and that's all.

(Colin Patterson, in response to David Hull's question: 'Did you have any formal training in taxonomy?', unpublished taped interview)

This book is about biological taxonomy and classification as seen through the lens of cladistics. We set out to address the question: How is taxonomy done?

One way to find out how taxonomy is done is to examine any leading taxonomy journal. Inspection of any issue of the journal *Phytotaxa*, for example, is revealing. Issue 231(3) was published on 23 October 2015. It has 7 Articles and 3 Correspondence pieces, roughly 100 printed pages. Ignoring the Correspondence, the Articles deal with the taxonomy of a wide range of botanical groups (as understood in the older 'non-animal-photosynthetic' sense): flowering plants, fungi, diatoms, ferns and so on. The contributions are concerned with describing (diagnosing) new taxa (most often at the species level), revising groups of species (some are substantial taxonomic revisions), creating identification keys, making appropriate nomenclatural acts or various combinations of these endeavours.

It is a similar experience if a recent issue of *Zootaxa* is examined. Issue 4034 (1) was published on 28 October 2015. It has 9 Articles and 2 Correspondence pieces, roughly 200 printed pages. Likewise, these contributions are primarily concerned with the same issues as the authors contributing to *Phytotaxa* but applied to animals.

The 16 Articles from both journals are all beautifully illustrated, most are clearly written and, even if they are considered by some to be minor contributions to science, add something significant to our growing knowledge of life on this planet. In short, among these 300 pages, there is a wealth of new information about life on Earth.

Considering the contents of these contributions in more detail, only one gives any account of the method used to identify 'key' characters or the principles behind the characterisation of any taxa, new or otherwise. By this, we do not mean the process with which the authors record their observations on specimens to uncover their characters (homologues) – these are usually explained in great detail. We mean how they establish the definitions (diagnoses) of the taxa they subsequently describe.[1] The exceptions were authors who used molecular data, either in addition to or as the sole source of their evidence; these studies gave very precise details of how their data were analysed – but it is (almost) impossible to use DNA data without doing so. In each case, however, a different kind of analysis was applied, and none offered any specific justification for their choice – they simply chose one or another computer program (or some combination) to process their data.[2]

If a method was chosen, then most, if not all, appear to have emerged as a result of, or a reaction to, either the numerical taxonomy of the 1950s and 1960s, the various developments related to Hennig's *Phylogenetic Systematics* in the 1960s, 1970s and 1980s, the necessity of dealing with vast amounts of molecular data from the 1990s onwards – or, as is more often the case, following a tradition of what we will call 'empiricism', learnt from, or handed down, from previous generations' study of a particular group of organisms.

The central aim of this book, then, is to explore the method, or methods, in taxonomy, specifically methods of discovery – which we see as the aim of 'empiricism' – and how those results are presented in a classification.

How Is Taxonomy Done?

Bolstered by the age of bioinformatics, the dawning of cyber-taxonomy and the creation of mega-journals to cope with the vast amounts of taxonomic work that require rapid publication, unknown organisms are now being discovered, described and classified at an astonishing rate. One might, for example, consider the success of the mega-journal *Zootaxa* (Zhang 2006). Launched in 2001, by 2005

---

[1] Only one contribution assembled a data matrix and used PAUP to find a tree using parsimony for the basis of their classification (Tahseen & Mustaqim 2015)

[2] There were four contributions that used molecular data.

'*Zootaxa* [had] published 1103 issues in a total of 31 038 printed pages, with contributions from 1619 authors from around the world describing 2337 new taxa' (Zhang 2006). That is a period of four years – and that commentary is over a decade old. A more recent contribution – 'Describing unexplored biodiversity: *Zootaxa* in the International Year of Biodiversity' – noted:

> In the International Year of Biodiversity (2010), *Zootaxa* published 1,582 papers (including 92 monographs) in 405 issues, with a total of 32,330 pages. These papers included descriptions of 3,951 new taxa, of which 3,664 are of the species-group, 268 of the genus-group and 19 of the family-group. It is estimated that the total new animal species described in 2010 is most likely to be between 15,000 and 20,000, and *Zootaxa* has thus contributed 18 to 24% of the total. (Zhang 2011)

A massive achievement indeed: but one might scour those many pages and find little discussion concerning how taxonomy is actually done.

In some respects, however, the question is actually easy to answer. Consider this example: Imagine a dataset acquired from 100 representatives (specimens) each of a harvestman (e.g., *Wintonia scabra*, https://en.wikipedia.org/wiki/Opiliones), a huntsman (e.g., *Delena cancerides*, https://en.wikipedia.org/wiki/Delena_cancer ides), a redback (e.g., *Latrodectus hasseltii*, https://en.wikipedia.org/wiki/Red back_spider) and a funnelweb spider (e.g., *Hadronyche versuta*, https://en .wikipedia.org/wiki/Hadronyche_versuta). From this dataset two characters are studied for each set of 100 specimens: (1) the presence of modified pedipalps (palps) and (2) spinnerets. The tarsus on the palp is developed in adult male spiders to transfer sperm. These characters are often referred to as homologues as they are names for parts of the organism, or more properly, the parts of the specimens (Chapter 7). Examining the funnelweb spider, the modified palp is different from those of the huntsman and the redback spider, and missing altogether in the harvestman. These data are informative as to their relationships: any particular funnelweb specimen is more closely related to all other funnelweb specimens than it is to any of the other organisms in this set. This detail can be written so:

W. scabra, L. hasseltii, D. cancerides (*H. versuta* specimen #1, *H. versuta* specimen #2 ... #100)

The round brackets group those that are more closely related to each other than to anything else. So in this example, all 100 specimens of *Hadronyche versuta* are represented by enclosing them within a pair of round brackets.

All modified palps found in spiders are considered to be the same organ, so a further statement can be made: all spiders are more closely related to each other, as they have modified palps, than they are to harvestmen, who do not. All palps are not necessarily the same, and modified versions of the character (homologue) are numbered individually in square brackets:

Character 1 (Ch. 1): no palps (palps [1], palps [2], palps [3])

This could be simplified as: Ch. 1: (0 (1, 2, 3)). Again, the rounded brackets group those items more closely related to each other than to anything else, in this case the set of homologues known as palps.

Taxon names can be substituted for the character:

W. scabra (*L. hasseltii, D. cancerides, H. versuta*)

The various types of spinnerets are also found in all spiders but not in harvestmen, yielding two characters with an identical set of relationships:

Character 2 (Ch. 2): no spinnerets (spinnerets [4], spinnerets [5], spinnerets [6])
Simplified as: Ch. 2: (0 (4, 5, 6))

Taxon names can be substituted for the character:

W. scabra (*L. hasseltii, D. cancerides, H. versuta*)

Two further characters can now be considered: the presence of (3) orthognaths (downward pointing fangs) and (4) labidognaths (inward pointing fangs). These yield the following relationship:

Character 3 (Ch. 3): no orthognath (orthognath [7], orthognath [8])
Simplified is: Ch. 3: (0 (7, 8))

Taxon names can be substituted for the character:

W. scabra, L. hasseltii, D. cancerides (*H. versuta*)

and

Character 4 (Ch. 4): no labidognath (labidognath [9], labidognath [10])
Simplified is: Ch. 4: (0 (9, 10))

Taxon names can be substituted for the character:

W. scabra, H. versuta (*L. hasseltii, D. cancerides*)

All four characters together yield the following:

W. scabra (*H. versuta* (*L. hasseltii, D. cancerides*))

This final statement of relationships among these animals is found merely by the addition of these 4 characters with 10 homologues ([1] – [10]):

Ch. 1: (0 (1, 2, 3)) + Ch. 2: (0 (4, 5, 6)) + Ch. 3: (0 (7, 8)) + Ch. 4: (0 (9, 10)) = [0] ([3, 6, 7, 8] ([1, 4, 9], [2, 5, 10]))

The taxon names can be substituted for the character, which may be appended to each name:

W. scabra[0] (*H. versuta*[3 6 7 8] (*L. hasseltii1*[4 9], *D. cancerides*[2 10]))

More data can be added in the form of a new specimen. For example, the trapdoor spider *Misgolas rapax* (https://en.wikipedia.org/wiki/Sydney_brown_ trapdoor_spider) has three more homologues that are part of the palp [11], spinneret [12] and orthognathus fangs [13]. Added to the existing set of relationships reveals it to be more closely related to the funnelweb than to any other taxon:

W. scabra[0] (*H. versuta*[3 6 7 8], *M. rapax*[11 12 13](*L. hasseltii*[1 4 9], *D. cancerides*[2 5 10]))

This set of relationships conveys various details of the organisms themselves, the characters that define them and how closely related one is to another.

This particular notation above might appear a little cumbersome, so a simpler way would be to relate the whole as a written classification as follows:

Arachnida
    Assamidae
                    Wintonia scabra
    Araneae (spiders)
        Mygalomorphae
            Hexathelidae
                Hadronyche versuta
            Idiopeae
                Misgolas rapax
        Araneomorphae
            Theridiidae
                Latrodectus hasseltii
            Sparassidae
                Delena cancerides

It should be easily appreciated that such a classification is inclusive, meaning that if another species of *Latrodectus* is found, say *Latrodectus antheratus*, then we would expect it to share features with other organisms in *Latrodectus*, Theridiidae, Araneomorphae, Araneae and so on.

Of course, when *Latrodectus antheratus* was found, a further interesting question can be addressed: *Latrodectus hasseltii* has only been found in Australia[3]; *Latrodectus antheratus* has only been found in Paraguay and Argentina. Why? And what of other species in the genus *Latrodectus*? Some estimate the number to be more than 30 (www.wsc.nmbe.ch/genus/3502).

Of course, with respect to features that define the taxonomic hierarchy, there is no certainty that all will be present in all organisms included in any one particular category. As William Bateson remarked:

> Nevertheless, if I may throw out a word of counsel to beginners, it is: Treasure your exceptions! When there are none, the work gets so dull that no one cares to carry it further. Keep them always uncovered and in sight. Exceptions are like the rough brickwork of a growing building which tells that there is more to come and shows where the next construction is to be. (Bateson (1908) 1928, p. 324)

Investigating further character systems, probing further the structure of organisms, searching for more specimens – in museums or in the field (Chapter 7) – is the stuff of taxonomy, and characters *are* found that do not correspond to known taxa, and known taxa *are* found to be not so clearly defined after all, and known taxa *are* found to be composed of unrelated organisms that need separating – this, then, is the key: *the scheme above is one of relationships, rather than of 'things' and the endeavours of taxonomists are to determine what exactly is and what exactly is not related in some specific way.*

One might read this account of the relationships among harvestmen and spiders and search for some particular theory or theories that guide, or have guided, the exploration. But there is none, save that observations (homologues) yield conclusions (taxa): what we understand as a version of *empiricism* is the version that probably all working taxonomists have adopted in one form or another, whether precisely stated or otherwise.

In terms of methodology, many books that deal with taxonomy never describe how one would go about discovering taxa, what that process might be, how the day to day activities of a taxonomist might work out. An exception is Richard E. Blackwelder's book simply entitled *Taxonomy* (1967) – but this is now a half a century old. To mention a few more recent books that specifically address taxonomy, *Describing Species* (Winston 1999) is a very readable and useful account of what one might need to tackle when undertaking a taxonomic study, with chapters on nomenclature, literature, material required, etc. Only at the very end of the book, in its final chapter, is there any mention of method at all, and this is found in a chapter entitled 'Further Studies in Systematics', as if discovery were achieved only *after* one learnt something about species, even genera, and then it is as if the

---

[3] Some 'alien' specimens have been found in Southeast Asia, Japan and New Zealand.

only kinds of explorations relate to phylogeny. *The New Taxonomy* (Wheeler 2008) is a collection of essays related to, or about, taxonomy, but it too lacks any chapter on how it is actually done; *Descriptive Taxonomy* (Watson et al. 2015), another excellent collection of essays relating to taxonomy, also has no guide as to method. One might read all these books with profit but still be left with one vital question: how is taxonomy done?

If the question is re-phrased as 'how is classification done?', then one inevitably encounters a central issue of the cladistic period: 'It is informative that Hennig (1966), the father of phylogenetics, devoted about two thirds of his book to issues related to classification, and that resistance to his ideas centred on questions of classification (e.g. Mayr 1974)' (Wheeler 2004, p. 576). More specifically, it centred on the influence cast by the *Modern Synthesis* (e.g., Huxley 1942; Mayr 1942) and its offspring *The New Systematics* (Huxley 1940), beginning in the 1940s and still with us today in one form or another (Pigiucci & Müller 2010[4]; see also the forward by Pigiucci & Müller in Huxley 2010). We tackle this period from the perspective of discovering relationships in Part II (Chapters 3–6).

To facilitate our discussion, definitions used in this book are briefly outlined in the next section. This is simply to provide early guidance to our usage as each term is explained and discussed more fully in Chapter 7.

Terms of Classification

Taxonomy is the study of classification (from the Greek τάξις *taxis*, meaning 'arrangement', and νομία *nomia*, meaning 'method'). Biological classification and taxonomy are discussed further in Chapter 2.

Cladistics: The goal of cladistics is to discover natural classifications hence it might be prudent to define it as: *an approach that aims to find natural taxa within a natural classification.* This is the subject matter of the entire book.

Cladogram is a hierarchical branching diagram, which includes all species, fossil and Recent. This is discussed in detail in Chapter 7.

Cladistic parameter: All cladograms have a branching aspect. This is its cladistic parameter and is the parameter used to deal with any Natural Classification. This is discussed in detail in Chapter 7.

Monophyly and aphyly: Monophyly refers to those groups of organisms captured by the cladistic parameter; aphyly refers to those groups not so captured.

---

[4] Very little is said of taxonomy, systematics, classification or cladistics in this edited volume – what is noted is not particularly complementary (e.g, Callebaut 2010, p. 456). Although Koonin provides a broad view of a 'postmodern synthesis of evolutionary biology', it is a rather conventional approach to the subject of classification (Koonin 2011).

Trees: Cladograms are not *trees*. In the following chapters in most cases the word 'tree' will refer to a *phylogeny, phylogram* or *phylogenetic (phyletic) tree*. A tree is a special kind of cladogram. This is discussed in detail in Chapter 7.

Homologue: A homologue is a part of an organism, usually abstracted from a part of a specimen, or specimens. All organisms might be viewed as being made of parts, such that each organism can be broken into its structural components. Homologues, however, are not just parts – they are parts that imply relationships. Homologues are the primary evidence on which all subsequent taxonomy is based. Character and homologue are interchangeable, with the taxonomist using character more frequently.

Homology is the actual relation derived from the homologues.

Relationship: The representation of a particular kind of comparison, such that minimally of three things, two will be more closely related to each other than either are to the third. This statement is economically represented by A(BC). One might see BC as the *homologues*, and A(BC) as *homology*; the latter understood to be equivalent to *taxon* and *monophyly*, as well as the cladistic parameter.

Heterobathmy literally means 'different steps on a stair' (from the Greek *bathmos,* a step, stair or grade). A cumbersome word, but when applied to taxonomy, it translates into finding the correct 'step' on the taxonomic hierarchy for each character observed. That is, asking the question: where does this character fit?

How This Book Is Structured

This book is structured in a reasonably conventional way, composed of a number of chapters. Each has been written so that hopefully it will 'stand-alone' as a statement on the particular subject addressed. We have tried to avoid repetition, but some is inevitable. Each chapter has illustrative examples, taken either from our own work or from the literature. Alongside the cited references, each chapter includes a list of *Further Reading*. In some cases, we have provided commentary on our choices. The list of items is included for those who wish to pursue any particular chapter's subject in greater depth – or even to explore some ideas alternative to our own. Naturally, our recommendations (and comments) reflect our own bias. For this we crave your indulgence but hope that we have been broad-minded enough to cover a range of material. We have dispensed with the traditional end-of-book glossary. While we accept that a glossary provides neat, concise definitions, it usually excludes the author's reasons for adopting that particular use.

It is important for us to stress that we specifically did not want to produce a cookbook. If anything, we have attempted to produce a book that explains the ingredients, offers routes to sourcing the best products and makes suggestions for possible recipes. It is up to you, dear reader, to make your own meal.

References

Adanson, M. 1763. *Familles des Plantes.* Vincent, Paris.

Bateson, BD. 1928. *William Bateson, F. R. S., Naturalist: His Essays & Addresses, Together with a Short Account of His Life.* Cambridge University Press, Cambridge.

Blackwelder, RE. 1967. *Taxonomy.* John Wiley & Sons, New York.

Cain, AJ. 1959. The post-Linnaean development of taxonomy. *Proceedings of the Linnean Society of London* 170: 234–244.

Callebaut, W. 2010. The dialectics of dis/unity in the evolutionary synthesis and its extensions. In: Pigiucci, M & Müller, GB (eds), *Evolution: The Extended Synthesis.* MIT Press, Cambridge, MA, pp. 443–481.

Felsenstein, J. 1982. Numerical methods for inferring evolutionary trees. *Quarterly Review of Biology* 57: 379–404.

Huxley, JS. 1940. *The New Systematics.* Oxford University Press, Oxford.

Huxley, JS. 1942. *Evolution: The Modern Synthesis.* George Allen & Unwin, London.

Huxley, JS. 2010. *Evolution: The Modern Synthesis: The Definitive Edition.* Forward by Massimo Pigiucci & Gerd B. Müller. MIT Press, Cambridge, MA.

Koonin, EV. 2011. *Logic of Chance: The Nature and Origin of Biological Evolution.* FT Press, New Jersey. [Updated edition].

Mayr, E. 1942. *Systematics and the Origin of Species.* Columbia University Press, New York.

Mayr, E. 1974. Cladistic analysis or cladistic classification? *Zeitschrift für zoologische Systematik und Evolutionsforschung* 12: 95–128.

Nelson, GJ. 1979. Cladistic analysis and synthesis: principles and definitions, with a historical note on Adanson's *Familles des Plantes. Systematic Zoology* 28: 1–21.

Pigiucci, M. & Müller, G. (eds) 2010. *Evolution: The Extended Synthesis.* MIT Press, Cambridge, MA.

Simpson, GG. 1961. *Principles of Animal Taxonomy.* Columbia University Press, New York.

Tahseen, Q. & Mustaqim, M. 2015. A taxonomic review of the genus *Goffartia* Hirschmann, 1952 (Rhabditida: Diplogastridae) with a note on the relationship of congeners. *Zootaxa* 4034: 70–86.

Watson, MF., Lyal, C. & Pendry, CA. (eds) 2015. *Descriptive Taxonomy: The Foundation of Biodiversity Research.* Cambridge University Press, Cambridge, UK.

Wheeler, QD. 2004. Taxonomic triage and the poverty of phylogeny. *Philosophical Transactions of the Royal Society of London, Series B*, 359: 571–583.

Wheeler, QD. (ed.) 2008. *The New Taxonomy.* Systematics Association Special Volume Series. CRC Press, Boca Raton, FL.

Winston, JE. 1999. *Describing Species, Practical Taxonomic Procedure for Biologists.* Columbia University Press, New York.

Zhang, Z.-Q. 2006. The making of a mega-journal in taxonomy. *Zootaxa* 1358: 67–68.

Zhang, Z.-Q. 2011. Describing unexplored biodiversity: *Zootaxa* in the International Year of Biodiversity. *Zootaxa* 2768: 1–4.

Further Reading

Blackwelder, RE. 1967. *Taxonomy*. John Wiley & Sons, New York.

A little outdated but useful for the discussion of what was called 'Omnispective taxonomy', an attempt to be wholly empirical. The book's merits are in the critique of the then prevailing views of Mayr. '*The Omnispective System* ... [is the] modern form of classical taxonomy, [and] has been in use for two hundred years and has produced 99% of the revisionary and monographic work in taxonomy' (Blackwelder, RE. 1977. Twenty five years of taxonomy. *Systematic Zoology* 26: 107–137).

Michel, E. (ed.) 2016. Anchoring biodiversity information: From Sherborn to the 21st century and beyond. *Zookeys* 550.

While focusing on Charles Davies Sherborn (1861–1942), 'The papers in this volume fall into three general areas. In the first section, seven papers present different facets of Sherborn as a man, scientist and bibliographer, and describe the historical context for taxonomic indexing from the 19th century to today. In the second section, five papers (with a major appendix) discuss current tools and innovations for bringing legacy information into the modern age. The final section, with three papers, tackles the future of biological nomenclature, including innovative publishing models and the changing tools and sociology needed for communicating taxonomy'.

Polaszek, A. 2010. *Systema Naturae 250 – the Linnaean Ark*. CRC Press, Boca Raton, FL.

Watson, MF., Lyal, C. & Pendry, CA. (eds) 2015. *Descriptive Taxonomy: The Foundation of Biodiversity Research*. Cambridge University Press, Cambridge, UK.

Wheeler, QD. (ed.) 2008. *The New Taxonomy*. Systematics Association Special Volume Series. CRC Press, Boca Raton, FL.

Polaszek, Watson et al. and Wheeler are all useful collections of papers on various aspects of the taxonomic enterprise, although, as we noted above, not much space is devoted to how taxonomy is actually done.

Winston, JE. 1999. *Describing Species, Practical Taxonomic Procedure for Biologists*. Columbia University Press, New York.

A useful book focused on the practice of species-level taxonomy.

2

Classification

The safest general characterization of the European philosophical tradition is that it consists of a series of footnotes to Plato.

(Whitehead 1929 [1978], p. 39)

It is a truism to note that classification is the oldest form of human understanding, but it has permeated every walk of life, not just in the sciences, but all kinds of endeavour.

Although this book is about biological classification, to introduce the topic we offer two non-biological examples, both taken from novels.

Our first example is from the Australian Richard Flanagan's novel *The Narrow Road to the Deep North*, which won the Man Booker Prize in 2014. Flanagan's novel is primarily about suffering and survival, surviving the enforced building of the Thailand–Burma Railway (the 'Death Railway') during World War II; the survival of the Australian prisoners of war who built it. There is a passage in the book that summarised the kinds of men left after building ceased:

> For good reason, the POWs refer to the slow descent into madness that followed simply with two words: *the line*. Forever after, there were for them only two sorts of men: the men who were *on the line*, and the rest of humanity, who were not. (Flanagan 2013, p. 26)

Flanagan's novel deals with the immense psychological disruption felt by humans under adverse conditions but the prose above captures the essence of classification, regardless of the reasons, the process of how we divide the world up to forge greater meaning, and how we go about naming those divisions.

Flanagan left his two divisions unnamed, just identified as 'two sorts of men'. His division, his classification, is a variation on similar sentiments expressed long ago in Plato's *The Statesman*, the concern for distinguishing Greeks from the rest of humanity (Plato 1848):

> What most Greeks do is make the division by separating Greeks from all the rest: they use the single term 'barbarian' for all other categories of people . . . and then expect there to be a single category . . . just because they've used a single term. (Plato in Annas & Waterfield 2010, p. 11)

Unlike Flanagan, Plato's visitor names his 'two sorts of men' – the group he belongs to, the Greeks, and the other sort, the Barbarians, the 'not-Greeks', namely, everyone else. But Plato's message is simple: just because something has a name ('Barbarian') or even a definition ('of incomprehensible speech') does not mean it is actually anything at all, apart from a convenient catch-all category – in this case a catch-all for not-Greeks. And Flanagan, possibly recognising this too, follows his sentences above with:

> Or perhaps only one sort: the men who *survived the line*. (Flanagan 2013, p. 26)

When the words of Plato's visitor are expanded, the issue comes into sharper focus:

> . . . the kind of a mistake a man would make who, seeking to divide the class of human beings into two, divided them into Greeks and barbarians. This is a division most people in this part of the world make. They separate the Greeks from all other nations making them a class apart; thus they group all other nations together as a class, ignoring the fact that it is an indeterminate class made up of people who have no intercourse with each other and speak different languages. Lumping all this non-Greek residue together, they think it must constitute one real class because they have a common name 'barbarian' to attach to it. (Skemp 1952, 262d, after Platnick 1989, p. 20)

One might understand the goal of taxonomy to be simple: eliminate the 'Barbarians', of which protists, invertebrates, reptiles, fishes, dinosaurs, etc. are but a few.

Our second example comes from the American novelist Don DeLillo, from his 8th book *White Noise*, published in 1985 (DeLillo 1985). The book's plot is not necessary for the points we want to make. What is of interest is this piece of dialogue:

> "A dog is a mammal."
> "So's a rat," Denise said.
> "A rat is a vermin," Babette said.
> "Mostly what a rat is," Heinrich said, "is a rodent."
> "It's also a vermin."
> "A cockroach is a vermin," Steffie said.
> "A cockroach is an insect. You count the legs is how you know."
> "It's also a vermin."
> "Does a cockroach get cancer? No," Denise said. "That must mean a rat is more like a human than it is like a cockroach, even if they're both vermin, since a rat and a

human can get cancer but a cockroach can't."

"In other words," Heinrich said, "She's saying that two things that are mammals have more in common than two things that are only vermin."

"Are you people telling me," Babette said, "that a rat is not only a vermin and a rodent but a mammal too?"

Snow turned to sleet, sleet to rain.

The dialogue from a postmodern novel says more about classification and its puzzles than many contemporary scientific texts.

Given all the various names included in this passage, one really needs a piece of paper to write them down to help make sense of them: dog, mammal, rat, rodent, vermin, cockroach, insect, human.

A simple list of those names looks like this:

Dog

Rat

Vermin

Cockroach

Mammals

Human

Rodent

Not that useful, unless you already know what these categories are or what they belong to: that you know what they mean.

Some of these names refer to particular biological species. All known species have proper names. By this, we mean all identified species have a two-part Latin binomial ('two-term') or binominal ('two-name') – the two parts are the genus name and the species name, a naming convention first introduced by Carolus Linnaeus (Carl von Linne, 1707-1778) in his *Species Plantarum* (Linnaeus 1753) and *Systema Naturae* (Linnaeus 1758-1759):

Dog	*Canis familiaris*[1]
Rat	*Rattus rattus*
Vermin	
Cockroach	*Periplaneta americana*
Mammals	
Human	*Homo sapiens*
Rodent	

---

[1] We will assume that the domesticated dog was the animal in mind.

Other names in the list are collections of species. Some of these have proper names as well, but with just a single part:

Dog	*Canis familiaris*
Rat	*Rattus rattus*
Vermin	
Cockroach	*Periplaneta americana*
Mammals	Mammalia
Human	*Homo sapiens*
Rodent	Rodentia

Mammalia and Rodentia are taxonomic groups of species, the former with something like 6000 known species, the latter half that amount, with around 3000 known species. To further understand these names, the list could be re-organised. All the creatures mentioned in this list are animals and belong to one of two taxonomic groups, insects or mammals: the cockroach is an insect, the human, rat and dog are mammals. The list could be subdivided even further because the rat (*Rattus rattus*) is a rodent (a member of Rodentia) as well; the human is a different kind of mammal, a primate; and the dog is a Carnivora ('a flesh-eater'). So an indented list would make even greater sense for these names:

Animals
    Insects: Cockroach [*Periplaneta americana*]
    Mammals:
        Rodent: Rat [*Rattus rattus*]
        Primate: Human [*Homo sapiens*]
        Carnivora: Dog [*Canis familiaris*]

So something – a rat, for example – really can be two things at the same time: a mammal and a rodent. In fact it can be many things all at once: an animal, a mammal and a rodent as well as a particular kind of rat:

Animalia
        Chordata
                Mammalia
                        Rodentia
                                Muroidea
                                        Muridae
                                            Murinae
                                              Rattus
                                                      Rattus rattus

Combined together, all these various names above make up the total classification, and each category should allow us to say something about all of its included members. We expect, for example, there to be some common properties for the genus *Rattus*, which are shared among the 70-odd different species. One estimate of the Murinae is that it includes more than 120 genera and nearly 600 species. One might assume that each genus has characters of its own to distinguish it from all the others. This kind of classification is said to be a *natural classification*, one that reflects (or attempts to reflect) the properties of the included organisms relative to all other organisms and is expressed as a series of relationships.

One category used by de Lillo's characters is not included in our classification above: vermin. Vermin are usually considered to be pests, those organisms (usually animals) that are said to cause a nuisance. With some justification, one might consider all of the animals in our classification to be vermin, but whatever conclusion one comes to as to what constitutes vermin, the term leads to a *practical or artificial classification* rather than a natural classification: Vermin can be identified for their treatment and eventual eradication[2].

The differences between these kinds of classifications – practical (artificial) and natural – are crucial. It might not be too much of an exaggeration to suggest that all classifications today are mixtures of the practical and the natural, sorting out the mixture being the source of most problems. We explore some of these issues related to this from a variety of angles in the pages that follow.

Taxonomy

Taxonomy is defined as the study of classification (from the Greek τάξις taxis, meaning 'arrangement', and νομία nomia, meaning 'method'). It was first used by Candolle in his book *Théorie élémentaire de la botanique: ou, Exposition des principes de la classification naturelle et de l'art de décrire et d'étudier les végétaux,*

---

[2] It is worth noting that the word for vermin in German is *Ungeziefer*. This term was used to describe the state Gregor Samsa finds himself in the opening line of Kafka's *The Metamorphosis* [*Die Verwandlung*]: "One morning, when Gregor Samsa woke from troubled dreams, he found himself transformed in his bed into a horrible vermin [Als Gregor Samsa eines Morgens aus unruhigen Träumen erwachte, fand er sich in seinem Bett zu einem ungeheueren Ungeziefer verwandelt]". Note that in the original German Kafka uses *Ungeziefer*, while the English translations often use *cockroach*, *bug* or the more general term *insect*, instead of vermin. Again this is an example of how an artificial classification may convey a practical sense to the term in mind or, in this case, an image. Kafka does not say what type of *Ungeziefer* Samsa is, rather it is up to us to imagine what the vermin is, in the same way the terms *monster* or *ghost* conjure up certain images in the mind. None of these terms is meant to reflect an actual natural or specific thing.

published in 1813 (this book is discussed in more detail in the section 'Natural and Artificial Classifications').

There are numerous definitions of taxonomy, with almost every textbook on the subject differing in one way or another. Taxonomy is linked with, even considered synonymous with, *classification*, *systematics* and, in more recent times, *phylogeny* – with all these terms themselves being variously understood. Here, as we stated above, we treat them all as synonyms – with the proviso that the last term in our list, phylogeny, requires further discussion (Chapters 3 and 7).

The primary task of the taxonomist is to provide a hierarchical classification that represents as accurately as possible current knowledge of the organisms of interest.

The names in a classification represent *taxa* (singular: *taxon*). For example, the jellyfish *Malo kingi* belongs to all the following taxa: Animalia, Cnidaria, Cubozoa, Carybdeida, Carukiidae, *Malo*. Is the species *Malo kingi* also a taxon? We believe it is, but this requires further discussion (Chapter 3).

Taxonomy need not just refer to the study of organisms. Indeed, there have been recent studies published on office chairs (Olivares 2011) and mythical creatures such as vampires (Woerner 2010). In this book we deal only with biological classification.

Classifications are often divided into *artificial classifications* and *natural classification* both based on an understanding of the construction and structure of any particular animal or plant: the study of their parts. We mentioned some briefly above with reference to *vermin*. The difference between the two kinds of classification was summarised above, but many years ago Thomas Henry Huxley (1825–1895) offered these thoughts:

> The differences between 'artificial' and 'natural' classifications are differences in degree, and not in kind. In each case the classification depends upon likeness; but in an artificial classification some prominent and easily observed feature is taken as the mark of resemblance or dissemblance; while, in a natural classification, the things classified are arranged according to the totality of their morphological resemblances, and the features which have been ascertained by observation to be the indications of many likenesses or unlikenesses. And thus a natural classification is a great deal more than a mere index. It is a statement of the marks of similarity of organization; of the kinds of structure which, as a matter of experience, are found universally associated together; and, as such, it furnishes the whole foundation for those indications by which conclusions as to the nature of the whole of an animal are drawn from a knowledge of some part of it. (Huxley 1902)

There is much in Huxley's statements to disagree with, but for the moment we can use it as a guide. *Artificial classification* is given above in the plural as there can be many kinds - we mentioned 'practical classification' as one kind; *natural classification* is given in the singular as there is but one. The two forms of classification need not be identical or even similar, but both are equally useful *when applied to their stated purpose*. As noted above, a core problem for most classifications of any particular taxonomic group of organisms is that they will be a mixture of the artificial and the natural. By that we mean the characters that are attributed to them are a mix of those that might be used for *identification* and those used for determining *relationships* (we return to this below).

The study of any organisms' structure and relationships, once known as *Comparative Biology*, is the process of teasing out one kind of character from the other, the artificial from the natural. We deal with characters in more detail in Chapter 3.

Biological Classification

At an early age most of us quickly become familiar with a variety of plants and animals, usually those we encounter in our day to day lives, depending on where we live: ducks, elephants, dogs, palms, roses and so on. Most of the names first learnt are *common names*, which are often descriptive: there are dabbling ducks, diving ducks, perching ducks and so on.

A recent editorial in the taxonomic journal *Zootaxa* commented on the kinds of diversity still being found and recorded, the kinds of animals that appear in the pages of popular nature magazines such as *National Geographic*, as well as in the general press:

> . . . a spectacular pink millipede *Desmoxytes purpurosea* . . . ; a rare frog *Philautus maia* . . . ; one of the most venomous snakes in the world *Oxyuranus temporalis* . . . and a dangerous box jellyfish *Malo kingi* . . . ; the longest insect *Phobaeticus chani* . . . ; a pygmy seahorse *Hippocampus satomiae* . . . ; the world's smallest snake *Leptotyphlops carlae* . . . ; a deep-reef fish *Chromis abyssus* . . . ; a carnivorous sponge *Chondrocladia (Meliiderma) turbiformis* . . . ; a pollinating cricket *Glomeremus orchidophilus* . . . ; a new duiker *Philantomba walteri* . . . (Zhang 2011, p. 4, ellipses represent authors and dates)

Phytotaxa, the botanical sister journal to *Zootaxa*, which only began publishing in 2009, documents botanical diversity in the same way:

> . . . a new pitcher plant from the Philippines, *Nepenthes attenboroughii* . . ., . . . a previously undescribed yam which is harvested locally in Madagascar, *Dioscorea orangeana* . . . the "suicide palm", *Tahina spectabilis* . . . and a caffeine-free coffee from Cameroon, *Coffea charrieriana* . . . (Christenhusz et al. 2011, p. 1, ellipses represent authors and dates)

Consider the paragraphs above but from the perspective of just the common names:

... a spectacular pink millipede; a rare frog; [a] venomous snake; a box jellyfish;
an insect; a seahorse; a deep-reef fish; a sponge; a cricket; a duiker ...
    a pitcher plant; a yam; a palm ...

Many of these names are familiar and in common usage: snakes, frogs and
jellyfish, pitcher plants and palms. Some may be less familiar, such as duikers (a
small mammal, an antelope, with 20+ species occurring in sub-Saharan Africa)
and yams (edible tubers, with 6000+ species in the genus *Discorea*, mostly occurring in the southern hemisphere but often found in greengrocers in the northern
hemisphere), depending on where one lives or was brought up.

Taxonomists deal with more precision than simply noting that there are snakes
and frogs and jellyfish. As noted above, they refer to the organism's two-part
scientific name such as *Oxyuranus temporalis* (a snake), *Philautus maia* (a frog)
and *Malo kingi* (a jellyfish), as there are different types (or kinds) of snakes, frogs
and jellyfish. An organism's scientific name adds clarity where common names
may confuse.

Example 1 Robins: *Erithacus rubecula*; *Turdus migratorius*: Regardless of where
one lives, many people are aware that there is a bird called a robin. There is a robin
in Europe and a robin in the USA – the two birds are different animals.[3] So what is
a robin? The European robin is called *Erithacus rubecula* and the American robin
is called *Turdus migratorius*. Birds in the genus *Turdus* are commonly referred to
as thrushes; birds in the genus *Erithacus* are commonly referred to as old-world
flycatchers (Table 2.1).

Our two robins (*Erithacus rubecula* and *Turdus migratorius*) are both Passer-
iformes (perching birds) and belong in the family Muscicapidae (often referred to
as 'typical flycatchers'), and in the family Turdidae (commonly known as the
thrushes) respectively. Both birds share the properties of a few more inclusive
categories such as animals and chordates, as well as birds. They differ in the
characters of their genus: *Erithacus* (of which there are only three species) and
Turdus (of which there are more than 60 species).

The two robins are, then, different animals – but to an almost greater extent, they
are actually the same. Nevertheless, the name robin – a widely used common
name – would serve to mislead when used colloquially and out of context.

Example 2: The orange or red roughy: *Hoplostethus atlanticus* is a large deep-sea
fish, referred to as a deep-sea perch. What is known of this fish? Initial exploration
of online resources yields data on its physical appearance, life-history, conservation status and much more. The many online pages contain much duplicated
information, mostly concerning how to eat the fish and, somewhat ironically,

---

[3] There are other robins: an Australian 'Scarlet robin' (*Petroica multicolor*), a 'Pekin robin' (*Leio-
thrix lutea*) and magpie robins (*Copsychus*). For more on why robins are robins, see Moss (2017).

Table 2.1 Classification of two robins

	Erithacus rubecula Robin (Europe)	*Turdus migratorius* Robin (USA)
Kingdom	Animalia	Animalia
Phylum	Chordata	Chordata
Class	Aves	Aves
Order	Passeriformes	Passeriformes
Family	Muscicapidae	Turdidae
Genus	*Erithacus*	*Turdus*

whether we should or not – its conservation status is placed as 'Conservation Dependent' (Near Threatened, Environment Protection and Biodiversity Conservation (EPBC) Act) – so we probably shouldn't eat it.

Inspection of the name *Hoplostethus atlanticus* on FishBase (an information resource for recent fishes) provides further data such as its 'Size/Weight/Age', 'Length at first maturity', 'Environment', 'Climate/Range' and 'Distribution' alongside lists of further sources of information on its structure and DNA, its vulnerability, even its price. One thing seems certain: Much is known about the orange roughy (*Hoplostethus atlanticus*).

A different source of information for *Hoplostethus atlanticus* reveals knowledge of another kind. The entry for *Hoplostethus atlanticus* in the World Register of Marine Species (WORMS, www.marinespecies.org/index.php), for example, is prefaced with a series of taxonomic names beginning with Animalia, the Kingdom to which this fish belongs, and ending in *Hoplostethus*, its genus name. A similar list is given in the equally useful Integrated Taxonomic Information System (www.itis.gov/), where the names are represented as an indented table rather than a list, as in Table 2.2. The names provide a guide to the more general properties of *Hoplostethus atlanticus*, in a similar fashion to the robins in Table 2.1.

The table of indented names (Table 2.2) is another typical representation of biological classification: a hierarchical arrangement of names, with the assumption being that the properties at one level are included in the next: all chordates are animals, all bony fishes are vertebrates (as well as chordates), all spiny-ray finned fishes are bony (as well as being vertebrates and chordates), and so on. Thus, *Hoplostethus atlanticus* is a spiny-ray finned fish (Actinopterygii); a bony fish (Osteichthyes).

Given this classification, the red roughy can be compared to the two robins: both are vertebrates (Vertebrata), chordates (Chordata) and animals (Animalia).

Table 2.2 Classification of *Hoplostethus atlanticus*

Kingdom	Animalia	Animals
Phylum	Chordata	Chordates
Subphylum	Vertebrata	Vertebrates
Class	Actinopterygii	Bony fishes
Subclass	Neopterygii	
Infraclass	Teleostei	
Superorder	Acanthopterygii	Spiny-ray fishes
Order	Beryciformes	
Suborder	Trachichthyoidei	
Family	Trachichthyidae	Roughies, slimeheads
Genus	*Hoplostethus*	*Hoplostethus*
Species	*atlanticus*	*atlanticus*

From Integrated Taxonomic Information System Report for *Hoplostethus atlanticus*.
www.itis.gov/servlet/SingleRpt/SingleRpt?search_topic=TSN&search_value=166139#null

The taxonomic names relate to the properties of the organisms themselves, properties often referred to as their characters, such that vertebrates have backbones and spinal columns; Actinopterygian fishes have fin rays (fins supported by bony spines called lepidotrichia); Trachichthyidae are called slimeheads as they have a network of muciferous canals in their heads; and so on. So if one is presented with a specimen that has fins with fin rays, then it is an Actinopterygian; if it has a network of muciferous canals in its head then it is a Trachichthyidae, a slimehead. There is – or should be – a neat symmetry between the classification and the properties of organisms. This isn't always the case and will be discussed further in Chapters 5 and 8.

But more importantly, taxonomic names give precise details of any organism's relationships known at that time. One cannot *necessarily* determine the characters of any organism just from its classification – but one can be absolutely certain as to its relationships.

Example 3: We referred in passing to a species of jellyfish called *Malo kingi*. From Wikipedia we learn the following:

> *Malo kingi* is an Irukandji jellyfish named after victim Robert King, a tourist from the United States who died from its sting. It was first described to science in 2007, and is one of two species in genus *Malo*. It has some of the world's most

potent venom, even though it is no bigger than a human thumbnail. As an Irukandji, it can cause Irukandji syndrome, characterized by severe pain, vomiting, a rapid rise in blood pressure, and death. (https://en.wikipedia.org/wiki/Malo_ kingi, accessed 28 May 2018)

Naturally these details are fascinating – if not a little frightening. Two questions might arise: How do I identify *Malo kingi*? And what are its relationships to other organisms?

To identify *Malo kingi* (which might be of some considerable importance) it would be best to examine the original paper in which it was described so as to learn something about the animal:

> This species [*Malo kingi*] differs from all other cubozoans in having halo-like rings of tissue encircling the tentacles, with club-shaped Type 4 microbasic mastigo-phores inserted end-on around the periphery of the rings. It further differs from its only current congener, *Malo maxima*, in having: a much smaller, more rounded body; different tubule winding patterns in the undischarged nematocysts; spines confined to the distal-most portion of the shaft in the discharged nematocysts; relatively broader pedalia; somewhat less well defined rhopalial horns; and perra-dial lappets with a greater number of nematocyst warts and nipple-like terminal extensions. (Gershwin 2007, p. 55)

As is usual, original descriptions are couched in the technical language of animal classification – the parts being named as: mastigophores; nematocysts; pedalia; rhopalial horns; perradial lappets – but with some help, those of us not familiar with the features of box jellyfish might be able to recognise the animal when necessary (see Chapter 7) – but first one would need to learn something about box jellyfish structure.

Another phrase stands out: *Malo kingi* 'is one of two species in genus *Malo*'. We might consult details of the second species, *M. maxima*. Wikispecies directs us to its description, where we learn that *M. maxima* is actually the first species described in what was at first a monotypic genus (Gershwin 2005, p. 1). In the same account another new species, *Carukia shinju*, was described: 'Both species are thought to be dangerous to humans, causing Irukandji syndrome' (Gershwin 2005, p. 1). So for purposes of safety we need to be able to identify species in the genera *Malo* and *Carukia*. This recalls our second question: What are the relationships of these organisms? How are these two genera – *Malo* and *Carukia* – related to each other?

At present, the genus *Malo* belongs to the taxonomic family Carukiidae, which has four genera[4]: *Malo* (four species), *Carukia* (two species), *Gerongia* (one species) and *Morbakka* (two species). The family belongs to the order Carybdeida,

---

[4] Here we follow Bentlage & Lewis's (2012) study.

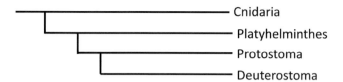

Figure 2.1 Cladogram showing the position of Cnidaria, which includes jellyfishes.

which has five families: Alatinidae, Carukiidae, Carybdeidae, Tamoyidae and Tripedaliidae. The order Carybdeida belongs to the class Cubozoa, a taxonomic group that includes all the box jellyfish (aptly called 'sea wasps'), of which *Malo kingi* is but one. In turn, the box jellyfish belong to the phylum Cnidaria, which includes all other kinds of jellyfish (true jellyfish, stalked jellyfish), the hydras, corals and sea anemones. A summary classification for *Malo kingi* is thus:

Regnum: Animalia
    Phylum: Cnidaria
        Classis: Cubozoa
            Ordo: Carybdeida
                Familiae: Carukiidae
                    Genus: *Malo*
                        Species: *Malo kingi*

The summary classification tells us of its relationships, not necessarily its properties. So given this classification and those above, the red roughy, robins and box jellyfish can be compared one to another even if the comparison might seem relatively trivial: they are all animals (Animalia). A listing of the properties of any of the box jellyfishes will be extremely useful but need not be part of a natural classification. A scheme for identification is artificial in the sense used here: it has a specific purpose other than the determination of relationships.

Any hierarchical listing of a classification can be represented as a tree diagram, as branching trees are, to a certain extent, always hierarchical (Figure 2.1). Ernst Haeckel (1834–1919) instigated the use of tree diagrams to represent summaries of what became known as *phylogenetic relationships*, a specific kind of relationship among organisms (see Chapter 3). Haeckel's illustration of the relationships of the Cnidaria, as he understood them, is reproduced as our Figure 2.2 (from Haeckel 1866, fig. III).

A general question might be: what do these names mean, beyond being simple labels for the various organisms that inhabit this planet? We have suggested that the names are representative of the organism's relationships. We tackle this in detail later (Chapter 3), but first we need to outline a little more about kinds of classifications.

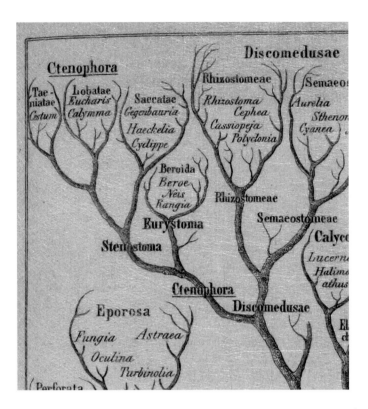

Figure 2.2 Haeckel's illustration of the relationships of Cnidaria (from Haeckel 1866, fig. III).

Natural and Artificial Classifications

Linnaeus may be viewed as having placed the study of systematic botany on a new and modern basis … The central problem, for him and later workers as well, is the meaning and significance of artificial and natural systems.

(Nelson & Platnick 1981, p. 88)

The *Théorie élémentaire de la botanique* was 'the first work in which the soul of the natural and artificial method had been laid bare'.

(Croizat 1945, p. 64).

The 3rd edition of Jean Baptiste Lamarck's (1744–1829) *Flore Française* was published in 1805. Augustin Pyramus de Candolle (1778–1841) was its sole author even though the work was credited to both Lamarck and Candolle, the former having been an active author of the first two editions. It was here Candolle wrote his first account of taxonomy in the section *Principes élémentaires de botanique* (Lamarck & Candolle 1805). After moving to Montpellier, Candolle began work on the more comprehensive *Théorie élémentaire de la botanique: ou, Exposition des principes de la classification naturelle et de l'art de décrire et d'étudier les végétaux* (*Elementary Theory of Botany*) (Candolle 1813).

Rather than dwell on the details of Candolle's arguments in *Théorie élémentaire de la botanique*, the basic outline will be presented below. Our efforts are to present as much as possible in Candolle's own words (albeit translated into English).

Candolle first acknowledged Linnaeus' role:

> Linnaeus was the first to distinguish carefully between the artificial method and the natural method ... he was the first to give examples of one and the other.

Candolle then deals with business:

> Rational classifications have a real relation with the objects to which they apply. These are the only ones that merit consideration.

The idea of a rational classification should be obvious; they should have some empirical content – but it is not enough to simply say that rational classifications have empirical content as the extremely odd (and fictional) classification much discussed by Jorge Luis Borges (1899–1986), apparently derived from 'a certain Chinese Encyclopedia from the *Celestial Emporium of Benevolent Knowledge*', is based on evidence of sorts. Borges' classification includes 14 categories:

1. Those that belong to the Emperor
2. Embalmed ones
3. Those that are trained
4. Suckling pigs
5. Mermaids
6. Fabulous ones
7. Stray dogs
8. Those included in the present classification
9. Those that tremble as if they were mad
10. Innumerable ones
11. Those drawn with a very fine camelhair brush
12. Others
13. Those that have just broken a flower vase
14. Those that from a long way off look like flies

Each of the 14 categories has a connection with evidence (data) of sorts for it could be established as factual that the Emperor does indeed have a collection of items that belongs to him; that we can and indeed do find stray dogs; and that it is more than possible that quite a few items 'from a long way off look like flies', when viewed appropriately. Not all 14 have this empirical connection as 'Those included in the present classification', by definition, include everything included (an entry designed to show a paradox), and 'Innumerable ones' depends on how many are too many to count. Of more significance is that the categories do and can contradict each other in ways that are insoluble. A stray dog, for example, may very well be fabulous, may well 'have just broken a flower vase' and from a long way off may look like a fly.

The 14 categories might also fit every possible animal, real or imaginary. Take Cerberus, the 'hound of hades' who guards the underworld. He may be classified in at least eight of the 14 categories. Given the contradictory nature of nearly all of the 14 categories when applied to anything, one might, then, describe this classification as irrational – great fun, but almost useless.[5]

Let us return to Candolle. Of rational classifications, he noted that there is a 'great diversity in the principles of the various authors . . . This diversity stems from the special purpose for which each is proposed'.

The idea that classifications are many and may have different purposes is central:

> Some authors wanted to study plants with respect to other areas of human understanding, and so they have classed plants in relation to the usages, to the properties, to their countries, etc. These classifications I call *practical* [italics ours].

Such classifications are common. For example, if we divide all plants into those that are poisonous and those that are not, it is a practical division. Its intent is to allow its users a way to avoid being poisoned. Books on poisonous plants are numerous but may be of limited value. A book on the poisonous plants of the British Isles would be more useful if you lived in Britain; or if you lived in South Africa, a book on the poisonous plants of South Africa. The addition of a geographical perspective identifies with greater accuracy the target audience: it would be of limited value having a book on the poisonous plants of the British Isles if you lived in South Africa (and wanted to avoid being poisoned). Whatever the case may be, the assumption is that the plants added to these kinds of books are already known by someone, even if it is only the compiler of the book on poisonous plants. No unique discovery, with respect to the plant itself, is expected from the user. This relates to Candolle's next point:

---

[5] But not completely useless: see the chapter entitled 'Classifying' in Michel Foucault's book *Les Mots et les Chose* (Foucault 1966; *The Order of Things*, 1970).

> Others have as their essential goal to give to persons who know nothing of the
> names of plants an easy way to discover the names . . . by inspection of the plant
> itself. These classifications have been given the name of *Artificial Methods*
> [italics ours].

Artificial methods, or classifications, should allow a person to find the name of any
known plant by examining key characters of any specimens. Keys are the most
well-known of the artificial methods.

> Finally there are those persons who want to study plants, either in themselves, or in
> their real relations among themselves, and to class them so that those plants most
> closely related in the order of nature are also those most closely related in our books.
> These classifications have received the name of *Natural Methods* [italics ours].

This is the method for those who wish to study nature to make discoveries, this
is the basis for natural classification. We will examine the nature of those discoveries later. It is the scientific aspect of taxonomy that is embedded in natural
classification. Candolle then states what should by now be obvious:

> These three sorts of classification [practical, artificial, natural] follow entirely differ-
> ent laws and rules. However, they have often been and still are often confounded.

One further aspect of Candolle's summary is important: He explains the difference
between a system and a method. Above we have just been dealing with methods but
for Candolle, a system is a key or classification based on a particular organ (a leaf,
the flower, etc.), a method is a key or classification based on all the organs of a plant
(or animal). Above it was clear that methods could be subdivided into artificial and
natural, depending on their purpose: artificial classifications are primarily intended
for identifying plants, natural classifications are primarily for expressing, in Candolle's words, 'real relations'. Artificial and natural classifications are both useful,
once we know their purpose.

For natural classifications, however, we are left pondering Candolle's notion of
real relationships. Much (and a very worthy) effort has been spent discussing the
meaning of 'relationships', which we attempt elsewhere. For now, it is worth
addressing a different question and ignore the meaning of 'real relationships' for
the moment: how do we *discover* relationships? It might seem odd, even absurd, to
ask how we find 'relationships' prior to having some idea of what they might be.
Here we can do no better than refer to the words of Polly Winsor:

> To us it may seem paradoxical that naturalists should use the word "related"
> without agreeing on its meaning, but actually this tolerance enabled them to
> make progress as a scientific community. (Winsor 2009, pp. 43–44).

We might also dwell on the words of the philosopher William Whewell
(1794–1866), taken from *The Philosophy of the Inductive Sciences*, published in

1840, summarising, in part, Candolle's views on scientific classification, 'a depart-ment of the philosophy of natural history which has been termed by some writers (as Decandolle) *Taxonomy*, as containing the *law of Taxis* (arrangement) . . . and . . . By some Germans this has been denominated *Systematik* . . .'.

Here we return to some words written by Whewell we noted earlier:

> The basis of all Natural Systems of Classification is the Idea of Natural Affinity. The Principle which this Idea involves is this:–Natural arrangements, obtained from *different* sets of characters, must *coincide* with each other.[6] (Aphorism C in Whewell 1840a, p. xxxiv; Whewell 1840b, p. 18; Whewell 1847, p. 463)

'Different sets of characters, must *coincide* with each other': Whewell captures the basis of the taxonomic method when the focus is natural classification and captures the basis of cladistic analysis as outlined in this book.

Artificial Classification: The word 'artificial' implies a sense of inferiority, as if any particular artificial classification is significantly less worthy than *the* natural classification. Encouraging this view are common definitions of the word artificial: 'false', 'misleading' or 'unnatural'. It is usually these interpretations that have been carried into the present when used in reference to classification. Yet a more appropriate definition of artificial in the context of biological classification would be 'man-made', imposed. Most things 'man-made' have a purpose or an intended purpose – later in this book we explore a number of areas of contemporary classification in which artificial classifications pose as natural (Chapter 8).

Artificial classifications are designed with a purpose in mind; in biological classifications that purpose relates to the identification of organisms. As Candolle noted:

> For their unique purpose and their unique result, artificial systems [classifica-tions] have, as we have seen, to make it possible to learn with more or less ease, the names of the species to which the system are applied (Candolle 1813, p. 44, translated).

Artificial classifications, then, are specifically designed for the identification of known organisms by someone unfamiliar with them. Many popular biology books are designed with this in mind, and many products of the taxonomic enterprise have identification as their primary purpose such as the many field guides pro-duced to allow non-experts to identify organisms they may be unfamiliar with.

Taxonomic keys are another prime example of artificial classifications. These tools can use any characteristic that might help establish identity, in the sense of separating one organism (species) from another. In this sense, they are, of course, extremely useful.

---

[6] Aphorism C, p. 18.

Example 4: A Taxonomic key. A taxonomic key is a useful tool for identifying taxa based on detailed characteristics, some of which may only be visible under a microscope. In fact, many entomological and botanical papers that describe new species come with a taxonomic key. The basis of the key is simple: it is a system by which the presence and/or absence of characters leads the user through a series of steps. In order to understand a key it is best to build one to understand its inner workings.

Consider five objects: a spoon, a cup, a small glockenspiel, a computer mouse and a pencil. You can arrange these into a useful key using a series of characteristics: one set to group the objects and another set to identify them. The first set contains shared characteristics. For example, the glockenspiel and the spoon are made mostly out of metal. The second set contains novel or unique characteristics, such as a 'thing that has tuned keys'. The spoon and glockenspiel are both made of metal, but the glockenspiel is distinguished from the spoon because it has tuned keys. If we do this for all five objects, we end up with the following key:

1a	Things made of metal	2
1b	Things made of non-metal	3
2a	Thing has tuned keys	Glockenspiel
2b	Thing has no buttons	Spoon
3a	Thing is made of wood	Pencil
3b	Thing is not made of wood	4
4a	Thing contains plastic	Computer mouse
4b	Thing has no plastic	Cup

Note that the key is clearly artificial as it contains non-homologous characteristics and no biological entities. The same may be true for keys of living organisms, such as flies, but it does not mean that every characteristic is artificial. In fact, if the monophyly of a taxon and its homologous characteristics are known, a completely natural key could, in principle, be constructed. But is it a natural *classification*? Let us return to the natural spider classification in Chapter 1. When we add a new specimen, such as the trapdoor spider *Misgolas rapax*, the natural classification accommodates it, as *Misgolas rapax* has a pedipalp, spinneret and orthognathus fangs. In fact *all* trapdoor spiders have pedipalps, spinnerets and orthognathus fangs, including ones that have yet to be discovered, meaning that natural classifications are predictive. Taxonomic keys, such as the one above, cannot accommodate any new objects. Take the addition of a computer keyboard, for example. It is made of metal and plastic, meaning that it fits in both 1a and 4a. If we do not know what a keyboard is, we will not be able to identify it using this key. We would need someone familiar with keyboards to draw up a new taxonomic key. In this sense taxonomic keys are designed to *identify* known objects of a finite quantity using any artificial and/or natural characteristics. In this sense, taxonomic keys are not natural classifications because they cannot predict as yet unknown objects or taxa. Rather they are either artificial or natural *systems* that identify known objects.

Natural Classification: We noted above for Whewell that the 'basis of all Natural Systems of Classification is the Idea of Natural Affinity'. The scientific problem of biological classification is discovering the symmetry between the various properties of any particular organism with the entire range of organisms that possess those same properties. To discover this symmetry is to discover an organism's relationships (taxa) and hence its place in the *natural classification*. Simple examples are: every organism found with a backbone or spinal column is considered to be a vertebrate; every organism with a flower is considered to be an angiosperm: evidence (backbone or spinal column, flower) is linked directly with the relationships (vertebrate, angiosperm). We might also note here that certain predictions are possible from these observations. One might be that an organism with a spinal column and leaves will never be found – this combination of characters would be irrational in the sense used above for Borges' classification. We address the topic of relationships in more detail in Chapter 3.

This may at first seem an unusual way to view natural classification, as the common perception today might directly equate natural classification with the results of whatever process or processes have brought any particular organisms into being: in short, the process of evolution and its results as expressed graphically in phylogenetic trees.

After the publication of Charles Darwin's (1809–1882) *Origin of Species*, natural classification was assumed to reflect, at least partially, the actual genealogy of organisms. Darwin expressed it as follows:

> Naturalists try to arrange the species, genera, and families in each class, on what is called the Natural System. But what is meant by this system? Some authors look at it merely as a scheme for arranging together those living objects which are most alike, and for separating those which are most unlike; or as an artificial means for enunciating, as briefly as possible, general propositions,—that is, by one sentence to give the characters common, for instance, to all mammals, by another those common to all carnivora, by another those common to the dog-genus, and then by adding a single sentence, a full description is given of each kind of dog. The ingenuity and utility of this system are indisputable. But many naturalists think that something more is meant by the Natural System; they believe that it reveals the plan of the Creator; but unless it be specified whether order in time or space, or what else is meant by the plan of the Creator, it seems to me that nothing is thus added to our knowledge. Such expressions as that

famous one of Linnæus, and which we often meet with in a more or less concealed form, that the characters do not make the genus, but that the genus gives the characters, seem to imply that something more is included in our classification, than mere resemblance. I believe that something more is included; and that propinquity of descent,–the only known cause of the similarity of organic beings,–is the bond, hidden as it is by various degrees of modification, which is partially revealed to us by our classifications. (Darwin 1859, p. 413)

We will return to Darwin's *Origin* momentarily, but here we note that he inadvertently set in motion a research programme to discover the *genealogy* of organisms, a task vigorously embraced by Ernst Haeckel, the results of his labours initially depicted in his now famous oak-like tree diagrams.

Haeckel had many critics, early commentary coming from zoologist Louis Agassiz (1807–1873):

when Haeckel sought to found a whole system of classification on the idea of the transformation of beings through successive changes, from generation to generation, he did not attempt to prove that any of these beings descended from any other; he did not add to the knowledge that we possessed before of the affinities of animals; he simply monopolized the affinities after those who had established them; he made of them so indications of reproductive links between the beings which possesses them, and, assuming that these affinities are more or less distinct, he prepared genealogical trees which are no more than a new expression of previously acquired knowledge. (Agassiz 1869, p. 375, translation after Morris 1997)

A more succinct summary of Haeckel's efforts was provided by palaeontologist Colin Patterson (1933–1998):

Haeckel's diagram expressing these ideas ... is now of interest chiefly for its overtly tree-like form, rather than for its content. So far as fishes are concerned, Haeckel adopted a somewhat outdated pre-Darwinian taxonomy to the tree format, replacing 'affinity' by 'ancestry'. (Patterson 1977, p. 591)

Agassiz and Patterson are separated by nearly a century, but in between a number of commentators made much the same point. A direct expression of these ideas is found in Edward Stuart Russell's (1887–1954) book, *Form and Function*, for example, where, according to Russell, Haeckel's trees 'are nothing more than graphic representations of the ordinary systematic relationships of organisms, with a few hypothetical ancestral groups thrown in' (Russell 1916, p. 251). Be that as it may, these words – Agassiz (1869), Russell (1916), Patterson (1977) – address issues pertaining to phylogeny. What of taxonomy?

Here we can return to Darwin's and Polly Winsor's historical discoveries: 'The importance of the achievements of taxonomy for the discovery and proof of evolution is hard to exaggerate ...' (Winsor 2013, p. 72). Winsor then cites another passage from the *Origin*,

It is a truly wonderful fact—the wonder of which we are apt to overlook from familiarity—that all animals and all plants throughout all time and space should be related to each other in groups, subordinate to groups, in the manner which we everywhere behold—namely, varieties of the same species most closely related, species of the same genus less closely and unequally related, forming sections and sub-genera, species of distinct genera much less closely related, and genera related in different degrees, forming sub-families, families, orders, sub-classes, and classes. (Darwin 1859, p. 128)

That 'truly wonderful fact' was based on the Linnean hierarchy, painstakingly put together by scores of naturalists, each working away at some part of that vast taxonomic hierarchy. As Winsor has pointed out, Darwin began to make tree-like (or coral-like) sketches in his notebooks from around 1837, when he first pondered evolution as an explanation for the Linnean hierarchy. It was sometime later, between 1855 and 1857, that he considered the principle of divergence. Winsor goes on: '... with respect to the role of taxonomy, the principle of divergence is of minor importance, it is the relationship between the Linnaean hierarchy and branching evolution that is primary' (Winsor 2009, p. 44). Just to emphasise this in modern-day language: the Linnean hierarchy is discoverable, it is an empirical generalisation, the principle of divergence is simply a model.

Much has been written about the hierarchy of life recently, some of it critical. *The New Scientist,* for example, in 2009 had a cover story 'Why Darwin was Wrong about the Tree of Life'. The writer, Graham Lawton, noted that:

The tree-of-life concept was absolutely central to Darwin's thinking, equal in importance to natural selection, according to biologist W. Ford Doolittle of Dalhousie University in Halifax, Nova Scotia, Canada. Without it the theory of evolution would never have happened. The tree also helped carry the day for evolution. Darwin argued successfully that the tree of life was a fact of nature, plain for all to see though in need of explanation. (Lawton 2009)

The gist of the article was that new evidence, particularly from microbial systematics, was showing anything but a tree-like structure, with most schemes, if represented accurately, being more or less reticulated networks. Popular commentary claimed that 'The tree of life is being politely buried', and '...our whole fundamental view of biology needs to change'. But what is (or might be) vanishing is the 'principle of divergence' model as a general concept and not the taxonomic hierarchy, the classification: 'Without it the theory of evolution would never have happened'.

Groups and Taxa: Taxonomic groups, such as genera, families and orders, are part and parcel of any classification, yet they tell us little about taxonomic relationship. In a system in which there are a finite number of known objects (e.g., individual organisms), we speak of groups, namely 'a number of things placed together as the result of deliberate arrangement or composition' (OED Online 2018). In this sense, a group is part of a system, whether it is natural or artificial. Taxa, however, are the relationships between an unknown number of *all* individual organisms that have ever lived or will ever live. In this sense, a taxon is part of a natural and artificial classification. The difference between a system and a classification is one of usage: a system is there to provide a way to identify known organisms within a taxonomy, whereas a classification is there to determine the relationships between organisms:

	System		Classification	
	Artificial	**Natural**	**Artificial**	**Natural**
Taxon	–	–	Aphyletic[7]	Monophyletic
Group	Group	Group	–	–
Predictive	No	No	No	Yes
ID characteristics	Yes	Yes	No	Yes

So the problem of taxonomy is: how does one discover taxa?

One might see the problem of taxonomy as relatively simple: how to identify and eliminate the artificial taxa, how to find the 'Barbarians'. This will be the subject of Chapter 13.

Summary

Natural classification provides the evidence for evolution, usually understood as providing the basis for concluding the shared or common history of all organisms. All birds, for example, can be assumed to have evolved from some common precursor. But a better way, a more empirical way, of putting it is: all members of natural taxon are more closely related to each other than they are to any other taxon. It is this inclusiveness that is evidence for evolution, rather than any appeal to largely (invisible) ancestors.

Taxonomy is the science of classification; artificial classifications are found through imposition – they are *created*, they are abstractions primarily designed to identify known organisms and are usually based on subjectively chosen characters; natural classifications are *discovered* – they are concerned with the discovery of real facts about the world, based on observable characters.

---

[7] See Chapter 6 for a discussion of these terms.

References

Agassiz, L. 1869. De l'espèce et de la classification en zoologie. In: *Le Darwinisme – Classification de Haeckel*, part 3. Ballière, Paris, pp. 375–391.

Annas, J. & Waterfield, R. 2010. *Plato: The Statesman*. Cambridge Texts in the History of Political Thought, Cambridge University Press, Cambridge, UK.

Bentlage, B. & Lewis, C. 2012. An illustrated key and synopsis of the families and genera of carybdeid box jellyfishes (Cnidaria: Cubozoa: Carybdeida), with emphasis on the "Irukandji family" (Carukiidae). *Journal of Natural History* 46: 2595–2620.

Candolle, A., de. 1813. *Théorie élémentaire de la botanique: ou, Exposition des principes de la classification naturelle et de l'art de décrire et d'étudier les végétaux.* Déterville, Paris.

Candolle, AP. de & Sprengel, K. 1820. *Grundzüge der wissenschaftlichen Pflanzenkunde.* C. Cnobloch, Leipzig.

Candolle, AP. de & Sprengel, K. 1821. *Elements of the Philosophy of Plants: Containing the Principles of Scientific Botany.* W. Blackwood, Edinburgh.

Christenhusz, MJM., Baker, W., Chase, MW., Fay, MF., Lehtonen, S., Van Ee, B., Von Konrat, M., Lumbsch, T., Renzaglia, KS., Shaw, J., Williams, DM. & Zhang, Z-Q. 2011. The first anniversary of *Phytotaxa* in the International Year of Biodiversity. *Phytotaxa* 15: 1–8.

Croizat, L. 1945. History and nomenclature of the higher units of classification. *Bulletin of the Torrey Botanical Club* 72: 52–75.

Darwin, C. 1859. *On the Origin of Species by Means of Natural Selection, or, the Preservation of Favoured Races in the Struggle for Life.* John Murray, London.

DeLillo, D. 1985. *White Noise.* Picador, New York.

Flanagan, R. 2013. *The Narrow Road to the Deep North.* Random House.

Foucault, M. 1966. *Les Mots et les Chose,* Editions Flammarion (Reprint edition, 1998) [Translated into English as *The Order of Things*, 1970, Routledge, Reprint, 2001].

Gershwin, L. 2005. Two new species of jellyfishes (Cnidaria: Cubozoa: Carybdeida) from tropical Western Australia, presumed to cause Irukandji Syndrome. *Zootaxa* 1084: 1–30.

Gershwin, L. 2007. *Malo kingi*: a new species of Irukandji jellyfish (Cnidaria: Cubozoa: Carybdeida), possibly lethal to humans, from Queensland, Australia. *Zootaxa* 1659: 55–68.

Haeckel, E. 1866. *Generelle Morphologie der Organismen: Allgemeine Grundzüge der organischen Formen-Wissenschaft, mechanisch begründet durch die von C. Darwin reformirte Decendenz-Theorie.* G. Reimer, Berlin.

Huxley, TH. 1902. Biology, I. Morphology (Part 8): Artificial and natural classification in taxonomy. In: *Encyclopaedia Britannica* (10th ed.). Adam & Charles Black and The Times, London.

Lamarck, JBPA. de M. de & Candolle, AP. de.1805. *Flore francqise, ou descriptions succinctes de toutes les plantes qui croissent naturellement en France, disposées selon une nouvelle méthode d'analyse, et précédées par un exposé des principes élémentaires de la botanique.* 3rd ed. Desray, Paris.

Lawton, G. 2009. Why Darwin was wrong about the tree of life. *The New Scientist,* 21 January.

Linnaeus, C. 1753. *Species Plantarum . . .* 2 volumes. Laurentius Salvius, Stockholm.

Linnaeus, C. 1758-1759. *Systema Naturae . . .* 10th ed., vol 1 [1758], vol 2 [1759]. Laurentius Salvius, Stockholm.

Morris, PJ. 1997. Louis Agassiz's additions to the French translation of his *Essay on Classification. Journal of the History of Biology* 30: 121-134.

Moss, S. 2017. *The Robin: A Biography.* Square Peg, London.

Nelson, G. & Platnick. NI. 1981. *Systematics and Biogeography: Cladistics and Vicariance.* Columbia University Press, New York.

OED Online. group, n. March 2018. Oxford University Press. /www.oed.com/view/Entry/81855?rskey=8Ln3Xd&result=1& isAdvanced=false (accessed May 28 2018).

Olivares, J. 2011. *A Taxonomy of Office Chairs.* Phaidon Press Ltd, New York.

Patterson, C. 1977. The contribution of paleontology to teleostean phylogeny. In: Hecht, MK., Goody, PC. & Hecht, BM. (eds), *Major Patterns in Vertebrate Evolution.* Plenum, New York, pp. 579-643.

Platnick, NI. 1989. Cladistics and phylogenetic analysis today. In: Fernholm, B., Bremer, K. & Jörnvall, H. (eds), *The Hierarchy of Life.* Elsevier, Amsterdam, pp. 17-24.

Plato. 1848. *The Statesman.* Loeb Classical Library No. 164. Harvard University Press, Cambridge, MA.

Russell, ES. 1916. *Form and Function: A Contribution to the History of Animal Morphology.* University of Chicago Press, Chicago [Reprint, 1982].

Skemp, JB. 1952. *Plato's Statesman.* Yale University Press, New Haven, CT.

Whewell, W. 1840a. *The Philosophy of the Inductive Sciences, Founded upon Their History.* J.W. Parker, London.

Whewell, W. 1840b. *Aphorisms Concerning Ideas, Science & the Language of Science.* Harrison and Co, London.

Whewell, W. 1847. *The Philosophy of the Inductive Sciences, Founded upon Their History,* 2nd ed. J.W. Parker, London.

Whitehead, AN. 1929 [1978]. *Process and Reality. An Essay in Cosmology.* Corrected Edition, edited by David Ray Griffin & Donald W. Sherburne. The Free Press, New York.

Williams, DM., Ebach, MC. & Wheeler, QD. 2010. Beyond belief. In: Williams, D.M. & Knapp, S. (eds), *Beyond Cladistics.* University of California Press, Berkeley, pp. 169-197.

Winsor, MP. 2009. Taxonomy was the foundation of Darwin's evolution. *Taxon* 58:43-49.

Winsor, MP. 2013. Darwin and taxonomy. In: Ruse, M. (ed.), *The Cambridge Encyclopedia of Darwin and Evolutionary Thought.* Cambridge University Press, Cambridge, pp. 72-79.

Woerner, M. 2010. *Vampire Taxonomy: Identifying and Interacting with the Modern-Day Bloodsucker.* Penguin, New York.

Zhang, Z-Q. 2011. Describing unexplored biodiversity: Zootaxa in the International Year of Biodiversity. *Zootaxa* 2768: 1-4.

Further Reading

Agassiz, L. 1859. *Essay on Classification.* Longman, Brown, Green, Longmans & Roberts, and Trübner & Co, London.

In spite of its age, Agassiz's *Essay* is still worth reading today. It first appeared in 1857 as a chapter in the first volume of *Contributions to the Natural History of the*

United States. Its publication in book form, two years later, could not have been more unfortunately timed, coinciding with the publication of Darwin's *Origin.*

The *Essay* has been reprinted on several occasions, notably in 1962 by The Belknap Press of Harvard University Press, with an introduction by Edward Lurie, Agassiz's first biographer (1960, *Louis Agassiz: A Life in Science,* University of Chicago Press, reprinted in 1988; this biography is still worth reading). The 1962 reprint was republished in 2004 by Dover Press. Both the 1859 and 1962 editions are available via the Biodiversity Heritage Library (BHL), as is the earlier version published in *Contributions to the Natural History of the United States* (volume 1, 1857).

Of further interest is a French language edition published near the end of Agassiz's life (Agassiz 1869. *De l'espèce et de la classification en zoologie,* translated by Felix Vogeli, Bailière, Paris; also available at BHL). This edition includes an additional chapter that never appeared in the English editions where Agassiz discusses his objections to Darwin's views on evolution and where he re-casts some of Haeckel's phylogenetic diagrams (for further discussion see Morris, PJ., 1997. Louis Agassiz's additions to the French translation of his Essay on Classification. *Journal of the History of Biology* 30: 121–134, which includes some passages translated into English; Williams & Ebach (2007), cited below, comment on the relevance of some of these passages).

Two good books dealing with aspects of Agassiz's intellectual development, both written by Winsor, are: 1976. *Starfish, Jellyfish, and the Order of Life: Issues in Nineteenth Century Science.* Yale University Press, New Haven CT; 1991. *Reading the Shape of Nature: Comparative Zoology at the Agassiz Museum.* University of Chicago Press, Chicago; also Irmscher, C. 2013.

Louis Agassiz: Creator of American Science. Houghton Mifflin Harcourt, Boston, New York.

Baehni, C. 1957. Les grands systèmes botaniques depuis Linné: a propos du 250e anniversaire de la naissance de Linné et de Buffon. *Gesnerus: Swiss Journal of the History of Medicine and Sciences* 14 (heft 3–4): 83–93.

Candolle, A-P. de. 1844. *Théorie élémentaire de la botanique…* 3rd ed. Roret, Paris.

Candolle's *Théorie …* is still one of the best introductions to the distinction between artificial and natural classifications, being 'the first work in which the soul of the natural and artificial method had been laid bare' (Croizat 1945, p. 64). An English translation was published in 1821 (Candolle and Sprengel 1821), derived from an earlier German translation (Candolle and Sprengel 1820) – Candolle was unhappy with both. William Whewell (Whewell, W. 1840. *The Philosophy of the Inductive Sciences.* J.W. Parker, London; Whewell, W. 1847. *The Philosophy of the Inductive Sciences, Founded upon Their History.* J.W. Parker, London) summarised some of Candolle's ideas and a brief account is given in Williams et al. (2010) and in our Chapter 7. No accurate (or acceptable) English translation exists of *Théorie élémentaire de la botanique*; one is much needed.

Funk, H. 2014. Describing plants in a new mode: the introduction of dichotomies into sixteenth-century botanical literature. *Archives of Natural History* 41: 100–112.

Griffing, LR. 2011. Who invented the dichotomous key? Richard Waller's watercolors of the herbs of Britain. *American Journal of Botany* 98 (12): 1911–1923.

Guyénot, E. 1941. *Les sciences de la vie aux XVIIe et XVIIIe siècles. L'idée d'évolution.* Paris.

Lecointre, G. & Le Guyader, H. 2007. *The Tree of Life: A Phylogenetic Classification.* Belknap Press, Cambridge, MA and London.

Lecointre & Le Guyader produced a comprehensive account of the diversity and classification of all organic life, including short summaries of each taxon recognised. The version cited here is derived and updated from the French edition of *Classification phylogénétique du vivant*, first published in 2001 (the 2nd edition was published in 2002, the 3rd in 2006). The English language translation is of the 3rd French edition.

Lefèvre W. 2001. Natural or artificial systems? In: Lefèvre W. (ed.), Between Leibniz, Newton, and Kant. *Boston Studies in the Philosophy and History of Science* 220: 191–209.

Ogilvie, BW. 2006. *The Science of Describing: Natural History in Renaissance Europe.* University of Chicago Press, Chicago.

Ong, WJ. 1958 [2004]. *Ramus, Method, and the Decay of Dialogue: From the Art of Discourse to the Art of Reason.* Harvard University Press, Cambridge, MA.

Pavord, A. 2005. *The Naming of Names: The Search for Order in the World of Plants.* Bloomsbury, New York.

Perrier, JOE. 1884. *La philosophie zoologique avant Darwin.* Félix Alcan, Paris.

This is an excellent account of developments in evolutionary ideas in France written by a scientist who worked at the *Muséum national d'Histoire naturelle*. Although a little dated, it is still a useful book to consult providing insight into the complexities of the comparative biology of a previous time. It is available in an English translation: *The Philosophy of Zoology Before Darwin: A translated and annotated version of the original French text by Edmond Perrier* (2009, translated by Alex McBirney, with annotations by Stanton Cook and Gregory Retallack, Springer).

Scharf, ST. 2009. Identification keys, the "Natural Method"" and the development of plant identification manuals. *Journal of the History of Biology* 42: 73–117.

This is an excellent account of developments in the understanding and construction of artificial and natural methods in classification. Scharf writes: 'British botanists, however, continued to use Linnaeus's sexual system almost exclusively for another two decades. Their reluctance to use other methods or systems of classification can be attributed to a culture suspicious of innovation, anti-French sentiment and the association of all things Linnaean with English national pride, fostered in particular by the President of the Linnean Society of London, Sir James Edward Smith. The British aversion to using multiple plant identification technologies in one text also helps explain why it took so long for English botanists to adopt the natural method, even after several Englishmen had tried to introduce it to their country'.

Shatalkin, AI. 2010. *Taxonomy. Foundations, Principles and Rules.* KMK Scientific Press, Moscow[In Russian].

Stagg, BC. & Verde, MF. 2019. A comparison of descriptive writing and drawing of plants for the development of adult novices' botanical knowledge. *Journal of Biological Education* 53: 63–78.

We include this as it directs attention to the ways in which some students learn about animals and plants, a factor that is of significance when developing useful artificial classifications specifically for taxon identification.

Stevens, PF. 1994. *The Development of Biological Systematics: Antoine-Laurent de Jussieu, Nature, and the Natural System.* Columbia University Press, New York.

This is a comprehensive account of developments in biological systematics up to 1859 with reference to, and a comprehensive account of, the Natural System. It is still a remarkably useful and accessible account, primarily focusing on developments in botanical classification.

Voss, EG. 1952. The history of keys and phylogenetic trees in systematic biology. *Journal of the Scientific Laboratories of Denison University* 43: 1–25.

Williams, DM. & Ebach, MC. 2008. *Foundations of Systematics and Biogeography*. Springer-Verlag New York Inc., New York.

PART II

Systematics: Exposing Myths

Taxonomy must be scientific. It must require for its devotees a training as rigid as that required by professional workers in morphology, physiology or ecology. Species-making by taxonomic tyros must be abandoned ... The recent ebullitions of the taxonomic radicals have evoked in botanists in general successively dissatisfaction, contempt and rage. These things will not be endured much longer; a little more and the sinning taxonomists will be 'cast out into the outer darkness where there shall be wailing and gnashing of teeth'.

(Cowles 1908, pp. 270–271)

Prior to the mid-1930s many aspects of taxonomy, its philosophy as well as its methods, were considered to be outmoded, redundant, for the most part archaic and due for retirement. In the eyes of general biologists, the subject was more than a little suspect: a discipline, if it was even considered such, that occupied a great deal of the time of myopic old men, hidden away in dusty rooms, usually in some old museum[1], counting the hairs on the legs of fleas, measuring the scales on a reptile's body or trying to figure out which way up some smudge on a rock might be[2].

---

[1] Ecologists, such as Henry Chandler Cowles (1896–1939), quoted above in the epigraph, were particularly affronted by the species descriptions and designations made by taxonomists who were considered stuck in herbaria and museums: 'A theoretical plant species may be produced in the laboratory, but the real species that make up the vegetation of the world are developed and must be studied out of doors' (Cowles 1908, p. 266). For early ecologists, taxonomy was, in part, seen as deprived of any method (experimentation) or skill, and species descriptions were seen as purely speculative. Indeed, 'One of the noblest aims of ecology is the destruction of many of the "species" of our manuals' (Cowles 1908, p. 266), by perhaps abandoning the Linnean system or allowing for trinomials.

[2] The original reference is lost to us but many have referred to fossil remains as a 'smudge on a rock'.

An influential attempt at rehabilitation began with Simpson's *The principles of classification and a classification of mammals* (Simpson 1945). In it he wrote of that same perception:

> A generation or so ago taxonomy had fallen into disrepute. It was believed to be a blind alley, a pursuit for hack workers only, or a phase happily outgrown. 'Taxonomist' was a term of reproach, and the bright young men and the older students who had, or thought they had, a truly broad, scientific, and modern viewpoint carefully avoided the subject.

He went on:

> ... 'taxonomy' seemed to them to mean little more than sorting dead specimens according to a few rather superficial morphological characters, writing out labels, and then stowing the specimens away in drawers.

And later:

> A revival of taxonomy is now occurring as these facts are being recognized - or, one should say, as they are being more widely and again recognized, for there have always been some taxonomists with a broader viewpoint, and few of the early naturalists had the narrow concept of taxonomy that finally brought it into disrepute. For instance, one of the most fundamental goals of modern biological research is comprehension of the way in which species arise, and we now find classifiers, experimental biologists, geneticists, ecologists, physiologists, statisticians - even sociologists, psychoanalysts, and many other apparently most nontaxonomic students - working together on this subject with full recognition of the fact that the common element in their problem is fundamentally taxonomic. (Simpson 1961, p. 1)

At the end of this passage, Simpson refers, in a footnote, to Julian Huxley's *The New Systematics* (Huxley 1942), a collection of papers that primarily focused on species and how to study their origin ('... one of the most fundamental goals of modern biological research is comprehension of the way in which species arise'), inspired, so it was claimed, by Darwin's *On the Origin of Species* (Darwin 1859) and, perhaps more directly, by Ernst Mayr's first book *Systematics and the Origin of Species, from the Viewpoint of a Zoologist* (Mayr 1942, reprinted in 1999).

Simpson was in the process of outlining the role of palaeontology, and hence, as he understood it, phylogeny, in the methods of taxonomy and went on to write a book on the subject, *Principles of Animal Taxonomy* (Simpson 1961).

As Nelson and Platnick noted of Simpson's 1961 book, it

> ... stemmed from his long experience and well-known expertise in paleontology, but perhaps more than anything else merely summarized the practice of the last 100 years. In doing so, his book marked and perhaps hastened the end of evolutionary theory as traditionally applied to classification. (Nelson & Platnick 1981, p. 131)

From the palaeontological aspect, a fresh look at systematics (and phylogeny) occurred, initially, but not entirely, inspired by the various English translations of the work of Willi Hennig, especially *Phylogenetic Systematics* (Hennig 1966). Of Hennig, Nelson and Platnick wrote:

> Hennig's critique consequently began to look less like a revolutionary manifesto, and more like an incisive exposition of the best of past taxonomic practice and theory. Indeed, it now seems an attempt, generally successful, to tighten the entire theoretical structure of evolutionary taxonomy. And the question is moot whether Simpson's or Hennig's is really the more traditional exposition of it. (Nelson & Platnick 1981, p. 131)

Note that Nelson and Platnick refer to 'evolutionary taxonomy', a subset, if you will, of biological taxonomy. It is the latter we are interested in (not the subset) – and it is the latter we refer to as *cladistics*.

Many further critiques of taxonomy arose in the context of the *Modern Synthesis* (a partisan account of the history of that 'movement' can be found in Mayr & Provine 1980), of which *The New Systematics* was an offshoot.

Much of the *Modern Synthesis* represented a neontological critique and might be summed up by one of Ernst Mayr's early essays 'Darwin and the evolutionary theory in biology' (included in *Evolution and Anthropology: A Centennial Appraisal*, Mayr 1959; it has since been reprinted in Mayr 1976 and in Sober 1994, with the new title 'Typological Thinking versus Population Thinking'[3]), during the Darwinian centennial year of 1959[4]. Many of Mayr's pronouncements on taxonomy, such as the ills of typology and the related issue of essentialism in taxonomy, have since been found wanting, much of it exposed via the work of historian Polly Winsor and, more recently, that of Joeri Witteveen (see also Provine, 2005, on some of Mayr's fanciful input into genetics). We discuss these matters more fully in Chapter 4.

Traditionally phylogenies are associated with, or represented by, tree-like diagrams, possibly because the first phylogenies were represented as such (Haeckel 1866) – indeed, the 100 years referred to by Nelson and Platnick above, from Haeckel to Hennig (1866–1966), eventually evolved into what has since been called 'tree-thinking'. The idea behind 'tree-thinking', where Robert O'Hara riffs on Mayr's 'population thinking', has encouraged simple classifications and simple

---

[3] On the 'Typological versus Population Thinking' essay (Mayr 1976): 'When one of Mayr's early historical treatments of the typology/population distinction (Mayr, 1959[b]) was reprinted in a collection of some of his key writings (Mayr, 1976[a]), he reminisced that it had constituted "the first full presentation" and "full articulation" of what this dichotomy consists in (Mayr, 1976[b], p. 26)' (Witteveen 2015).

[4] A celebratory year best captured in the three volumes of *Evolution after Darwin* (Tax 1960a, b and Tax & Callender 1960, all now available online, and another partisan review in Smocovitis 1999).

cladograms to be interpreted as actual phylogenies. Taking a further step away, Morrison has suggested that phylogenies are best treated as networks (Morrison 2016). We will deal with these issues later in Chapter 3 and more fully, with reference to taxonomic groups, in Chapter 6.

Thus, in the context of taxonomic myths, we deal with *Relationship Diagrams* (Chapter 3), *Essentialism and Typology* (Chapter 4), *Monothetic and Polythetic Taxa* (Chapter 5) and *Non-taxa or the Absence of -Phyly: Paraphyly and Aphyly* (Chapter 6). Of course, Simpson's cartoon description that ''taxonomy' seemed . . . to mean little more than sorting dead specimens according to a few rather superficial morphological characters' is itself a myth – not only is it a betrayal of taxonomy's rich past, it is, and has been, a damaging pronouncement that often gets repeated in more modern times.

Another Darwinian centennial celebration occurred in the 1980s, its vast output summarised by Wassersug and Rose (1984). While at that time many taxonomists were concerning themselves with further tightening the theoretical structure of 'evolutionary taxonomy' (mostly based upon these myths), seeing the Emperor (Mayr?) if not fully clothed then nearly so (de Queiroz 1988), while for others his nakedness was all too obvious (Nelson 1971)[5].

References

Cowles, HC. 1908. An ecological aspect of the conception of species. *The American Naturalist* 47: 265–271.

Darwin, C. 1859. *On the Origin of Species by Means of Natural Selection, or, the Preservation of Favoured Races in the Struggle for Life.* John Murray, London.

de Queiroz, K. 1988. Systematics and the revolution. *Philosophy of Science* 55: 238–259.

Haeckel, E. 1866. *Generelle Morphologie der Organismen: Allgemeine Grundzüge der organischen Formen-Wissenschaft, mechanisch begründet durch die von C. Darwin reformirte Decendenz-Theorie.* 2 volumes. G. Reimer, Berlin.

Hennig, W. 1966. *Phylogenetic Systematics.* University of Illinois Press, Urbana [Reprinted 1979, 1999].

Huxley, J. 1942. *Evolution. The Modern Synthesis.* George Allen & Co, London.

Mayr, E. 1942. *Systematics and the Origin of Species.* Columbia University Press, New York.

Mayr, E. 1959. Darwin and the evolutionary theory in biology. In: *Evolution and Anthropology: a Centennial Appraisal.* The Anthropological Society of Washington, Washington, DC, pp. 1–10 [reprinted in Mayr, E. 1976. *Evolution and the Diversity of Life: Selected Essays.*

---

[5] In 2009 there were further Darwin centennials: the 150th anniversary of the publication of *On the Origin of Species* and the 200th anniversary of Darwin's birth. The number of publications celebrating these two anniversaries was awe-inspiring. For the interested a compilation can be found here: http://darwin-online.org.uk/2009.html. Over 200 items are listed.

Harvard University Press, Cambridge, MA, pp. 26–29].

Mayr, E. 1976. Typological thinking versus population thinking. In: *Evolution and the Diversity of Life*. Harvard University Press, Cambridge, MA, pp. 26–29.

Mayr, E. & Provine, W. (eds) 1980. *The Evolutionary Synthesis*. Harvard University Press, Cambridge, MA.

Morrison, DA. 2016. Genealogies: pedigrees and phylogenies are reticulating networks not just divergent trees. *Evolutionary Biology* 43: 456–473.

Nelson, GJ. 1971. "Cladism" as a philosophy of classification. *Systematic Zoology* 20: 373–376.

Nelson, GJ. & Platnick, NI. 1981. *Systematics and Biogeography: Cladistics and Vicariance*. Columbia University Press, New York.

Provine, WB. 2005. Ernst Mayr, a retrospective. *Trends in Ecology and Evolution* 20: 411–413.

Simpson, GG. 1945. The principles of classification and a classification of mammals. *Bulletin of the American Museum of Natural History* 85: 1–350.

Simpson, GG. 1961. *Principles of Animal Taxonomy*. Columbia University Press, New York.

Smocovitis, VB. 1999. The 1959 Darwin Centennial Celebration in America. *Osiris* 14: 274–323.

Sober, E. (ed.). 1994. *Conceptual Issues in Evolutionary Biology*. The MIT Press. Bradford Books, Cambridge, MA, London, pp. 157–160.

Tax, S. 1960a. *Evolution after Darwin*. Vol. 1. *The Evolution of Life*. University of Chicago Press, Chicago.

Tax, S. 1960b. *Evolution after Darwin*. Vol. 2. *The Evolution of Man*. University of Chicago Press, Chicago.

Tax, S. & Callender, C. (eds.). 1960. *Evolution after Darwin*. Vol. 3. *Issues in Evolution*. University of Chicago Press, Chicago.

Wassersug, RJ. & Rose, MJ. 1984. A reader's guide and retrospective to the 1982 Darwin centennial. *The Quarterly Review of Biology* 59: 417–437.

Witteveen, J. 2015. "A temporary oversimplification": Mayr, Simpson, Dobzhansky, and the origins of the typology/population dichotomy (part 1 of 2). *Studies in History and Philosophy of Biological and Biomedical Sciences* 54: 20–33.

3

Relationship Diagrams

Nowadays some students receive the impression that little of value was understood about systematics before the revolution begun by Willi Hennig. Such views seriously distort history ... With respect to the actual living things around us, however, the shape of nature is not so very different from what it was 150 years ago. What has been lost is our familiarity with it.

(Winsor 2009, pp. 2, 43)

In the preceding chapters we discussed the classification and relationships of a few animals and plants. We discussed these in relation to what can be referred to as *derivative cladograms* sensu Nelson: 'a graphic representation of a hierarchical classification' (Nelson 1979, p. 5; see Chapter 7). In some we provided a written classification. For the box jellyfish *Malo kingi* (see Chapter 2), for example, an indented written classification, with ranks, would look something like this:

Kingdom (regnum): Animalia
      Phylum: Cnidaria
            Class (classis): Cubozoa
                  Order (ordo): Carybdeida
                        Family (familia): Carukiidae
                              Genus: *Malo*
                                    Species: *Malo kingi*

The indented list is an obvious hierarchy and corresponds to the various taxa to which *Malo kingi* belongs. The series of ranked names simply provides an indication of the relative hierarchical level of each taxon. For example, the family (Familia) name Carukiidae is recognisable to zoologists as a family as it ends in –idae (it would end in –aceae for botanical groups).

The names are referred to as *ranks* such that there is a rank Kingdom, Phylum, Class, Order, etc. There are further ranks at other levels (e.g., Domain, Phylum, Subspecies, etc.) not included in our example. All are merely conventions for indicating a particular level in relation to other levels. If one sees a family name then that will include genera and the genera will include species. There is no convention as to how large these taxa may be, such that a genus can include from 1 to 1000s of species.

Somewhat surprisingly, there are still misunderstandings concerning the *meaning* of ranks, as if a particular rank, the genus, say, has some specific definition common to all genera. They, of course, do not. There is no equivalence between a genus of diatoms and a genus of trilobites, save both are monophyletic, or at least presumed to be so. We will not pursue this issue further (but see Laurin 2010). Instead, we refer the reader to the recent debate between Lambertz and Perry (2015, 2016) and Giribet et al. (2016). The latter authors, in words we cannot improve upon, summed up the situation thus:

> The debate over whether Chordata is a phylum or a superphylum is ultimately futile because it does not alter anything about our knowledge or interpretation of the phylogeny and evolution of chordates. *No single character or combination of characters makes a taxon a phylum or a class*; nor is there a prize for erecting a phylum—although it is true that novel body plans receive especial attention. (Giribet et al. 2016, p. 427, our italics)

Regardless of how any classification has been arrived at, the underlying assumption is that each taxon is associated with a particular set of properties of the organisms concerned (their parts, the characters or *homologues*, see Chapter 7). Some of those parts are visible to the naked eye (some parts of the phenotype, e.g., bones, petals, etc.), some are not (some parts of the phenotype, e.g., cell structure; and all parts of the genotype, e.g., DNA, amino acids, etc.) – but these properties (parts) are accessible and can be studied. Information on the parts themselves is not necessarily embedded in the classification, which strictly speaking represents just the relationships – a classification is a hierarchy, a hierarchy is a set of relationships.

Platnick succinctly captured some advantages of ranked classifications:

> Using the Linnaean hierarchy, when I identified the spider in John's garden as a salticid, *I was asserting that John's spider is more closely related to any single species currently included within the Salticidae than it is to any single species that is currently excluded from that family*. In other words, if my identification, and the current classification, are both correct, then John's spider is more closely related to salticid species #1 than it is to any of the 32,752 spider species currently excluded from the Salticidae. It is also more closely related to salticid species #2 than it is to any non-salticid spider. (Platnick 2009, quoted in Giribet et al. 2016, p. 429, our italics)

If the taxonomic ranks are removed, the series becomes a simple list of indented names – but the name-endings still indicate the rank so the names remain useful:

Animalia
    Cnidaria
        Cubozoa
            Carybdeida
                Carukiidae
                    Malo
                        Malo kingi

Most biologists operate with just the names, the rank being implicitly understood.

As noted above, each taxon name summarises its 'content', so to speak. The genus *Malo* has four known species: *M. bella*, *M. filipina*, *M. maxima* and *M. kingi*[1]. The family name summarises the included genera. For Carukiidae there is *Malo* (with four species), *Carukia* (with two species), *Gerongia* (with one species) and *Morbakka* (with two species). The order Carybdeida summarises the included families. For Carybdeida there are five. Thus, each taxon name is shorthand for those species contained within. For the non-specialist (as we both are with respect to jellyfishes), the classification informs about the relationships directly.

Any series of taxon names can be converted into a diagram. Almost every biological classification ever produced is to some extent hierarchical, and so the diagram is often given as a branching tree-like structure and so is a *derivative cladogram*. A summary diagram (*a derivative cladogram*) for the family-level relationships within Cubozoa was provided by Bentlage and Lewis (2012, p. 2597, their figure 1, reproduced here as our Figure 3.1).

To be sure, classifications and cladograms do look like trees, at least in the sense they branch – but this is a direct consequence of their hierarchical nature and a direct consequence of how each character's distribution might be determined. The legend for Bentlage and Lewis's figure 1 (2012, p. 2597) reads, in part, 'The families of the Carybdeida can be identified using the following characters. Alatinidae (C): rhopaliar niche ostium T-shaped. Carukiidae (D): rhopaliar niche ostium with rhopaliar horns, stomach lacking gastric phacellae ...' (Bentlage & Lewis 2012, p. 2597). This illustrates the concept of *heterobathmy*, where characters fit at certain levels in the classification (for further commentary see Chapter 7).

From a mathematical point of view, some have considered cladograms (and classifications) as actual examples of trees, in the sense defined by the strictures of graph theory, rather than a scheme of relationships, and have used the tree-image as a model of evolutionary change (some examples are given in Chapter 7).

There are a number of accounts that summarise the vast range of diagrams that have been used to represent organism relationships but, at least as far as we are concerned, their generality is most often misinterpreted, or not recognised at all.

---

[1] As we noted in Chapter 1, we follow Bentlage and Lewis's account of these animals.

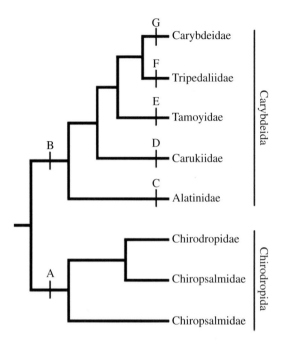

Figure 3.1 Summary diagram (*derivative cladogram*) for family-level relationships within Cubozoa (after Bentlage & Lewis 2012, p. 2597, their figure 1, with permission).

A Short Historical Digression

There seems little point offering yet another review of the various kinds of 'ladder-map-tree' images that have been used in systematic biology to depict taxon relationships. These have been documented in a number of books and many papers (see Box 3.1 and the notes in our reference list).

Box 3.1   Diagrams of Relationships: 'La Scalla, la mappa, l'albero'

The use of diagrams to depict biological diversity (biological relationships) goes back to at least the twelfth century, if not earlier (Ragan 2009). The diagrams are of different kinds, not all are specifically tree-like, branching. Giulio Barsanti, for example, wrote of three kinds of image, which, according to him, succeeded one another chronologically: the ladder, the map and the tree ('La Scalla, la mappa, l'albero', Barsanti 1988). The latter he understood as being 'more complex ... [and] conceived by Pallas (1766),[1,] becoming 'the "genea-logical tree" only with great difficulty in 1801[2] [1809]' with Lamarck[3] (Barsanti 1988, pp. 56–57; see also Barsanti 1992; Bigoni & Barsanti 2011; and Hellström 2012). Pallas suggested a tree-like representation for the relationships of organisms: 'On the other hand, the whole system of organic bodies may be well represented by the likeness of a tree that immediately from the root divides both the simplest plants and animals [...]' (Pallas 1766,

Box 3.1 *(cont.)*

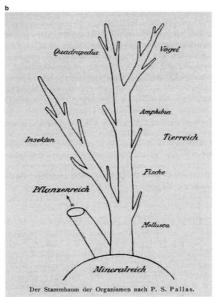

Figure B3.1.1 (a) The extravagant tree-like image drawn by Carl Edward von Eichwald (1795–1876) after Pallas (Eichwald 1829, p. 41); (b) Thienemann's simpler version (Thienemann 1910, p. 251).

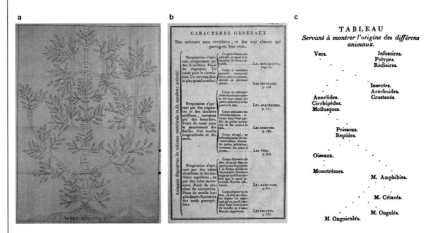

Figure B3.1.2 (a) The first published *tree* of taxon relationships from Augustin Augier de Favas (1758–1825) from his 'Arbre botanique' (Augier 1801); (b) published in the same year as Lamarck's *Système des animaux sans vertèbres* (Lamarck 1801), his first account of the invertebrates, bringing their included number of classes to seven, the chart reproduced here – it is more like a key, an artificial classification; (c) Lamarck's oft-reproduced diagram from *Philosophie zoologique* (Lamarck 1809).

Box 3.1 *(cont.)*

p. 23, translation from Ragan 2009, p. 5). Pallas published no diagrams of his own but, according to Ragan, Carl Edward von Eichwald (1795–1876) rendered his idea graphically with an extravagant tree-like image (Eichwald 1829, p. 41, our Figure B3.1.1a, reprinted in Gould 1997, p. 33, Ragan 2009, figure 4, and Mindell 2013, p. 480, and further discussion in Archibald 2014), or, in Ragan's apt turn of phrase, 'a bunch of asparagus-shoots' (Ragan 2009, p. 5). Later Thienemann represented Pallas's idea in a more conventional tree-like graphic (Thienemann 1910, p. 251, reproduced as our Figure B3.1.1b).

These various kinds of representations were noted in Hellström's (2015) and Scharf's (2012) review of Pietsch's book *Trees of Life: A Visual History of Evolution* (Pietsch 2012). Earlier discussions note similar kinds of representation, albeit in slightly different forms (e.g., O'Hara 1988 and Stevens 1994). Rieppel, for example, discusses 'The series, the network, and the tree' (his 'series' being equivalent to Barsanti's 'ladder') and writes of 'the metaphor of order in nature' (Rieppel 2010). Hellström recognised a different trio of diagrams, some from outside the biological literature: the 'taxonomical' tree, the genealogical tree and the cosmological tree (Hellström 2012).

Often acknowledged as the first published *tree* of taxon relationships, Augustin Augier de Favas (1758–1825) published his 'Arbre botanique' (Augier 1801, reproduced here as Figure B3.1.2a) in the same year Lamarck published his *Système des animaux sans vertèbres* (Lamarck 1801), the latter being Lamarck's first account of the invertebrates, bringing their included number of classes to seven (Lamarck 1801, p. 50, the table is reproduced here as Figure 3.1.2b, see Newth 1952[4]). Lamarck's oft-reproduced diagram appeared a few years later in his *Philosophie zoologique* (Lamarck 1809, reproduced as our Figure B3.1.2c). Lamarck's 1801 table is more like a key, an artificial classification for documenting characters for ease of identification (Lamarck 1801, p. 50; for more on Lamarck's taxonomy, as opposed to his views on 'transmutation–evolution', see Barsanti 2000[5]).

Moving from one kind of diagram to another did not necessarily follow the neat sequence of ladder to map to tree, for there are more recent representations of each.[6] The general point we wish to make is, regardless of motivation behind the image, or what meaning was intended by its use, the resulting table or diagram, however represented, can be viewed as having a *cladistic aspect*, and all specify, however vaguely in some cases, a scheme of relationships (for a further example, see Box 3.3; for a discussion of unrooted 'trees', see Box 3.4).

References

Archibald, JD. 2014. *Aristotle's Ladder, Darwin's Tree: The Evolution of Visual Metaphors for Biological Order*. Columbia University Press, New York.

Augier, A. 1801. *Essai d'une Nouvelle Classification des Végétaux*. Bruyset, Lyons.

Barsanti, G. 1988. Le immagini della natura: scale, mappe, alberi 1700–1800. *Nuncius* 3: 55–125.

Box 3.1 *(cont.)*

Barsanti G. 1992. *La scala, la mappa, l'albero: immagini e classificazioni della natura fra sei e ottocento*. Sansoni, Florence.

Barsanti, G. 2000. Lamarck: taxonomy and theoretical biology. *Asclepio* 52(2): 119–132.

Bigoni, F. & Barsanti, G. 2011. Evolutionary trees and the rise of modern primatology: the forgotten contribution of St. George Mivart. *Journal of Anthropological Sciences* 89: 1–15.

Egerton, FN. 1985. Review of: Zoological Philosophy: An Exposition with Regard to the Natural History of Animals by J. B. Lamarck. *Isis* 76: 422–423.

Eichwald, E. 1829. *Zoologia specialis quam expositis animalibus tum vivis, tum fossilibus potissimum Rossiae in universum, et Poloniae in species, in usum lectionum publicarum in Universitate Caesarea Vilnensi habendarum. Parsprior. Propaedeuticam zoologiae atque specialem Heterozoorum expositionem continens.* Josephus Zawadzki, Vilnae.

Gould, SJ. 1997. Redrafting the tree of life. *Proceedings of the American Philosophical Society* 141: 30–54.

Hasegawa, M. 2013. Phylogeny mandala: a method for illustrating the biodiversity. *Biostory* 20: 74–83.

Hasegawa, M. 2017. Phylogeny mandalas for illustrating the Tree of Life. *Molecular Phylogenetics and Evolution* 117: 168–178.

Hellström, NP. 2012. Darwin and the tree of life: The roots of the evolutionary tree. *Archives of Natural History* 39: 234–252.

Hellström, NP. 2015. Review of J. David Archibald, Aristotle's ladder, Darwin's tree. The evolution of visual metaphors for biological order, Columbia University Press, 2014. *Archives of Natural History* 42(2): 378–379.

Lamarck, JB. 1801. *Systême des animaux sans vertèbres, ou Tableau général des classes, des classes, des ordres et des genres de ces animaux.* Déterville, Paris.

Lamarck, JB. 1809. *Philosophie zoologique*. Chez Dentu...; et L'Auteur, au Muséum d'Histoire Naturelle (Jardin des Plantes), Paris.

Lamarck, JB. 1984 [1914]. *Zoological Philosophy: An Exposition with Regard to the Natural History of Animals*. Translated by Hugh Elliot. Introductory essays by David L. Hull and Richard W. Burkhardt, Jr. University of Chicago Press, Chicago, London.

Minaka, N. & Sugiyama, K. 2012. *Phylogeny Mandala: Chain, Tree, and Network*. NTT Publishing, Tokyo.

Mindell, DP. 2013. The tree of life: metaphor, model, and heuristic device. *Systematic Biology* 62: 479–489.

Newth, DR. 1952. Lamarck in 1800. *Annals of Science* 8: 229–254.

O'Hara, RJ. 1988. Homage to Clio, or, toward an historical philosophy for evolutionary biology. *Systematic Zoology* 37: 142–155.

Pallas, PS. 1766. *Elenchus zoophytorum sistens generum adumbrationes generaliores et specierum cognitarum succinctas descriptiones cum selectis auctorum synonymis.* F. Varrentrapp, Hagæ.

Pietsch, TW. 2012. *Trees of Life: A Visual History of Evolution.* Johns Hopkins University Press, Baltimore.

Ragan, MA. 2009. Trees and networks before and after Darwin. *Biology Direct* 4(43) [38 pages with reviews].

Box 3.1   *(cont.)*

Rieppel, O. 2010. The series, the network, and the tree: changing metaphors of order in nature. *Biology and Philosophy* 25: 475–496.

Roveda, L. 2005. Lamarck et l'art des distinctions. *Revue d'histoire des sciences* 58(1): 145–168.

Scharf, S. 2012. Review of J. David Archibald, Aristotle's ladder, Darwin's tree. The evolution of visual metaphors for biological order, Columbia University Press, 2014. *Isis* 103: 773–774.

Stevens PF. 1994. *The Development of Biological Systematics: Antoine-Laurent de Jussieu, and the Natural System*. Columbia University Press, York.

Thienemann, A. 1910. Die Stufenfolge der Dinge, der Versuch eines natürlichen Systems der Naturkörper aus dem achtzehnten. Jahrhundert. Eine historische Skizze. *Zoologische Annalen (Würzburg)* 1910(3): 185–274.

Further Reading

A great deal has been written on the diagrams that have been used to represent biological diversity. Below we offer a few titles that may be of interest to those wishing to follow up on the subject.

Archibald, JD. 2014. *Aristotle's Ladder, Darwin's Tree: The Evolution of Visual Metaphors for Biological Order*. Columbia University Press, New York.

Barsanti, G. 1992. *La scala, la mappa, l'albero: immagini e classificazioni della natura fra sei e ottocento*. Sansoni, Florence.

Bouquet, MR. 1994. Family trees and their affinities: the visual imperative of the genealogical diagram. *Journal of the Royal Anthropological Institute* n.s. 2: 43–66.

Daugeron, B. 2009. L'usage méthodique de la carte en botanique. Classer en histoire naturelle au XVIIIe siècle. *Comite Francais de Cartographie* 199: 97–104.

Delisle, RG. 2007. *Debating Humankind's Place in Nature 1860–2000: The Nature of Paleoanthropology*. Pearson Prentice Hall, Upper Saddle River, NJ.

Eco, U. 2014. *From the Tree to the Labyrinth: Historical Studies on the Sign and Interpretation*. Harvard University Press, Cambridge, MA.

Fisler, M., Crémière, C. & Lecointre, G. 2014. Chapitre 2. Qu'est-ce qu'un arbre des idées ? Explicitation des notions d'arbre et de phylogénie et histoire des représentations de l'arbre. In: *Apparenter la pensée: Vers une phylogénie des concepts savants*. Editions Matériologiques, Paris, pp. 103–144.

Klapisch-Zuber, C. 2004. *Stammbäume: Eine illustrierte Geschichte der Ahnenkunde*. Knesebeck, Munich.

Lecointre, G. 2011. Que représénte-t-on avec un arbre? *Biosystema* 28: 9–40.

This paper covers a number of other topics related to its title. It is part of an excellent collection, all contributions are worth reading (*Biosystema* 28, 2011, 'L'arbre du vivant existe-t-il?', eds V. Malécot, N. Léger & P. Tassy).

Lima, M. 2014. *The Book of Trees – Visualizing Branches of Knowledge*. Princeton Architectural Press, New York.

Minaka, N. 2006. *Tree-Thinking in Historical Sciences [Keitouju-shikou no sekai]*. Koudansha Publ., Tokyo.

Box 3.1 *(cont.)*

Minaka, N. 2018a. *Phylogenetic Thinking. A History and Philosophy of Systematic Biology in the Twentieth Century*. Keiso Shobo Publishing Co., Tokyo. [English translation due shortly].

Minaka, N. 2018b. *Systematic Thinking. Diagrams in Taxonomy and Phylogeny*. Shunjusha Publishing Co., Tokyo.

Nelson, GJ. & Platnick, NI. 1981. *Systematics and Biogeography: Cladistics and Vicariance*. Columbia University Press, New York.

Nelson and Platnick included a detailed account of the history of systematics entitled 'Systematic History: Kinds of Branching Diagrams' (their Chapter 2), beginning with Theophrastus (371–287 BC). Their account is novel by including representation of the many branching diagrams and written classifications of the past with an alternative diagram representing the 'structural components' of each classification. These diagrams are *derivative cladograms* representing the *cladistic parameter* of each diagram or classification (see Chapter 7).

Pietsch, TW. 2012. *Trees of Life: A Visual History of Evolution*. Johns Hopkins University Press, Baltimore.

Thienemann, A. 1910. Die Stufenfolge der Dinge, der Versuch eines natürlichen Systems der Naturkörper aus dem achtenzehnten. Jahrhundert. Eine historische Skizze. *Zoologische Annalen (Würzburg)* 1910(3): 185–274.

Vargas, P. & Kayman, MA. 2014. *The Tree of Life*. Sinauer Associates, an imprint of Oxford University Press, Oxford.

On Peter Simon Pallas (1741–1811)

Ragan, MA. 2009. Trees and networks before and after Darwin. *Biology Direct* 4(43): unpaginated.

On Jean-Baptiste Pierre Antoine de Monet, Chevalier de Lamarck (1744–1829)

There is a vast literature on Lamarck. Burkhardt (1977) and Corsi (1988) are good places to start, followed by the many important contributions of Goulven Laurent (1925–2008, www.annales.org/archives/cofrhigeo/goulven-laurent.html). Laurent's work is written in French but deserves translation into English to gain his work a wider audience.

Burkhardt, R. 1977. *The Spirit of System: Lamarck and Evolutionary Biology*. Harvard University Press, Cambridge, MA, London, UK[2nd edition with an additional essay 'Lamarck in 1995', published 1995].

Corsi, P. 1988. *The Age of Lamarck: Evolutionary Theory in France 1790-1830*. University of California Press, Berkeley.

Corsi, P., Gayon, J. Gohau, G. & Tirard, S. 2006. *Lamarck, philosophe de la nature*. [Preface by Armand de Ricqlès] Science, Histoire et Société. Presses Universitaires de France, Paris.

Jordanova, LJ. 1984. *Lamarck*. Oxford University Press, Oxford.

Laurent, G. 1997a. *Jean-Baptiste Lamarck: 1744-1829*. Comité des travaux historiques et scientifiques, CTHS.

Laurent, G. 1997b. Lamarck (1744–1829) et la paléontologie. *Eclogae Geologicae Helvetiae* 92: 115–121.

Box 3.1 *(cont.)*

Laurent, G. 2001. *La naissance du transformisme: Lamark entre Linné et Darwin*. Vuibert, Paris. [Reprinted in 2017]
Roveda, L. 2005. Lamarck et l'art des distinctions. *Revue d'histoire des sciences* 58(1): 145–168.

On Augustin Augier de Favas (1758–1825)

Hellström, NP., André, G. & Philippe, M. 2017a. Augustin Augier's botanical tree: transcripts and translations of two unknown sources. *Huntia* 16(1): 17–38.
Hellström, NP., André, G. & Philippe, M. 2017b. Life and works of Augustin Augier de Favas (1758–1825), author of "Arbre botanique" (1801). *Archives of Natural History* 44: 43–62.
Stevens, PF. 1983. Augustin Augier's 'Arbre botanique' (1801), a remarkable early botanical representation of the natural system. *Taxon* 32: 203–211.
Tassy, P. 1991 (1998, 2nd edition). *L'arbre à remonter le temps*. Christian Bourgois, Paris (2nd edition Diderot Multimedia, Paris).
Tassy presents a useful discussion on Augier (pp. 36–40) – the book is excellent. Even now, after more than 20 years an English translation would still be welcome (the 2nd edition differs only in having a new preface).

On Carl Edward von Eichwald (1795–1876)

Ragan, MA. 2009. Trees and networks before and after Darwin. *Biology Direct* 4(43): unpaginated.

[1] Peter Simon Pallas (1741–1811).
[2] The 1801 date refers to Lamarck's *Système des animaux sans vertébres*, for which see below.
[3] Jean-Baptiste Pierre Antoine de Monet, Chevalier de Lamarck (1744–1829).
[4] This translation is reprinted in Lamarck (1984 [1914]). A review of this book by Egerton (1985) gives some background to the translation of Lamarck 1800 [1801]. Egerton refers to it as 'Lamarck's first lecture on evolution'. See also Roveda (2005).
[5] Barsanti suggests that Lamarck's taxonomy can be understood as a precursor to numerical taxonomy as he, Lamarck, considered a range of characters from various sources (Barsanti 2000, p. 121 and footnote 3).
[6] More recently, Hasegawa described and illustrated some 'Phylogeny Mandalas', circular diagrams with accompanying specimen illustrations (Hasegawa 2013, 2017). Minaka may have been the first to use the term Mandala for these kinds of illustrated circular trees (Minaka & Sugiyama 2012).

The ebbing and flowing of complexity in these various diagrams is a feature of the chronology of tree-like diagrams (a glance at the many images in Pietsch (2012) and Archibald (2014), for example, would suffice to illustrate this point; see the diagrams in Box 3.2 for a specific example). The general view is that these diagrams are, for the most part, metaphors or models.

Box 3.2   A Few Diagrams from the Classifications of Diatoms (1823–1988)

Of diatoms, Darwin wrote in chapter 6 (*Difficulties of the Theory, Utilitarian Doctrine, How Far True: Beauty, How Acquired*) of the 4th edition of the *Origin*,

> Few objects are more beautiful than the minute siliceous cases of the diatomaceae: were these created that they might be examined and admired under the higher powers of the microscope? (Darwin 1866, p. 200)

His comments have their origin in correspondence with George Henry Kendrick Thwaites (1811–1882; Burkhardt & Smith 1987, pp. 131–133), an erstwhile diatomist who, in 1849, moved from Bristol, UK, to Ceylon (Sri Lanka), eventually becoming Director of the Peradeniya Botanical Gardens (1857–1880, Willis 1901). Still, Darwin's question was largely rhetorical. What of their classification?

Written Classifications (1823–1872): A good place to start is with Carl Adolph Agardh (1785–1859), a phycologist and, eventually, bishop of Karlstad, Sweden. In his *Systema Algarum*, Agardh published one of the first comprehensive classifications of diatoms (Agardh 1823–1828). He classified them (at that time just over 40 species were known) in one order of Algae (Ordo I, Diatomeae), placing them in nine genera: *Achnanthes*, *Frustulia*, *Meridion*, *Diatoma*, *Fragilaria*, *Melosira*, *Desmidium*, *Schizonema* and *Diatoma*, the latter genus being subdivided into five 'groups' (Agardh 1824 in 1823–1828; *Gomphonema*, another genus of diatoms, was placed separately in an appendix).

In the later *Conspectus Criticus Diatomacearum* he rearranged the genera (now with an increased number of species) into three families, Cymbelleae, Styllarieae and Fragilarieae, based on the shape of their valves (the silica casing that encloses the single cell): Styllarieae included genera with cuneate (wedged-shaped) valves; Cymbelleae included genera with cymbelloid valves; Fragilarieae included genera with rectangular valves, the latter being subdivided into two groups on the basis of their colony form. Agardh compared each of these three families with four different colony 'types': those with no obvious colony formation ('libera'); those attached by a stalk ('Stipitata'); those attached in chains ('In frondem composita'); and those in 'cymbelloid' chains ('Fila cymbellarum frondem formantia'). As a summary, Agardh (1832 in 1830–1832) presented a table contrasting valve shape with colony structure (reproduced as Figure B3.2.1a). His final classification, however, reflected frustule shape rather than colony formation. The classification derived from the table in Figure B3.2.1a can be represented as a derivative cladogram (see Chapter 7) with two branches and one node of two branches. The three each represent a family and the node represents the subdivisions of Fragilariae (Figure B3.2.1b). For Agardh, 'true' relationships could only be understood from the characters of the silica frustule, with colony formation providing only additional (and, for him, analogous) information.

Agardh was aware of the many issues in classification but the central question for him was: How does one discover characters of significance?

> that such a work as this cannot be free of hypotheses is self-evident. Hypotheses have always been necessary; they have never harmed the sciences, but on the contrary, have aided them even if they were found in the final analysis to be unsubstantiated. (Agardh 1829–1830; modified from a translation in Ott 1991, p. 304)

Box 3.2 *(cont.)*

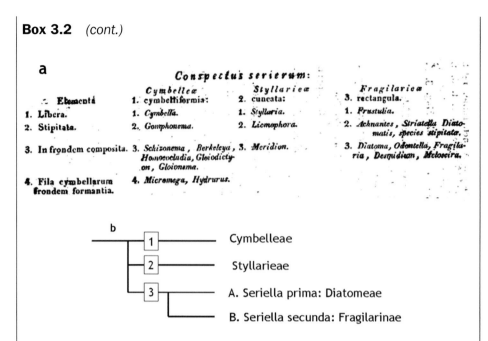

Figure B3.2.1 (a) Agardh's summary (1832 in 1830–1832) of diatom relationships; (b) a derivative cladogram with two branches and one node of two branches.

Agardh was clear what he was doing:

> I take the liberty to offer you a book on Algae ... [he wrote on 23rd September 1824 to Saint-Hilaire] ... The idea that I have followed is not so much like those in present day systems that squeeze plants into a tidy frame, but rather like yours, to arrange them one nearer the other according to their greatest affinity. (From Woelkerling & Lamy 1999, p. 78; their translation)

Agardh's classification was influential but the characters he chose to contrast – colony form and frustule shape – were to become problematic for future diatomists, as 'colony form' is actually dependent on a suite of interrelated characters (spines, pore fields, etc.) and frustule shape depends to a certain extent on subjective judgements.

Like Agardh, Friedrich Traugott Kützing (1807–1859) contrasted the siliceous characteristics of the diatom frustule with the mode of colony formation and similarly concluded that the former was of prime importance. Influential in this decision may have been that Kützing had discovered the siliceous nature of the valves (Kützing 1844a). He classified diatoms into three tribes, two of which were subdivided in a similar way: Tribes I and II were each split into two orders, 'Astomaticae' and 'Stomaticae' – he considered these two tribes to be subdivided on the basis of whether they do ('Stomaticae') or do not ('Astomaticae') possess some kind of 'opening', misinterpreted by Kützing as a kind of stomach (Kützing 1844b); the third tribe was also subdivided into two orders but with different names, Disciformes and Appendiculatae (Figure B3.2.2a). Kützing's classification can be represented with a derivative cladogram of three nodes (Figure B3.2.2b).

Box 3.2 *(cont.)*

Figure B3.2.2 (a) Kützing's summary classification (Kützing 1844b); (b) Kützing's classification as a derivative cladogram with three nodes.

In 1846 Giuseppe Giovanni Antonio Meneghini (1811–1899) wrote a lengthy critique of Kützing's work, drawing attention to the unevenness of the characters that described many of his taxa (Meneghini 1846, translated into English in 1853). He revised Kützing's classification, primarily at the generic level. Kützing and Meneghini's understanding of

Box 3.2   *(cont.)*

what characters were important differed: Meneghini (1846, 1853) suggested that *combinations* of characters should be used to 'define' groups and did not think relative importance could be attributed to any particular character system:

> what value these characters have ... I do not believe that we can decide in the actual state of science. (Meneghini 1853, p. 405)

Thus, Meneghini's alternative suggestions implied that the discovery of important classificatory (taxonomic) characters was irrelevant, and a combination was required. Nevertheless, Meneghini did concede that importance might be given to '... conformity of the two primary surfaces [of the valves]' (Meneghini 1853, p. 384). In the introduction to his study, Meneghini discussed his views on natural groups and the meaning of taxonomic categories:

> The words Animal and Plant, like words in common use, as Species, Genus, Order, Class, Kingdom, do not denote any existing thing in particular. To the naturalist there exist individuals only. (Meneghini 1853, p. 346)

Meneghini's views might be considered to share certain similarities with what has been understood as phenetics, if that understanding is taken to mean a belief that there is not one natural system of classification but many artificial ones – and the act of classification being one of imposition rather than discovery.

In his *A Synopsis of the British Diatomaceae* (1853 and 1856 in Smith 1853–1856), William Smith divided diatoms into two tribes, of which one was subdivided into four subtribes, the other five. Each tribe and subtribe was characterised by some properties of the entire colony. For example, subtribe 3 (of Tribe 1) was characterised by '[C]onnecting membrane evanescent, or obsolete; frustules after self-division united in a compressed filament' (Smith 1853, p. 7) and his subtribe 4 (of Tribe 1) was characterised by '[C]onnecting membrane subpersistent; frustules after self-division united in a zig-zag chain' (Smith 1853–1856, p. 7) (Figure B3.2.3a, b).

William Smith acknowledged that his classification may not be 'natural' (in some unstated sense) but maintained that the arrangement was intended, '... to aid identification of species by a statement of the most obvious characters' (Smith 1856, p. 39). The implication being that 'the most obvious characters' were invariant and, as a consequence, 'important', a finding seemingly at odds with conclusions reached by earlier authors, as colony structure was not proving useful in predicting further characters of *any* taxa and has since become minimally important for identification, as Agardh (1832 in 1830–1832) suggested some years before. William Smith's approach to classification highlights a distinction between classifications created for utility (artificial) and those meant to represent nature (natural).

Hamilton Lanphere Smith's (1819–1903, Professor of Natural Philosophy and Astronomy at Kenyon College in Ohio, USA) approach to classification had a lasting impact. Convinced that the presence or absence of a raphe – a slit in the silica valve, often associated with enabling the ability of movement – coupled with basic valve symmetry, would prove to be decisive, he distinguished between those diatoms that possessed a raphe and those that did not, along with a third group that appeared to have a raphe but

Box 3.2   *(cont.)*

a

C<small>LASS</small> CRYPTOGAMIA.

S<small>UB-CLASS</small> ALGÆ.   N<small>ATURAL</small> O<small>RDER</small> DIATOMACEÆ.

Plant a F<small>RUSTULE</small> ; consisting of a unilocular or imperfectly septate cell invested with a bivalve siliceous epidermis. G<small>EMMIPAROUS INCREASE</small>, by S<small>ELF-DIVISION</small> ; during which process the cell secretes a more or less siliceous C<small>ONNECTING</small> M<small>EMBRANE</small>. R<small>EPRODUCTION</small>, by C<small>ONJUGATION</small> and the formation of Sporangia.

T<small>RIBE</small> I. *Frustules naked ; not imbedded in gelatine nor enclosed in membranaceous tubes.*

    S<small>UB-TRIBE</small> 1. *Connecting membrane deciduous ; frustules solitary or during self-division in pairs, rarely in greater numbers, adherent or free, dispersed, or aggregated into a mucous stratum.*

        22 G<small>ENERA</small>. Epithemia, Eunotia, Cymbella, Amphora, Cocconeis, Coscinodiscus, Eupodiscus, Actinocyclus, Arachnoidiscus, Triceratium, Cyclotella, Campylodiscus, Surirella, Tryblionella, Cymatopleura, Nitzschia, Amphiprora, Amphipleura, Navicula, Pinnularia, Stauroneis, Pleurosigma.

    S<small>UB-TRIBE</small> 2. *Connecting membrane subpersistent ; frustules after self-division attached by a gelatinous cushion, or dichotomous stipes.*

        7 G<small>ENERA</small>. Synedra, Doryphora, Cocconema, Gomphonema, Podosphenia, Rhipidophora, Licmophora.

    S<small>UB-TRIBE</small> 3. *Connecting membrane evanescent, or obsolete ; frustules after self-division united into a compressed filament.*

        12 G<small>ENERA</small>. Meridion, Bacillaria, Himantidium, Odontidium, Denticula, Fragilaria, Eucampia, Achnanthes, Diadesmis, Rhabdonema, Striatella, Tetracyclus.

    S<small>UB-TRIBE</small> 4. *Connecting membrane subpersistent ; frustules after self-division united into a zigzag chain.*

        6 G<small>ENERA</small>. Diatoma, Grammatophora, Tabellaria, Amphitetras, Biddulphia, Isthmia.

    S<small>UB-TRIBE</small> 5. *Connecting membrane subpersistent as a siliceous annulus ; frustules after self-division united into a cylindrical filament.*

        3 G<small>ENERA</small>. Podosira, Melosira, Orthosira.

T<small>RIBE</small> II. *Frustules invested with a gelatinous or membranaceous envelope.*

    S<small>UB-TRIBE</small> 6. *Frond indefinite, mammillate ; frustules scattered.*

        1 G<small>ENUS</small>. Mastogloia.

    S<small>UB-TRIBE</small> 7. *Frond definite, compressed or globular ; frustules scattered.*

        2 G<small>ENERA</small>. Dickieia, Berkeleyia.

    S<small>UB-TRIBE</small> 8. *Frond definite, filamentous ; frustules in rows.*

        3 G<small>ENERA</small>. Encyonema, Colletonema, Schizonema.

    S<small>UB-TRIBE</small> 9. *Frond definite, filamentous ; frustules fasciculated.*

        1 G<small>ENUS</small>. Homœocladia.

b

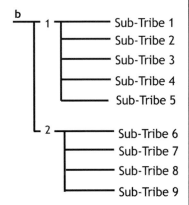

Figure B3.2.3 (a) William Smith's summary classification (Smith 1853, p. 7); (b) Smith's classification as a derivative cladogram with two nodes and nine branches.

Box 3.2 *(cont.)*

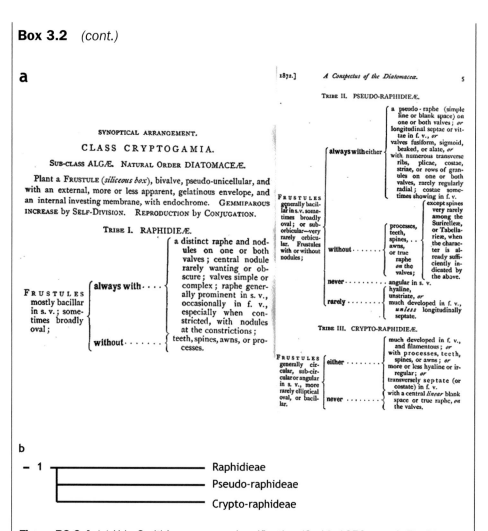

a

SYNOPTICAL ARRANGEMENT.

CLASS CRYPTOGAMIA.

SUB-CLASS ALGÆ. NATURAL ORDER DIATOMACEÆ.

Plant a FRUSTULE (*siliceous box*), bivalve, pseudo-unicellular, and with an external, more or less apparent, gelatinous envelope, and an internal investing membrane, with endochrome. GEMMIPAROUS INCREASE by SELF-DIVISION. REPRODUCTION by CONJUGATION.

TRIBE I. RAPHIDIEÆ.

TRIBE II. PSEUDO-RAPHIDIEÆ.

TRIBE III. CRYPTO-RAPHIDIEÆ.

b

– 1
— Raphidieae
— Pseudo-raphideae
— Crypto-raphideae

Figure B3.2.4 (a) H.L. Smith's summary classification (Smith 1872, pp. 4–5); (b) H.L. Smith's classification as a derivative cladogram with three branches.

on closer inspection did not. This latter 'raphe-mimic' Smith called a 'pseudo-raphe'. Smith also introduced the term 'crypto-raphe' to account for species that are bilaterally symmetrical (and the few that were radially symmetrical) which bore central nodules.

These three divisions came to be colloquially known as 'centric', 'araphid' and 'raphid' diatoms, the latter two together forming the pennate diatoms: 'These form the three Tribes of the Synopsis, and very seldom will any difficulty arise as to which tribe a diatom may belong to . . .' (Figure B3.2.4a, after Smith 1872, pp. 4–5). Smith thus helped create today's 'conventional wisdom' concerning the most decisive and important characters. Smith's classification can be represented by a derivative cladogram with three branches, each representing a family (Figure B3.2.4b).

Box 3.2   *(cont.)*

Diagrams with Classifications (1901–1988): With respect to characters and their significance, Constantin S. Merezhkowsky (1855–1921) suggested that functionality would explain a particular character's importance:

> What is the most important character which should serve as the base for a truly natural classification? I believe that it is the presence or absence of movement, which is merely dependent on the presence or absence of a slit in the walls of the frustule [the raphe]; this character should be taken into consideration before any other. (Merezhkowsky 1902, p. 65, an English summary of Merezhkowsky 1901, which includes both Russian and French text)

In his classification, Merezhkowsky divided the diatoms into 'two great groups', those that move, the 'mobilées' (Mobiles), and those that do not, the 'immobilées' (Immobiles) (Figure B3.2.5a); H.L. Smith's raphid diatoms were (more or less) equivalent to Merezhkowsky's mobilées, and his cryptoraphids and pseudoraphids were (more or less) equivalent to his immobilées.

From detailed studies of diatom auxospores and plastid morphology, as well as the usual features of the valve and frustules, Merezhkowsky presented a diagram intending to depict the evolutionary relationships of the major groups (Figure B3.2.5a, after Merezhkowsky (1903a); a slightly different diagram was presented in Merezhkowsky (1903b), opposite p. 204). In this diagram Merezhkowsky named several (hypothetical) ancestral taxa: *Archaideae*, the ancestor to raphid diatoms ('mobilées'); *Protonées* the ancestor to the *Archaideae*; *Copuloneis* the ancestor to the *Protonées* plus Tabellarioideae; and *Urococcus* ancestor to all diatoms (Figure B3.2.5a). In addition, some previously described taxa were also placed in an ancestral position: Melosireae, leading to (ancestral to) the Anaraphideae, Fragilarioideae leading to (ancestral to) *Copuloneis*, *Auricula* leading to (ancestral to) the Carinatae and *Libellus* leading to (ancestral to) Polyplacatae (Figure B3.2.5a).

In 1897 Hippolyte Peragallo published a preliminary discussion of diatom classification as a preamble to his *Diatomées Marines de France*, co-authored with his brother Maurice (H. & M. Peragallo 1897–1908). While being critical of all previous classifications, Hippolyte settled for the tripartite division of anaraphids, pseudoraphids and raphids (H. Peragallo 1897, pp. 16–17). He also included a genealogy diagram enclosing various diatom generic names in boxes, which were interconnected forming a semireticulated, tree-like structure: 'centric' diatoms at the base, raphid pennate diatoms at the tips, passing through various 'araphid' genera (Figure B3.2.5b).

Like Peragallo, Achilli Forti (1878–1937) extended and elaborated Merezhkowsky's classification, representing certain 'genealogical' aspects in a series of complex diagrams, linking genera via a number of reticulating lines in a network (Figure B3.2.5c, one of 11 (Forti, 1911)). Forti's diagrams are more like maps than trees.

A final diagram is that of Nikolaev, a classification of (mostly) fossil diatoms, supported by observations on many new fossil specimens (Nikolaev 1988, 1990). While dealing primarily with 'centric' diatoms (i.e., non-pennates, '... the diatom equivalent of

Box 3.2 *(cont.)*

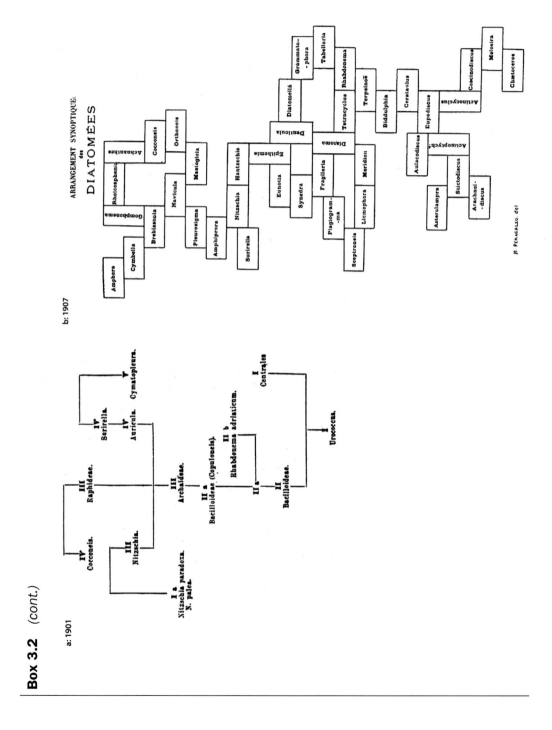

Box 3.2 *(cont.)*

Figure B3.2.5 Four diatom 'phylogenetic' trees, 1901–1988. (a) A diagram intending to depict the evolutionary relationships of the major groups (after Merezhkowsky 1903a; a slightly different diagram was presented in Merezhkowsky 1903b, opposite p. 204). (b) A genealogy diagram enclosing various diatom generic names in boxes, interconnected forming a semireticulated, tree-like structure (after H. Peragallo 1897, p. 16–7). (c) Achilli Forti extended and elaborated Merezhkowsky's classification, representing certain 'genealogical' aspects in a series of complex diagrams (one of 11, after Forti 1911). (d) A more recent diagram from Nikolaev, dealing (mostly) with fossils (after Nikolaev, 1984, with permission).

Box 3.2 *(cont.)*

'invertebrate''; Alverson & Theriot 2005), they place within the single class Centrophyceae six subclasses (after Nikolaev et al. 2001, pp. 38–41, Figure B3.2.5d).

The general significance of these diagrams is that regardless of the philosophy of the author, each represents a cladistic aspect to the development of diatom classification. By that we mean each has a series of defined relationships. One might see, then, that some common relationships remain regardless of whatever philosophical viewpoint was taken, even regardless of the kinds of data and their interpretation.

References

Agardh, CA. 1823–1828. *Systema Algarum*. Litteris Berlingianis, Lundae.

Agardh, CA. 1829–1830. *Lärobok i Botanik*. N.M. Thomsons Boktryckeri, Malmö.

Agardh, CA. 1830–1832. *Conspectus Criticus Diatomacearum*. Litteris Berlingianis, Lundae.

Alverson, AJ. & Theriot, EC. 2005. Comments on recent progress toward reconstructing the diatom phylogeny. *Journal of Nanoscience and Nanotechnology* 5: 57–62.

Burkhardt, F. & Smith, S. (eds) 1987. *The Correspondence of Charles Darwin*. Vol 3. Cambridge University Press, Cambridge, pp. 1844–1846.

Darwin, C. 1866. *On the Origin of Species by Means of Natural Selection, or, the Preservation of Favoured Races in the Struggle for Life*, 4th ed., with additions and corrections. John Murray, London.

Forti, A. 1911. Contribuzioni diatomologische. XII. Metodo di classificazione delle Bacillariee Immobili fondato sull'affinità morfologica dei frustuli ed in relazione con l'evoluzione dell'auxospora. *Atti del Real Istituto Veneto di Scienze, Lettere ed Arti* 71: 677–731.

Kützing, FT. 1844a. *Die Sophisten und Dialektiker die gefährlicihsten Feinde der wissenschaftlichen Botanik, &c*. Nordhausen, W. Köhne.

Kützing, FT. 1844b. *Die kieselschaligen Bacillarien oder Diatomeen*. Nordhausen, W. Köhne.

Meneghini, G. 1846. Sulla animaliti delle Diatomee e revisione organografica dei generi di Diatomee stabilite dal Kützing. *Atti del Real Istituto Veneto di Scienze, Lettere ed Arti* 1845: 1–191.

Meneghini, G. 1853. On the animal nature of diatomeae with an organographical revision of the genera established by Kützing. *Botanical and Physiological Memoirs (Ray Society)* II: 345–513.

Merezhkowsky, C. 1901. Étude sur l'endochrome des diatomées. *Mémoires de l'Académie Impériale des Sciences de St. Petersbourg*, ser. 8, 11(6): 1–40.

Merezhkowsky, C. 1902. On the classification of diatoms. *Annals and Magazine of Natural History*, ser. 7, 9: 65–68.

Merezhkowsky, C. 1903a. Les types des auxospores chez les diatomées et leur evolution. *Annales des sciences naturelles Botanique* 17: 225–262.

Merezhkowsky, C. 1903b. *K morfologii diatomovŷkh vodorovslei = Zur Morphologie der Diatomeen* [*Morphology of diatoms*]. Kasan, Imperatorskaya Universiteta, pp. 1–427.

Nikolaev, VL. 1984. To the construction of the system of diatoms (Bacillariophyta). *Botanicheskii Zhurnal* 69: 1468–1474. http://botjournal.ru/en/.

Nikolaev, VL. 1988. Systema klassa Centrophyceae (Bacillariophyta). *Botanischeskii Zhurnal* 73: 486–496.

Box 3.2 *(cont.)*

Nikolaev, VL. 1990. The system of centric diatoms. In: Simola, H. (ed.), *Proceedings of the 10th International Diatom Symposium*. O. Koeltz Scientific Books, Koenigstein, pp. 17–22.

Nikolaev, VL., Harwood, DM. & Samsonov, NI. 2001. [*Early Cretaceous Diatoms*]. Russian Academy of Sciences, Komarov Botanical Institute, St. Petersburg.

Ott, FD. 1991. Carl Adolph Agardh, Professor, Bishop. A translation of J.E. Areschoug's 1870 Memorial. *Archiv für Protistenkunde* 139: 297–312.

Peragallo, H. 1897. Diatomées marines de France. *Micrographie Préparateur* 5: 9–17.

Peragallo, H. & M. 1897–1908. *Diatomées Marines de France et des districts Maritimes Voisins*. Micrographie-éditeur, Grez-sur-Loing.

Smith, HL. 1872. Conspectus of the families and genera of the Diatomaceae. *The Lens* 1: 1–19, 72–93, 154–157.

Smith, W. 1853–1856. *A Synopsis of the British Diatomaceae*, volumes 1 & 2. Smith and Beck, London.

Willis, JC. 1901. The Royal Botanic Gardens of Ceylon, and their history. *Annals of the Royal Botanic Gardens, Peradeniya* 1: 1–15.

Woelkerling, WJ. & Lamy, D. 1999. *Non-geniculate coralline red algae and the Paris Muséum: Systematics and scientific history*. Paris Publications Scientifiques du Muséum, A.D.A.C., Paris.

Further Reading

Williams, DM. 2007. Classification and diatom systematics: the past, the present and the future. In: Brodie, J. & Lewis, J. (eds), *Unravelling the Algae*. CRC Press, pp. 57–91.

Williams, DM. & Kociolek, JP. 2011. An overview of diatom classification with some prospects for the future. In: Seckbach, J. & Kociolek, JP. (eds), *The Diatom World*. Series: Cellular Origin, Life in Extreme Habitats and Astrobiology, vol. 19. Springer, Dordrecht, pp. 47–91.

A fuller version of the account above can be found in these two publications.

'A Central Metaphor' (Gregory 2008)

Rather than begin at the beginning, wherever that might be (cf., Barsanti 1988, 1992; Ragan 2009; Bigoni & Barsanti 2011; Tassy 2011), we start, as many others do, with a comment on the famous and oft-reproduced 'tree-like' diagram Charles Darwin sketched in his B notebook in 1837. Much has been written about this sketch (see Box 3.3), an illustration that pre-dates the only diagram published in the *Origin* (Darwin 1859; Darwin did sketch a few genealogies in his notebooks, see Bredekamp 2005, Archibald 2014 and Pietsch 2012, for examples). As T. Ryan Gregory noted, 'Charles Darwin sketched his first evolutionary tree in 1837, and trees have remained a central metaphor in evolutionary biology up to the present' (Gregory 2008), while later Eldredge offered the view that Darwin's 1837 diagram is the '... closest equivalent to Einstein's handwritten $E = MC^2$ ... [when Darwin]

Box 3.3   Charles Darwin's 1837 'I Think' Tree As a Derivative Cladogram

Apart from overwrought comparisons to Einstein and his famous equation, more sober judgement suggested that Darwin's 1837 'I think' tree (Figure B3.3.1a, after Barrett et al. 1987, p. B44[1]) was derived if not directly then 'spiritually', so to speak, from

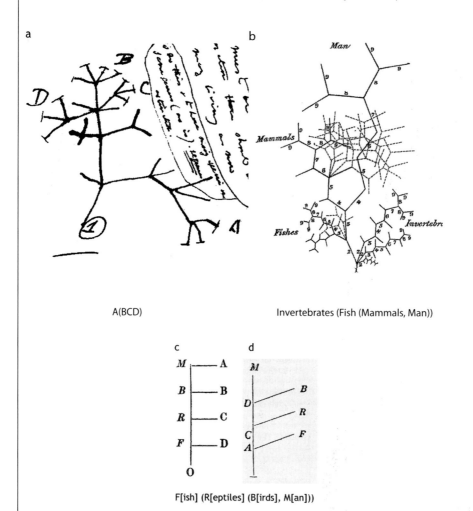

a

A(BCD)

b

Invertebrates (Fish (Mammals, Man))

c   d

F[ish] (R[eptiles] (B[irds], M[an]))

Figure B3.3.1 (a) Darwin's 1837 'I think' tree (see Barrett et al. 1987, p. B44) representing the relationships A(BCD); (b) Barry's diagrammatic representation of Karl Ernst von Baer's developmental scheme (Barry 1837, p. 346), depicting the relationships: Invertebrates (Fish (Mammals, Man)); (c) Barry's diagram as simplified by Carpenter (1841, p. 197) and (d) as simplified by Chambers (1844, p. 212), both depicting the relationships: F[ish] (R[eptiles] (B[irds], M[an])).

Box 3.3 *(cont.)*

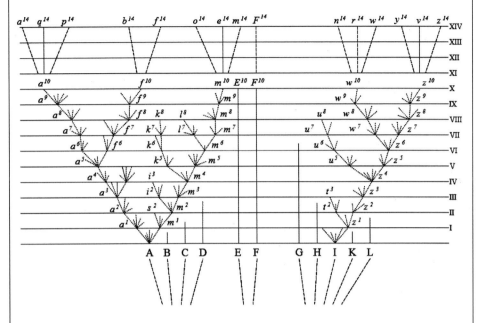

Figure B3.3.2 The only diagram from Darwin's *Origin* also said to be influenced by Barry's figure (Darwin 1859).

Martin Barry's (1802–1855, Figure B3.3.1b[2]) diagrammatic representation of Karl Ernst von Baer's (1792–1896) developmental scheme, with its branching motif (Barry 1837, p. 346, reproduced here as Figure B3.3.1d[3]). The diagrams were based on a 'table' in von Baer (1828, p. 225, reproduced here as our Figure B3.3.3a). Von Baer's table can be viewed as a hierarchical classification, a branching structure of diversification – a derivative cladogram, if you will (see below). The influence of Barry's diagram was extensive (Richards 1987; Camardi 2001; Torrens & Barahona 2013): it was simplified by William Carpenter in the 2nd edition of his *Principles of General and Comparative Physiology* (Carpenter 1841, p. 197, reproduced here as Figure B3.3.1c); Carpenter's diagram was used, and further interpreted, by Chambers in his *Vestiges of the Natural History of Creation* (Chambers 1844, p. 212, reproduced here as Figure B3.3.1d); the latter publication goaded Richard Owen to write to its author concerning 'the idea and diagram on page 212' (letter in Rev. R. Owen, 1894, p. 251) which, Owen claimed, embraced his own ideas on the relevance of development; and it has also been suggested that the published illustration in Darwin's *Origin* (Figure B3.3.2) was influenced by Barry's figure (e.g., Richards 1987, pp. 134, 169–170, footnote 187).

Art historian Horst Bredekamp finds the coral, rather than the tree, a more promising source of inspiration for the 1837 diagram. As it happens, combining the two metaphors – coral and tree – Darwin did note that 'The tree of life should perhaps be called the coral of

Box 3.3 *(cont.)*

a

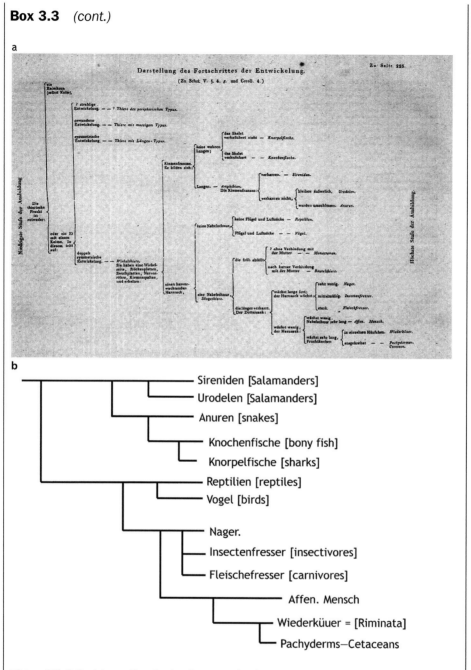

b

Sireniden [Salamanders]
Urodelen [Salamanders]
Anuren [snakes]
Knochenfische [bony fish]
Knorpelfische [sharks]
Reptilien [reptiles]
Vogel [birds]
Nager.
Insectenfresser [insectivores]
Fleischefresser [carnivores]
Affen. Mensch
Wiederküuer = [Riminata]
Pachyderms–Cetaceans

Figure B3.3.3 (a) von Baer's developmental scheme was based on this 'table' (von Baer 1828, p. 225); (b) Conventional derivative cladogram, similar to the diagrams of Carpenter and Chambers, of von Baer's classification with the relationships: Fish (Reptiles (Birds, Man)).

Box 3.3 *(cont.)*

life'. Bredekamp even located a particular specimen that may have inspired the 1837 illustration: the specimen turned out to be the rhodophyte *Bossea orbignyana* rather than a coral (Porter 1987, p. 194; Bredekamp 2005, translated into Italian and French, Bredekamp 2006, 2008, respectively; reviews in Maderspacher 2006; Weber 2008).

Amundson wrote of Barry's tree that 'it is impossible for a modern reader to resist seeing this as a phylogenetic tree' (Amundson 2005, p. 70). Maybe so – but Brower thought that Barry's tree was 'strongly suggestive of a cladogram … and nested taxonomic groups were scrutable to Linnaeus (1758), among others, from study of the hierarchical patterns of characters they exhibited' (Brower 2016, p. 106). To us, as well, it is impossible to resist seeing it as a *derivative cladogram*, for that is what it is, specifying the following relationships: Invertebrates (Fish (Mammals, Man)) (Figure B3.3.1b). Similarly, the diagrams of Carpenter and Chambers specify the following relationships: Fish (Reptiles (Birds, Man)) (Figure B3.3.1c, d) – and inspection of von Baer's classification (von Baer 1828, p. 225, Figure B3.3.3b) shows exactly those relationships.

In this respect, Darwin's 1837 diagram is less informative than either Barry's, Carpenter's or Chambers', and although it relates no specific organisms, it nevertheless specifies the following relationship: A(BCD) (Figure B3.3.1a).

References

Amundson, R. 2005. *The Changing Role of the Embryo in Evolutionary Thought: Roots of Evo-Devo.* Cambridge University Press, Cambridge, UK.

Baer, KE. von 1828. *Über Entwickelungsgeschichte der Thiere: Beobachtung und Reflexion.* Bei den Gebrüdern Bornträger, Königsberg.

Barrett, PH., Gautry, PJ., Herbert, S. Kohn, D. & Smith, S. (eds) 1987. *Charles Darwin's Notebooks, 1836-1844: Geology, Transmutation of Species, Metaphysical Enquiries.* British Museum (Natural History) & Cambridge University Press, Cambridge.

Barry, M. 1837. Further observations on the unity of structure in the animal kingdom, and on congenital anomalies, including "hermaphrodites"; with some remarks on embryology, as facilitating animal nomenclature, classification, and the study of comparative anatomy. *Edinburgh New Philosophical Journal* 22: 345–364.

Bredekamp, H. 2005. *Darwins Korallen. Frühe Evolutionsmodelle und die Tradition der Naturgeschichte.* Klaus Wagenbach Verlag, Berlin.

Bredekamp, H. 2006. *I coralli di Darwin. I primi modelli evolutivi e la tradizione della storia naturale.* Nuova Cultura, Torino.

Bredekamp, H. 2008. *Les coraux de Darwin – Prermier modèles évolutionniste et la tradition de l'histoire naturelle.* Les Presses du Réel, Dijon.

Brower, AVZ. 2016. Are we all cladists? In: Williams, DM., Schmidt, M. & Wheeler, QD. (eds), *The Future of Phylogenetic Systematics: the Legacy of Willi Hennig.* Cambridge University Press, Cambridge, UK.

Camardi, G. 2001. Richard Owen, morphology and evolution. *Journal of the History of Biology* 34: 481–515.

Carpenter, W. 1841. *Principles of General and Comparative Physiology.* 2nd ed. John Churchill, London.

Box 3.3 *(cont.)*

Chambers, R. 1844. *Vestiges of the Natural History of Creation*. John Churchill, London.

Darwin, C. 1859. *On the Origin of Species by Means of Natural Selection, or, the Preservation of Favoured Races in the Struggle for Life*. John Murray, London.

Maderspacher, F. 2006. The captivating coral – the origins of early evolutionary imagery. *Current Biology* 16(13): R476–R478.

Owen, R. 1894. *The Life of Richard Owen*. Vol. 1. J. Murray, London.

Porter, DM. 1987. Darwin's notes on *Beagle* plants. *Bulletin of the British Museum (Natural History) Historical Series* 14: 145–233.

Richards, E. 1987. A question of property rights: Richard Owen's evolutionism reassessed. *British Journal for the History of Science* 20: 129–171.

Richards, R. 1992. *The Meaning of Evolution*. University of Chicago Press, Chicago.

Torrens, E. & Barahona, A. 2013. Las Musas de Darwin tras el diagramma de 1859. [Darwin's muses behind his 1859 diagram]. *Arbor, Ciencia, Pensamiento y Cultura* 189–763 (Septiembre-Octubre): a072.

Voss, J. 2007. *Darwins Bilder: Ansichten der Evolutionstheorie 1837-1874*. Fischer Taschenbuch, Frankfurt.

Voss, J. 2010. *Darwin's Pictures: Views of Evolutionary Theory, 1837-1874*. Yale University Press, New Haven, CT.

Weber, TR. 2008. [Review of: Horst Bredekamp, *Darwins Korallen. Frühe Evolutionsmodelle und die Tradition der Naturgeschichte*. Berlin: Wagenbach Verlag, 2005. Pp. 111. ISBN 978-3803151735. €22.50 (hardback). Olaf Breidbach, *Visions of Nature: The Art and Science of Ernst Haeckel*. Munich: Prestel Verlag, 2006. Pp. 299. ISBN 978-3791336640. $100.00 (hardback)]. *British Journal for the History of Science* 41: 301–302.

Wilkins, J. 2008. Darwin. In: Tucker, A. (ed.), *A Companion to Philosophy of History and Historiography*. Wiley-Blackwell Publications, Chichester, pp. 404–415.

Further Reading

Darwin's various 'tree-of-life' sketches have been discussed and reproduced in many places, usually from the same original source and often alongside other unpublished tree-like sketches made by Darwin. The latter are usually cruder than those he published, but were, of course, never intended to be. Many of these can now be viewed via the Darwin Online website (http://darwin-online.org.uk/).

In addition, these references might be useful to examine the breadth of studies relating to Darwin's diagrams and the varying understanding of genealogy:

Bouzat, JL. 2014. Darwin's diagram of divergence of taxa as a causal model for the origin of species. *The Quarterly Review of Biology* 89: 21–38.

Brink-Roby, H. 2009. Natural representation: diagram and text in Darwin's *On the Origin of Species*. *Victorian Studies* 51: 247–273.

Helmreich, S. 2018. Ghost Lineages, Ghost Acres, and Darwin's "Diagram of Divergence of Taxa" in On the Origin of Species: Critical Commentary Inspired by Banu Subramaniam's Ghost Stories for Darwin. *Catalyst: Feminism, Theory, Technoscience* 4(2): 1-17.

Klapisch-Zuber, C. 1991. The genesis of the family tree. *I Tatti Studies in the Italian Renaissance* 4: 105–129.

Box 3.3   *(cont.)*

Priest, G. 2018. Diagramming evolution: the case of Darwin's trees. *Endeavour* 42(2–3): 157–171.

Weigel, S. 2003. Genealogie. Zu Ikonographie und Rhetorik einer epistemologischen Figur in der Geschichte der Kultur- und Naturwissenschaft. In: Schramm, H. (ed.), *Bühnen des Wissens. Interferenzen zwischen Wissenschaft und Kunst.* Dahlem University Press, Berlin, pp. 226–267. [In English: Genealogy: on the iconography and rhetorics of an epistemological topos. In: Alexandre, AF., Guerreiro, A. & Pombo, O. (eds), *Enciclopédia e Hipertexto.* Duarte Reis, Lisboa, 2006, pp. 1–21.]

[1] Darwin's first attempt at these kinds of diagram was in July 1837 (Barrett et al. 1987, p. B26).

[2] 'There is no direct evidence that Darwin took his inspiration from Barry, but the puzzle pieces fit snugly together, they look homologous' (Richards 1992, p. 111).

[3] Also in: Richards (1987, p. 135), Richards (1992, p. 109), Camardi (2001, p. 497), Amundson (2005, p. 70), Voss (2007, p. 137), Voss (2010, pp. 94–100), Wilkins (2008, p. 408), Brower (2016, p. 106).

writes ... 'I think', ... and sketches an abstract evolutionary tree' (Eldredge 2005, p. 1865, we return to Einstein below).

Gregory was not the first to invoke the notion of metaphor. The Oxford English dictionary informs us that a metaphor is 'a figure of speech in which a word or phrase is applied to something to which it is not literally applicable ... a thing regarded as symbolic of something else', and David Mindell reminds us that 'The price of metaphor is eternal vigilance' (attributed to Rosenblueth and Wiener by Lewontin 2001[2], in Mindell 2013).

Extravagant claims like Eldredge's to one side, Darwin's conviction concerning the facts of evolution, rather than any specific mechanism of diversification, came directly from the classificatory (Linnean) hierarchy, the scheme thrashed out by numerous taxonomists whose job it was to make sense of the vast amount of new material that was being collected from around the world and accumulating in museums and herbaria – as they still do. Darwin's 'principle of divergence', his way of explaining the cause of the taxonomic hierarchy, did not arise in Darwin's mind until much later, around 1855–1857 (Ospovat 1981, p. 170 *et seq.*). As historian Polly Winsor concluded, 'with respect to the role of taxonomy, the principle of divergence is of minor importance, it is the relationship between the Linnaean hierarchy and branching evolution that is primary' (Winsor 2009, p. 7).

---

[2] Like others, we tried without success to find the original source of this quotation and have come to the conclusion it should be attributed to Lewontin, as he has used it frequently but without citing the original source.

Alfred Russel Wallace (1823–1913) recognised the 'central metaphor' as it related to the taxonomic hierarchy but wrote of analogy[3] instead of metaphor:

> Returning to the analogy of a branching tree, as the best mode of representing the natural arrangement of species and their successive creation ... (Wallace 1855, p. 191; Wallace 1856, p. 205, reproduced as our Figure 3.2)

Just before Wallace, Charles Victor Naudin (1815–1899, Marza & Cerchez 1967) had written:

> Considered from this point of view, the plant kingdom would not be a linear series whose terms would increase or decrease in organizational complexity from one end to the other; it would not be like the disordered tangle of intersecting lines as in a geographical map, whose regions, different in shape and size, would meet as a greater or lesser number of points; it would be a like a tree whose roots are mysteriously hidden in the depths of time, which gives birth to a limited number of successively divided and subdivided stems. (Naudin 1852, p. 102, our translation, see also Padian 1999, p. 359)

Wallace took his lead from Hugh Edwin Strickland (1811–1853, see van Wyhe 2016[4]), who proposed that:

> The natural system may, perhaps, be most truly compared to an irregularly branching tree, or rather to an assemblage of detached trees and shrubs of various sizes and modes of growth. (Strickland 1841, p. 190)

Darwin pursued the comparison as well, writing, instead, of the 'tree' as simile[5]: 'The affinities of all the beings of the same class have sometimes been represented by a great tree. I believe this simile largely speaks the truth'. He continued:

> The green and budding twigs may represent existing species; and those produced during each former year may represent the long succession of extinct species ...

---

[3] 'A comparison between one thing and another, typically for the purpose of explanation or clarification' (Oxford English Dictionary).

[4] 'Strickland was no evolutionist and his proposal to graphically represent the varying degrees of similarities between groups of animals on tree-like diagrams had long been in the literature' (van Wyhe 2016, p. 60). Van Wyhe goes on to comment further on Wallace's views concerning the tree analogy: 'This passage [from Wallace] was perhaps inspired by the fine tree diagram of mollusca ... in Knight's *English cyclopaedia* (1854) which Wallace had with him at the time of writing the paper. The diagram is far more suggestive (at least to a modern reader) of a branching evolutionary tree than Wallace's Strickland-inspired diagrams in his 1856 paper ...' (Van Wyhe 2016, p. 60). Yet the illustration of Knight's 'tree' is clearly a written classification, although it does indeed have a tree-like structure (Knight 1854, p. 155). Van Wyhe, inadvertently perhaps, illustrates our point about classifications and tree-like diagrams.

[5] 'The use of an expression comparing one thing with another, always including the words "as" or "like"' (OED) 'A comparison of one thing with another, esp. as an ornament in poetry or rhetoric' (OED, accessed 28 May 2018).

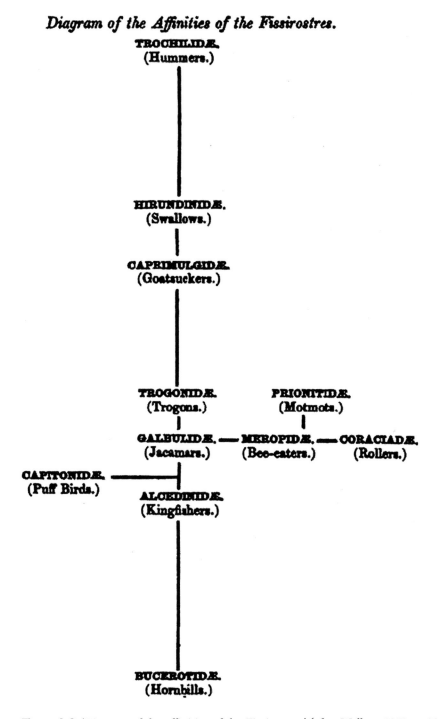

Figure 3.2 'Diagram of the affinities of the Fissirostres' (after Wallace 1856, p. 205).

> As buds give rise by growth to fresh buds, and these, if vigorous, branch out and
> overtop on all sides many a feebler branch, so by generation I believe it has been
> with the great Tree of Life, which fills with its dead and broken branches the crust
> of the earth, and covers the surface with its ever branching and beautiful ramifi-
> cations. (Darwin 1859, pp. 129–130)

Relevant to this is a passage from Polly Winsor:

> The rhetoric in current debates sometimes implies that only an evolutionist can
> construct a meaningful reference system for living things, but what this key quote
> highlights is that by Darwin's time, many European naturalists believed that
> organisms were all "related" and that their classifications should express these
> relationships. (Winsor 2009, p. 1)

The key Darwin quote referred to by Winsor occurs one page earlier than that
given above:

> It is a truly wonderful fact – the wonder of which we are apt to overlook from
> familiarity – that all animals and all plants throughout all time and space should
> be related to each other in group subordinate to group, in the manner which we
> everywhere behold – namely, varieties of the same species most closely related
> together, species of the same genus less closely and unequally related together,
> forming sections and sub-genera, species of distinct genera much less closely
> related, and genera related in different degrees, forming sub-families, families,
> orders, subclasses, and classes. (Darwin 1859, p. 128)

Rather than any model of evolution (e.g., such as the principle of divergence),
the hierarchy of life was largely an empirical discovery, one made time and
time again by a variety of taxonomists all with differing points of view, and
differing methodologies, and little to guide them but the accumulating specimens, the examination of their parts (characters, homologues) that yielded
groups (taxa) – that is, gathering evidence, yielding conclusions. Phylogeny, as
is generally understood today (but see Chapter 13), is derivative rather than
empirical.

Phylogeny

Phylogenetic trees began with Ernst Haeckel as he originally coined the word
phylogeny.[6] Haeckel was influenced by three sources: the single diagram in Darwin's *Origin* (Darwin 1859), the trees relating languages published by his friend

---

[6] Bigoni and Barsanti (2011) showed that St. George Mivart published the first evolutionary tree,
one representing the relationships among mammals (Mivart 1865, p. 592). Interesting though this
is, Haeckel coined the word phylogeny and so, in a strict sense, his must be considered the first.

August Schleicher (1821–1868) and the palaeontological trees of Heinrich Georg Bronn (1800–1862) (see Williams & Ebach 2008). At first many of Haeckel's diagrams did resemble actual trees, probably inspired by the European Oak, *Quercus robur* (Oppenheimer 1987, p. 127), but later they became more stylistic stick-like affairs (Haeckel 1896).

While the taxonomic (Linnean) hierarchy was Darwin's key to an evolutionary explanation, the tree as 'central metaphor' (or analogy, or simile, or 'evolutionary icon' even, Hellström 2011) eventually became 'a condensation of real events, rather than a metaphor' (Beer 2009, p. 33; 'The price of metaphor is eternal vigilance'), setting in motion, we suspect, the beginnings of the notion that is now referred to as 'tree-thinking', the literal interpretation of the branching representation of taxon relationships as a phylogenetic framework, rather than simply a metaphor derived from the classificatory hierarchy. Of course, 'tree-thinking' is but one interpretation of phylogeny (see Chapter 7).

Here we briefly return to Einstein. In the 2016 edition of the *American Journal of Undergraduate Research*, Moylan et al. published a paper entitled 'Einstein's 1905 Paper on $E=mc^2$' (Moylan et al. 2016). Their concluding Press Summary read as follows:

> It is a well-known and indisputable fact that Einstein's first paper on $E=mc^2$, published in 1905, is problematic in that it suffers from the error of circular reasoning. Despite this defect it is blindly referenced in many scientific articles as the official derivation of the famous formula. Furthermore, the arguments in that paper are still used today by some authors to purportedly derive $E=mc^2$. In view of this disturbing state of affairs, it seems worthwhile to have a clear and as elementary as possible explanation of the erroneous reasoning in Einstein's 1905 paper on $E=mc^2$. (Moylan et al. 2016, p. 10)

An error of circular reasoning: Nils Petter Hellström reviewed Pietsch's book on trees in biology, *Trees of Life: A Visual History of Evolution* (Hellström 2013, pp. 184–185). Commenting on the recent trend to publish circular trees, the kind of diagram now favoured by many molecular systematists, Hellström wrote,

> Pietsch has failed to discern such incongruities because he believes in an actual tree of life. Introducing the last diagram in his book, by Hillis, Zwickl and Gutell (2003), Pietsch writes that it was based on rRNA sequences 'from about three thousand species from throughout the tree of life' (p. 312). *Here the metaphor is the model for itself.* (Hellström 2013, p. 185, our emphasis)

We have already noted that the taxonomic hierarchy is an empirical discovery, emerging from decades of study by many comparative biologists. The interpretation of that hierarchy as a diverging evolutionary tree began early in the nineteenth century. Its acceptance as fact caused little violence to the taxonomic enterprise ('… systematists will be able to pursue their labours as at present',

Darwin 1859, p. 484) but its adoption as a vital part of many, if not all, current phylogeny reconstruction programs does indeed make *the metaphor a model for itself.* And so it is: another error of circular reasoning. To use the tree metaphor for phylogenetic reasoning is in error, as the tree shape is derived from the empirical discovery of taxonomists.

We remain agnostic with respect to whether evolution actually proceeds in a hierarchical fashion, a reticulate fashion, a mixture of both, neither or something else yet known to us (de Pinna 2014; see the section on Non-Trees below).

'Tree-Thinking' and 'Evolutionary Fundamentalism'

If we could just stop talking about 'accurate phylogenies' and 'common ancestors' as if they were real things instead of inferences from rigorous analyses of empirical data, *it would recast the debate between people with alternate belief systems as one between people with evidence, and people without.* That is the real challenge of tree-thinking.
(Andrew Brower 2006, Comment on Tree Thinking, Science e-letters: https://science.sciencemag.org/
content/310/5750/979/tab-e-letters, our emphasis)

The term 'tree-thinking' was first introduced by Robert O'Hara nearly 30 years ago (O'Hara 1988). Nearly a quarter of a century later it was the primary orientation of a book-length treatment, *Tree Thinking: An Introduction to Phylogenetic Biology* (Baum & Smith 2012), and five years after that discussed as a saviour to us all (Funk 2018[7]).

For O'Hara, 'tree-thinking', the literal interpretation of the branching representation of taxon relationships (cladograms), allowed certain kinds of evolutionary questions to be posed: 'The ability to analyze evolutionary "why" questions … comes from what I call "tree thinking" (after Mayr's "population thinking"). Tree thinking is absolutely necessary for answering almost all evolutionary "why" questions' (O'Hara 1988, p. 151).

O'Hara contrasted this way of thinking with what he called 'group-thinking', a facet of pre-evolutionary thinking:

> A pre-evolutionary perspective on diversity results in what may be called 'group thinking,' and state questions arise out of group thinking; it is tree thinking that allows one to convert a question of state into an evolutionary question of change. (O'Hara 1988, p. 151)

---

[7] 'In the first 15 years of the 21st century, tree-thinking pervaded the life sciences (e.g., Baum & Smith, 2012) and the public imagination, leading to, among other things, the emergence of Evolutionary Medicine, Evolutionary Ecology, and new tree-based Food Safety methods' (Funk 2018, p. 180).

That is because:

> One needs to know the chronicle of events in a situation before causation can be inferred with any confidence; in evolutionary terms, one needs to know phylogeny. (O'Hara 1988, p. 151)

It may seem a little more than odd that criticism was placed within the legacy of 'pre-evolutionary' classifiers, even though it was upon these scientists' work that Darwin based his own interpretations (O'Hara 1994, 1997). Interestingly, too, O'Hara understands his 'tree-thinking' as having the same significance, and same root, if you will, as Ernst Mayr's 'population thinking' (for more details see Chapter 4).

The most significant issue with respect to systematics (taxonomy) as a science was captured in Andrew Brower's phrase 'evolutionary fundamentalism':

> Baum's dictum (Baum & Offner, 2008) that 'one can develop a solid understanding of what a phylogenetic tree represents without knowing much about how scientists actually infer the structure of those trees' is a dangerous recipe for evolutionary fundamentalism – 'knowledge' based on belief without the critical capacity to trace its empirical source or understand why we believe what we believe. (Brower 2016, p. 384)

Or the view we follow in this book, that the cladogram is the result of empirical study. A study of the exchange between Weinstein (2016) and Brower (2016) might help to appreciate the differences:

> Among other things, you are a mammal. This is an assertion that no rational person is likely to dispute. Such a person might reasonably ask: 'What makes people mammals?' This question would likely be answered as follows: 'People have hair, make milk, have a single bone in their lower jaw, and so on.' But this answer is completely wrong. It is, in fact, the correct answer to the question, 'What characteristics can be used to recognize a mammal?' The correct answer to the question above is: 'People are mammals because they are descended from the most recent common ancestor of all mammals. (Weinstein 2016)

> Thus, some characters, such as backbones, mammary glands or feathers, are unique to particular clades and actually do diagnose or 'define' the taxa that possess them. Posing counterfactual hypotheticals such as milkless mammals or featherless birds is fine, in theory, but such creatures would either need to be recognized on the basis of other congruent characters, such as heterodont dentition (for the mammal) or some combination of the bird features mentioned above, or they would not be classifiable . . . common ancestry is an inference, not an observation, and hypotheses about common ancestry are results of systematics, not assumptions that justify it (Brower 2016).

One is a metaphysician, the other an empiricist.

One outcome of the 'tree-thinking' approach is to have prevented, or prevent, 'non-tree' interpretations of evolutionary relationships, which, as before, seem to be empirical discoveries (Booth et al. 2016, although see Chapter 7 for further explanation).

Non-trees

'Central metaphors' and metaphysics to one side, some taxon relationships have recently been portrayed in ways other than a simple bifurcating tree (an early example is in Hilario & Gogarten 1993). For example, there is the 'ring of life' (Figure 3.3a, after Rivera & Lake 2004), schemes with interlinking, anastomosing networks of the major eukaryote groups (Figure 3.3b; Doolittle 1999, 2000; Doolittle & Bapteste 2007), the construction of interconnecting networks relating various taxa (Figure 3.3c, Hertel et al. 2006), and so on. These arrays of non-trees have resulted primarily from the analysis of molecular data (genomic data), although the general argument was explored in a cladistic context via supposed problems with hybridisation in classification some three decades ago (e.g., Bremer & Wanntorp 1979 and Funk 1985, see Figure 3.4a and b, respectively; for a general review and history, see Kressing 2016).

Explanation for non-trees comes primarily from the process of Lateral (or Horizontal) Gene Transfer (LGT, HGT), being the horizontal transfer of a gene, or some genetic material, from one organism to another more distantly related organism by means other than direct ancestry (e.g., Dagan & Martin 2006). The idea was first proposed to support the theory of serial endosymbiosis (Margulis 1998; Lane & Archibald 2008), that idea having a relatively long history, dating from the work of Merezhkowsky (Figure 3.5a after Merezhkowsky 1910, p. 366). It is seen by some as a 'meta-revolution' in biology:

> The genomic revolution did more than simply allow credible reconstruction of the gene sets of ancestral life forms. Much more dramatically, it effectively overturned the central metaphor of evolutionary biology (and, arguably, of all biology), the Tree of Life (TOL), by showing that evolutionary trajectories of individual genes are irreconcilably different. Whether the TOL can or should be salvaged—and, if so, in what form—remains a matter of intense debate that is one of the important themes of this book. (Koonin 2011, p. viii)

LGT explains instances of what has been called *xenology* ('foreign genes', Gray & Fitch 1983, p. 64), 'a form of homology (inferred common ancestry) in which the sequence (gene) homology is incongruent with that of the organisms carrying the gene, and horizontal gene transfer or transfection is the assumed cause' (Patterson 1988, p. 612).

Xenology finds its closest morphological equivalent in *parallelism*, a term which still remains hard to define but can be simplified by associating it with incongruent homologies (similarities). Parallelism is a topic that is being discussed again (e.g.,

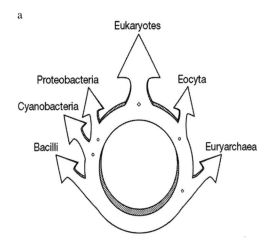

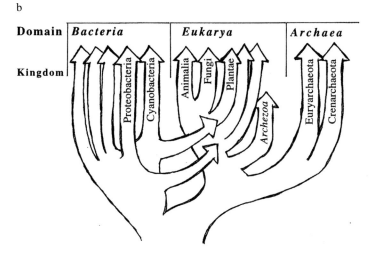

Figure 3.3 (a) The 'ring of life' (after Rivera & Lake 2004, p. 153, fig. 3, with permission); (b) interlinking, anastomosing networks of the major eukaryote groups (after Doolittle 1999, p. 2125, fig. 2); (c) interconnecting networks relating various taxa (after Hertel et al. 2006, p. 6, fig. 2, with permission).

Scotland 2011), although the notion never really disappeared: 'the significance of this similarity [parallelism] is thus dependent on the existence of *a relevant underlying process*' (Sanderson & Hufford 1996, p. 328, our emphasis). And even further back, George Gaylord Simpson wrote:

> In the most restricted sense virtually all evolution involves parallelism. Homologous genes tend to mutate in the same way (p. 9) ... Homology is always valid

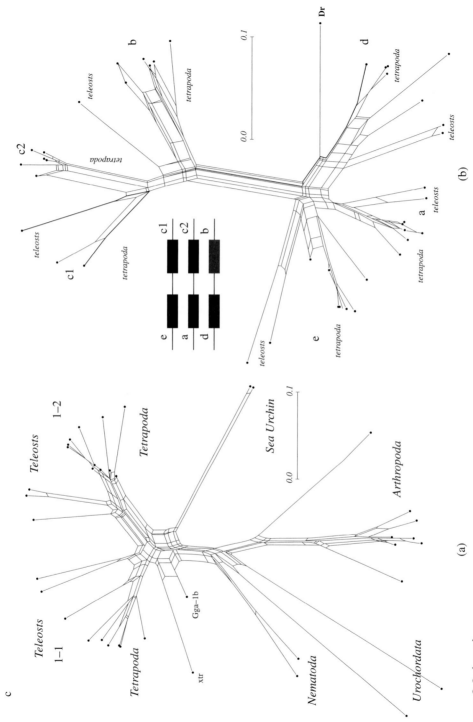

Figure 3.3 (cont.)

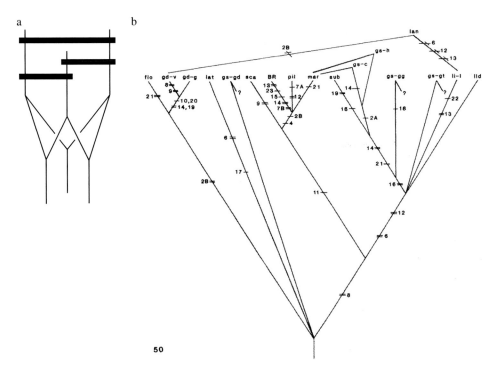

Figure 3.4 (a) Cladogram demonstrating conflicting characters when hybridisation occurs (after Bremer & Wanntorp 1979, p. 627, fig. 4, with permission); (b) a similar but more complex hybridisation cladogram '... of *Chrysopsis* and *Bradburia* with reticulations' (after Funk 1985, p. 698, fig. 50, reproduced with permission of the Missouri Botanical Garden Press, St. Louis).

> evidence of affinity. Parallelism is less direct and reliable, but it is also valid
> evidence within somewhat broader limits. It may lead to overestimates of degree
> of affinity, but it is not likely to induce belief in wholly false affinity (p. 10).
> (Simpson 1945, pp. 9–10)

Simpson turned out to be mistaken: similarities identified as parallelisms are simply incongruent characters. All the same, it has been argued that reticulate networks ('non-trees') allow incongruent 'homologies' to be accommodated on the same diagram relative to congruent homologies (e.g., Huson & Bryant 2006). The general idea is similar – not exactly, of course – to the quinarian diagrams of William Sharp Macleay (1792–1865, as in Figure 3.5b, after Swainson 1836–1837) and his circular systems, his attempt to represent analogies and affinities in one diagram. Perhaps the apparent death of circular systems was not severe enough.

To some, orthologous (homologous) genes do not support 'tree-thinking' (Bapteste et al. 2005) and general incongruence among gene trees presents problems

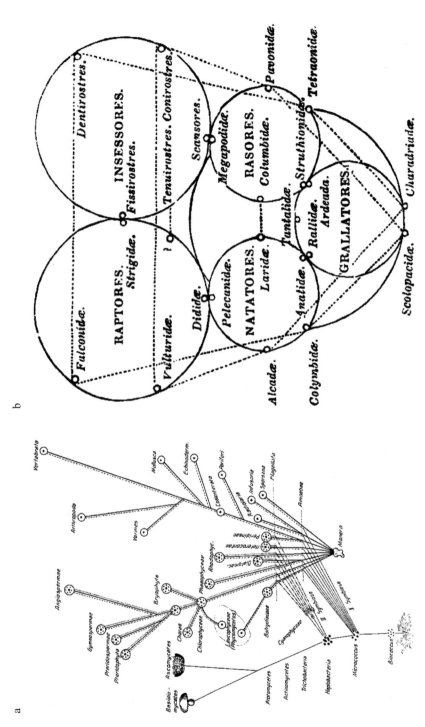

Figure 3.5 (a) The first diagram depicting serial endosymbiosis (after Mereschkowsky 1910, p. 366); (b) quinarian diagram of the relationships of birds (after Swainson 1836–1837).

Box 3.4   Unrooted 'Trees' and Networks

A history of unrooted trees is given by de Pinna et al. (2016) and of networks by Morrison (2016). Briefly, in phylogenetic research, an unrooted tree first appeared in Edwards and Cavalli-Sforza (1964, p. 75, figure 1, reproduced as our Figure B3.4.1) who based their analytical approach, in part, on earlier efforts by Robert Prim (b. 1921, a mathematician) on networks (Prim 1957). Farris developed the concept of networks and trees with respect to rooted and unrooted trees and pursued the parallel between his own Wagner networks and the earlier Prim networks (Farris 1970,[1] an example of results was given in his figure 3, reproduced as our Figure B3.4.2[2]). Wilkinson et al. provided a new terminology for the parts of unrooted trees (Wilkinson et al. 2007).

Of the ubiquity of unrooted trees since Farris, de Pinna et al. note:

> The tremendous advance of molecular data has also resulted in major expansions of the theoretical and methodological tools employed in each method. Still, throughout all such changes and trends, the use of unrooted trees has remained unscathed, a remarkable case of constancy amidst a history of major methodological change. The principle first enunciated for parsimony applies equally well for other kinds of analysis: the likelihood, posterior probability, and neighbour-joining distances of a tree are all independent of the placement of the root. This happens because the processes of change in core versions of all those methods are time-reversible or symmetrical. (de Pinna et al. 2016, p. 318)

They go on to note some exceptions, notably – and of significance for this book – *three-item analysis* and *traditional Hennigian phylogenetics* (see Chapter 7). And, we might add, the general practice of taxonomists through the ages (see Ebach & Williams, 2020).

We might discover why unrooted trees have 'remained unscathed' by inspection of Prim's 1957 paper where he states 'The basic problem considered is that of interconnecting a given set of terminals with a shortest possible network of direct links' (Prim 1957, p. 1389). Prim set out to track the shortest network (route) from one point to another, so as to get from one place to another efficiently (Prim 1957, figure 1, reproduced as our Figure B3.4.3).[3] Thus one can easily appreciate how the problem of getting from one geographical place to another efficiently (from Helena to Bismark, for example, our Figure B3.4.3) might apply analogously to the biological problem of getting from one character state to another, from character state A to character state B.

de Pinna et al. further note:

> The nearly universal application of unrooted trees throughout the explosive advance of phylogenetic reconstruction in recent decades is highly significant. Few other fundamental concepts in phylogenetics have remained similarly immune to change during that period. (de Pinna et al. 2016, p. 318)

Indeed. Yet, as we have noted elsewhere, this is based on the notion that going from character (state) A to character (state) B, B would need to be considered a modification or transformation of A, hence the more usual notation of A → B (Figure B3.4.4b) and with a 'time-reversible' element added, yields the notation A ⟵⟶ B (Figure B3.4.4c). This

Box 3.4 *(cont.)*

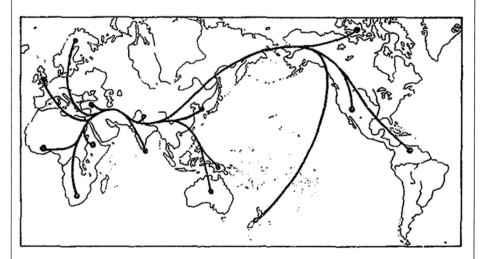

Figure B3.4.1 An unrooted tree (after Edwards & Cavalli-Sforza 1964, p. 75, figure 1, with permission).

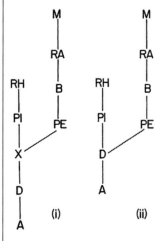

(i)   (ii)

Figure B3.4.2 Wagner network (after Farris 1970, fig. 3, with permission).

nevertheless betrays what might has been referred to as an 'Aristotelian' version of character change or character transformation (Platnick 1979; Platnick & Rosen 1987; see also Platnick & Nelson 1981 and Nelson & Platnick 1981, p. 238), with the implication that A disappears when it transforms to B. As Platnick put it:

> To view some character X as being composed of three states ... implies that the character states are alternatives, when they are actually additions ... In this case, character X is actually equivalent to state 1 (i.e., it defines a group, all the

Box 3.4 *(cont.)*

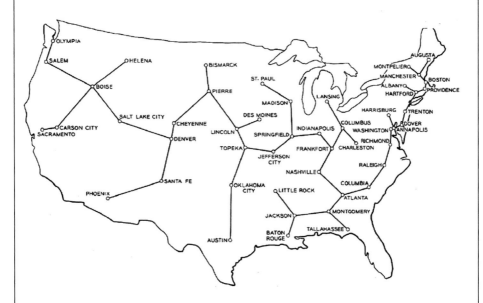

Figure B3.4.3 Prim network (after Prim 1957, fig. 1, reused with permission of Nokia Corporation and AT & T Archives).

members of which have state 1, either in its original or some modified form). (Platnick 1979, p. 543, his illustration is reproduced as our Figure B3.4.4a)

When considering any kind interpretation of transformation or any kind of instance of change, then the appropriate representation would be *a*(*bc*) (with *a* having character state A, *b* and *c* having character state B). Even *without* any notion of change, should one adopt that view, one might easily appreciate that the issue is still not one of representing a linear direction of change but of representing the relationship as implied by the given data, which is, almost by definition, hierarchical (Figure B3.4.4d). Thus, we offer two conclusions:

1. The Prim network/'travelling salesman'/unrooted tree approach to understanding *relationships* amongst organisms is simply misapplied mathematics, as is the graph theory derived from it, when used for systematics/taxonomy (for further comments on this see Chapters 7 and 8);

2. Unrooted trees (and networks) are almost entirely irrelevant to understanding *relationships* among organisms, and to creating classifications.

Of course, from a purely mathematical point of view, there is a relationship between unrooted and rooted trees (Dobson 1974; and for characters, McMorris & Zaslavsky 1981), at least a relationship between the numbers of rooted trees that can be derived

Box 3.4 *(cont.)*

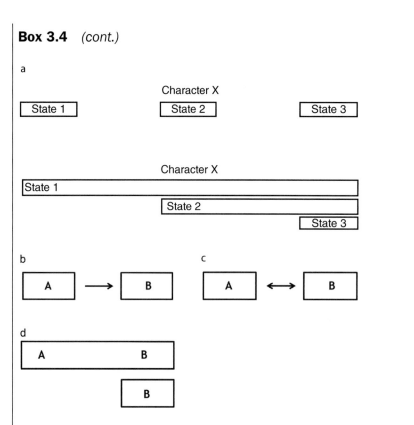

Figure B3.4.4 (a) Relations between characters and states (after Platnick 1979, p. 543, with permission); (b–d) relations between characters and states as linear series and as transformation series.

Table B3.4.1 The relationship between number of taxa, number of unrooted trees and number of rooted trees

Number of taxa	Number of unrooted trees	Number of rooted trees
3	1	3
4	3	15
5	15	105
6	105	945
7	945	10 395
8	10 395	135 135
9	135 135	2 027 025
10	2 027 025	34 459 425

Box 3.4 *(cont.)*

from any particular unrooted tree (Table B3.4.1, see Phipps 1976 and Felsenstein 1978, who both have presented many more numbers tabulated for different kinds of trees and classifications than we have; see also Felsenstein 2018, pp. 329–330, for a short historical account of these numbers and their generation). While superficially interesting, from the perspective of classification (even from the perspective of phylogeny for that matter), unrooted trees should be of no great importance or interest.

   Felsenstein offered three reasons why these numbers of trees might be of interest (Table B3.4.1). His third reason was that 'from time to time a taxonomist will propose a method of finding evolutionary trees in which one proposed step is examining all possible trees to see whether some criterion is satisfied. Enumeration of evolutionary trees may then be a powerful argument for adopting some procedure either less ambitious or more powerful' (Felsenstein 1978, p. 33). Oddly, by viewing systematic data as *relational* instead of *linear* it becomes both 'less ambitious', by not attempting to recreate, reconstruct or 'infer' phylogeny, and at the same time 'more powerful', by providing an exact method for biological classification.

References

Cook, W. 2011. *In Pursuit of the Traveling Salesman: Mathematics at the Limits of Computation*. Princeton University Press, Princeton, NJ.

de Pinna, M., Bockmann, FA. & Bagils, RZI. 2016. Unrooted trees discovered independently in philology and phylogenetics: a remarkable case of methodological convergence. *Systematics and Biodiversity* 14: 317–326.

Dobson, A. 1974. Unrooted trees for numerical taxonomy. *Journal of Applied Probability* 11: 32–42.

Ebach, M.C. & Williams, D.M. 2020. Ronald Brady and the cladists. *Cladistics* 36: 218–226

Edwards, AWF. & Cavalli-Sforza, LL. 1964. Reconstruction of evolutionary trees. In: Heywood, VH. & McNeill, J. (eds), *Phenetic and Phylogenetic Classification*. The Systematics Association Publication No. 6. The Systematics Association, London, pp. 67–76.

Farris, J.S. 1970. Methods for computing Wagner trees. *Systematic Zoology* 19: 83–92.

Felsenstein, J. 1978. The number of evolutionary trees. *Systematic Zoology* 27: 27–33.

Felsenstein, J. 2018. Anthony Edwards, Luca Cavalli-Sforza, and phylogenies. In: Winther, RG. (ed.), *Phylogenetic Inference, Selection Theory, and History of Science*, Cambridge University Press, Cambridge, UK, pp. 325–333.

McMorris, FR. & Zaslavsky, T. 1981. The number of cladistic characters. *Mathematical Biosciences* 54: 3–10.

Morrison, DA. 2016. Genealogies: pedigrees and phylogenies are reticulating networks not just divergent trees. *Evolutionary Biology* 43: 456–473.

Nelson, GJ. & Platnick, NI. 1981. *Systematics and Biogeography: Cladistics and Vicariance*. Columbia University Press, New York.

Box 3.4   *(cont.)*

Phipps, JB. 1976. The numbers of classifications. *Canadian Journal of Botany* 54: 686–688.

Platnick, NI. 1979. Philosophy and the transformation of cladistics. *Systematic Zoology* 28: 537–546.

Platnick, NI. & Nelson, G. 1981. The purposes of biological classification. *PSA: Proceedings of the Biennial Meeting of the Philosophy of Science Association, 1978*, volume 2 (Symposia and Invited Papers), pp. 117–129.

Platnick, NI. & Rosen, DE. 1987. Popper and evolutionary novelties. *History and Philosophy of the Life Sciences* 9: 5–16.

Prim, RC. 1957. Shortest connection networks and some generalizations. *The Bell System Technical Journal* 36: 1389-1401.

Wilkinson, M., McInerney, JO., Hirt, RP., Foster, PG. & Embley, TM. 2007. Of clades and clans: terms for phylogenetic relationships in unrooted trees. *Trends in Ecology & Evolution* 22: 114–115.

Further Reading

General

Platnick, NI. & Rosen, DE. 1987. Popper and evolutionary novelties. *History and Philosophy of the Life Sciences* 9: 5–16.

The authors note in the acknowledgements: 'Although the manuscript was completed early in 1980, circumstances beyond the control of the authors have prevented its timely publication. Because the piece has been widely circulated in manuscript, no attempt was made here to alter the 1980 document, beyond updating a few references then in press'. This paper is also relevant to Chapter 4.

Networks

Kikusawa, R. & Reid, LA. (eds) 2018. *Let's Talk about Trees: Tackling Problems in Representing Phylogenetic Relationships among Languages*. Senri Ethnological Studies 98. National Museum of Ethnology, Osaka.

Kutschera, U. 2011. From the scala naturae to the symbiogenetic and dynamic tree of life. *Biology Direct* 6: 33 [25 pages with reviews]

Minaka, N. 2016. Chain, tree, and network – The development of phylogenetic systematics in the context of genealogical visualization and information graphics. In: Williams, DM., Schmitt, M. & Wheeler, QD. (eds) *The Future of Phylogenetic Systematics: The Legacy of Willi Hennig*. Cambridge University Press, Cambridge, pp. 410–430.

Minaka, N. 2018. Tree and network in systematics, philology, and linguistics – Structural model selection in phylogeny reconstruction. In: Kikusawa, R. & Reid, LA. (eds), *Let's Talk about Trees: Tackling Problems in Representing Phylogenetic Relationships among Languages*. Senri Ethnological Studies 98. National Museum of Ethnology, Osaka, pp. 9–24.

Morrison, DA. 2014. Is the tree of life the best metaphor, model or heuristic for phylogenetics? *Systematic Biology* 63: 628–638.

> **Box 3.4** *(cont.)*
>
> Morrison, DA. 2016. Genealogies: pedigrees and phylogenies are reticulating networks not just divergent trees. *Evolutionary Biology* 43: 456–473.
>
> Wheeler, WC. 2015. Phylogenetic network analysis as a parsimony optimization problem. *BMC Bioinformatics* 16: 296.
>
> These references have been suggested not because we believe them to be of any great assistance to the development of cladistics (classification), but so that the reader can become more aware of why Networks, etc., are irrelevant to classification.
>
> [1] 'Prim Networks are quite crude approximations to Wagner Networks, but have the advantages that they can be computed exactly and very efficiently. Prim Networks may be most useful in evolutionary studies as tools for preliminary analyses of data, the more elaborate programs described below being reserved for refining final conclusions' (Farris 1970, p. 87).
>
> [2] The only difference is the addition of a 'hypothetical intermediate' (X) in the Wagner Network.
>
> [3] There is little need to follow this history too diligently but to simply note that Prim understood his problem to be similar to that of the *'Traveling Salesman Problem:* Find a closed path of minimum length connecting a prescribed terminal set' (Prim 1957, p. 1401). Both are problems dealing with getting from one place to another efficiently (see Cook (2011) for a recent overview).

for the effectiveness of these data, rather than provides alternative explanations for incongruence (LGT = parallelism).

The issue can be simplified: it is the conflict of characters (or genes, or taxa, in the case of biogeography) that confounds empirical discovery of the underlying taxonomic hierarchy, as it indeed confounded Macleay and others. As we noted above: in many DNA analyses, the program designed to find the result assumes a tree structure, the metaphor is the model for itself – some programs do not, as some find networks, or combinations of networks and trees. It seems that no one yet has discovered the exact empirical relationships between the evidence (DNA sequences) and the conclusions (taxa) – and programs to discover such relationships have their result built-in, so to speak (tree, non-tree). Nevertheless, from the point of view of classification, one needs to find the best fit of all congruent characters before any incongruent characters, those that do not conform or fit to the overall *regular* pattern of homologies, become obvious. As Brower concluded:

> It is no sin to popularize the idea of horizontal gene transfer as a complicating source of homoplasy for phylogenetic patterns [classifications], but this is neither new nor radical, except when wielded as a nihilistic critique of systematics. (Brower 2019)

Further comment can be found in Chapter 13.

1859 and Its Consequences

In spite of Darwin's statement to the contrary (see above), it is sometimes assumed that everything changed after 1859 with the publication of Darwin's *Origin* and it was recognised that '... community of descent is the hidden bond which naturalists have been seeking, and not some unknown plan of creation, or the enunciation of general propositions, and the mere putting together and separating objects more or less alike' (Darwin 1859, p. 404). 'Descent with modification' was the organising principle: that is, common origins and common ancestors. But these served as explanations not guides to *discovery*.

Be that as it may, doubts concerning Darwin's proposed mechanisms of evolutionary change soon emerged and a whole slew of alternatives were proposed. Matters coalesced into what might be called a post-Mendelian synthesis, characterised by 'population thinking'. A general consequence was to abandon the search for a natural classification, as if such a pursuit was not possible or ill-conceived, without some underlying changes to its basic philosophy. Cladistics, in the more general form we outline here – and understood as an attempt at reforming taxonomy – demonstrates that such a search can still be undertaken, albeit with a little more focus (Chapter 7), and the interference from essentialism and typology was at best minimal.

Summary

The hierarchical nature of classification was effectively an empirical discovery, a parameter of the data, rather than any underlying assumptions about causes (mechanisms), models (trees) or metaphors (trees, networks), hence the centrality of the *cladistic parameter* (Chapters 1 and 7), not just in the present day, but throughout the history of biological classification (Nelson & Platnick 1981; Williams & Ebach 2008).

Cladograms deal with evidence (homology) and their implications for taxon relationships to achieve a classification of organisms, rather than being a vehicle for various explanations of incongruence (non-homology).

We expect that classification and its study will outlive 'tree-thinking' – it will almost certainly outlive 'population thinking', if it has not done so already.

References

Archibald, JD. 2014. *Aristotle's Ladder, Darwin's Tree: The Evolution of Visual Metaphors for Biological Order*. Columbia University Press, New York.

Bapteste, E., Susko, E., Leigh, J., MacLeod, D., Charlebois, RL. & Doolittle, WF. 2005. Do orthologous gene phylogenies really support tree-thinking? *BMC*

Evolutionary Biology 5: 33; http://doi:10
.1186/1471-2148-5-33.

Barsanti, G. 1988. Le immagini della natura:
scale, mappe, alberi 1700–1800.
Nuncius 3: 55–125.

Barsanti, G. 1992. *La scala, la mappa,
l'albero: immagini e classificazioni della
natura fra sei e ottocento*. Sansoni,
Florence.

Baum, DA. & Smith, SD. 2012. *Tree
Thinking: An Introduction to
Phylogenetic Biology*. W. H. Freeman,
New York.

Beer, G. 2009. *Darwin's Plots: Evolutionary
Narrative in Darwin, George Eliot and
Nineteenth-Century Fiction*. 3rd ed.
Cambridge University Press,
Cambridge.

Bentlage, B. & Lewis, C. 2012. An illustrated
key and synopsis of the families and
genera of carybdeid box jellyfishes
(Cnidaria: Cubozoa: Carybdeida), with
emphasis on the "Irukandji family"
(Carukiidae). *Journal of Natural History*
46: 2595–2620.

Bigoni, F. & Barsanti, G. 2011. Evolutionary
trees and the rise of modern
primatology: the forgotten contribution
of St. George Mivart. *Journal of
Anthropological Sciences* 89: 1–15.

Booth, A., Mariscal, C. & Doolittle, WF.
2016. The Modern Synthesis in the light
of microbial genomics. *Annual Review
of Microbiology* 70: 279–297.

Bredekamp, H. 2005. *Darwins Korallen.
Frühe Evolutionsmodelle und die
Tradition der Naturgeschichte*. Klaus
Wagenbach Verlag, Berlin.

Bremer, K. & Wanntorp, H-E. 1979. Hierarchy
and reticulation in systematics. *Systematic
Zoology* 28: 624–627.

Brower, AV. 2006. Comment on Tree
Thinking, Science e-letters. www
.sciencemag.org/cgi/eletters/310/5750/
979#3414

Brower, AV. 2016. Tree-thinking. *Inference:
International Review of Science* 2(4).
https://inference-review.com/letter/
tree-thinking [with further response
from Weinstein]

Brower, AV. 2019. The Tree of Life:
metaphysics vs. metaphor. *Cladistics*.
35: 600–602.

Dagan, T. & Martin, W. 2006. The tree of one
percent. *Genome Biology* 7: 118.1–118.7.

Darwin, C. 1859. *On the Origin of Species by
Means of Natural Selection, or, the
Preservation of Favoured Races in the
Struggle for Life*. John Murray, London.

De Pinna, M. 2014. Species tot sunt diversae
quot diversas formas ab initio creativiv:
A dialogue on species. *Arquivos de
Zoologia* 45: 25–32.

Doolittle, WF. 1999. Phylogenetic
classification and the universal tree.
Science 284: 2124–2128.

Doolittle, WF. 2000. Uprooting the tree of
life. *Scientific American* Feb: 90–95.

Doolittle, WF. & Bapteste, E. 2007. Pattern
pluralism and the Tree of Life
hypothesis. *PNAS* 104: 2043–2049.

Eldredge, N. 2005. Darwin's *Other* Books:
"Red" and "Transmutation" Notebooks,
"Sketch," "Essay," and *Natural
Selection. PLoS Biology* 3(11): e382.
https://doi.org/10.1371/journal.pbio
.0030382

Funk, H. 2014. Describing plants in a new
mode: the introduction of dichotomies
into sixteenth-century botanical
literature. *Archives of Natural History*
41: 100–112.

Funk, VA. 1985. Phylogenetic patterns and
hybridization. *Annals of the Missouri
Botanical Garden* 72: 681–715.

Funk, VA. 2018. Collections-based science
in the 21st Century. *Journal of
Systematics & Evolution* 56: 175–193.

Giribet, G., Hormiga, G. & Edgecombe, GD.
2016. The meaning of categorical ranks

in evolutionary biology. *Organisms, Diversity & Evolution* 16: 427–430.

Gray, GS. & Fitch, WM. 1983. Evolution of antibiotic resistance genes: the DNA sequence of a kanamycin resistance gene from *Staphylococcus aureus*. *Molecular Biology & Evolution* 1: 57–66.

Gregory, TR. 2008. Understanding evolutionary trees. *Evolution: Education and Outreach* 1: 121–137.

Haeckel, E. 1896. *Systematische Phylogenie: Entwurf eines natürlichen Systems der Organismen auf Grund ihrer Stammesgeschichte.*Vol. 3: *Systematische Phylogenie der Wirbelthiere (Vertebrata)*. G. Reimer, Berlin.

Hellström, P. 2011. The tree as evolutionary icon: TREE in the Natural History Museum, London. *Archives of Natural History* 38: 1–17.

Hellström, P. 2013. *Review of: Pietsch, T. W. Trees of life. A visual history of evolution*. The Johns Hopkins University Press, Baltimore: 2012. *Archives of Natural History* 40: 184–185.

Hertel, J., Lindemeyer, M., Missal, K., Fried, C., Tanzer, A., Flamm, C., Hofacker, IL., Stadler, PF. and the Students of Bioinformatics Computer Labs 2004 and 2005. 2006. The expansion of the metazoan microRNA repertoire. *BMC Genomics* 7: 25.

Hilario, E. & Gogarten, JP. 1993. Horizontal transfer of ATPase genes – the tree of life becomes a net of life. *BioSystems* 31: 111–119.

Huson, DH. & Bryant, D. 2006. Application of phylogenetic networks in evolutionary studies. *Molecular Biology & Evolution* 23: 254–267.

Knight, C. (ed.) 1854. *The English Cyclopaedia: A New Dictionary of Universal Knowledge: Natural History*. Bradbury & Evans, London.

Koonin, EV. 2011. *Logic of Chance: The Nature and Origin of Biological Evolution*. FT Press, New Jersey [Updated edition].

Kressing, F. 2016. Lateral and vertical transfer in biology, linguistics and anthropology: an account of widely neglected ideas in the formation of evolutionary theories. *Evolutionary Biology* 43: 474–480.

Lambertz, M. & Perry, SF. 2015. Chordate phylogeny and the meaning of categorical ranks in modern evolutionary biology. *Proceedings of the Royal Society B: Biological Sciences* 282(1807): 2014–2327.

Lambertz, M. & Perry, SF. 2016. Again on the meaning of categorical ranks in modern evolutionary biology? *Organisms, Diversity & Evolution* 16: 723–725.

Lane, CE. & Archibald, JM. 2008. The eukaryotic tree of life: endosymbiosis takes its TOL. *Trends in Ecology and Evolution* 23: 268–275.

Laurin, M. 2010. The subjective nature of Linnaean categories and its impact in evolutionary biology and biodiversity studies. *Contributions to Zoology* 79: 131–146.

Margulis, L. 1998. *Symbiotic Planet: A New Look at Evolution*. Basic Books, New York.

Marza, VD. & Cerchez, N. 1967. Charles Naudin, a pioneer of contemporary biology (1815–1899). *Journal d'agriculture tropicale et de botanique appliquée* 14(10–11): 369–401.

Merezhkowsky, C. 1910. Theorie der zwei Plasmaarten als Grundlage der Symbiogenesis, einer neuen Lehre von der Entstehung der Organismen. *Biologische Centralblatt* 30: 353–367.

Mindell, DP. 2013. The tree of life: metaphor, model, and heuristic device. *Systematic Biology* 62: 479–489.

Mivart G. St. 1865. Contributions towards a more complete knowledge of the axial skeleton in the Primates. *Proceedings of the Zoological Society of London*, 33: 545–592.

Moylan, P., Lombardi, J. & Moylan, S. 2016. Einstein's 1905 Paper on E=mc2. *American Journal of Undergraduate Research* 13: 5–10.

Naudin, C. 1852. Considérations philosophiques sur l'espèce et la variété. *Revue Horticole* 1(4): 102–109.

Nelson, G. 1979. Cladistic analysis and synthesis: principles and definitions, with a historical note on Adanson's *Familles des Plantes. Systematic Zoology* 28: 1–21.

Nelson, GJ. & Platnick, NI. 1981. *Systematics and Biogeography: Cladistics and Vicariance.* Columbia University Press, New York.

O'Hara, RJ. 1988. Homage to Clio, or, toward an historical philosophy for evolutionary biology. *Systematic Zoology* 37: 142–155.

O'Hara, RJ. 1994. Evolutionary history and the species problem. *American Zoologist* 34: 12–22.

O'Hara, RJ. 1997. Population thinking and tree thinking in systematics. *Zoologica Scripta* 26: 323–329.

Oppenheimer, JM. 1987. Haeckel's variations on Darwin. In: Hoenigswald, HM. & Wiener, LF. (eds), *Biological Metaphor and Cladistic Classification.* University of Pennsylvania Press, Philadelphia, pp. 123–135.

Ospovat, D. 1981. *The Development of Darwin's Theory: Natural History, Natural Theology, and Natural Selection, 1838-1859.* Cambridge University, Cambridge, UK.

Oxford English Dictionary (OED) Online. simile, n. March 2018. Oxford University Press. www.oed.com/view/Entry/179881?rskey=i3BDH9&result=1&isAdvanced=false (accessed May 28 2018).

Padian, K. 1999. Charles Darwin's views of classification in theory and in practice. *Systematic Biology* 48: 352–364.

Patterson, C. 1988. Homology in classical and molecular biology. *Molecular Biology and Evolution* **5**: 603–625.

Pietsch, TW. 2012. *Trees of Life. A Visual History of Evolution.* Johns Hopkins University Press, Baltimore.

Platnick, NI. 2009. In: Knapp, S. & Wheeler, Q. (eds), *Letters to Linnaeus.* The Linnean Society of London, London, pp. 199–203.

Ragan, MA. 2009. Trees and networks before and after Darwin. *Biology Direct* 4: 43.

Rivera, MC. & Lake, JA. 2004. The ring of life provides evidence for a genome fusion origin of eukaryotes. *Nature* 431: 152–155.

Sanderson, MJ. & Hufford, L. (eds) 1996. *Homoplasy. The Recurrence of Similarity in Evolution.* Academic Press, San Diego.

Scotland, RW. 2011. What is parallelism? *Evolution & Development* 13: 214–227.

Simpson, GG. 1945. The principles of classification and a classification of mammals. *Bulletin of the American Museum of Natural History* 85: 1–350.

Strickland, HE. 1841. On the true method of discovering the natural system in zoology and botany. *Annals of Natural History* 6(36): 184–194.

Swainson, W. 1836-1837. *On the Natural History and Classification of Birds.* 2 volumes. Longman, Rees, Orme, Brown, Green, and Longman, London.

Tassy, P. 2011. Trees before and after Darwin. *Journal of Systematics and Evolutionary Research* 49: 89–101.

Wallace, AR. 1855. On the law which has regulated the introduction of new species. *Annals and Magazine of Natural History* (Ser. 2) 16: 184–196.

Wallace, AR. 1856. Attempts at a natural arrangement of birds. *Annals and Magazine of Natural History* (Ser. 2) 18: 193–216.

Weinstein, B. 2016. On being a fish. *Inference: International Review of Science* 2(3). https://inference-review.com/article/on-being-a-fish

Williams, DM. & Ebach, MC. 2008. *Foundations of Systematics and Biogeography.* Springer-Verlag New York Inc., New York.

Winsor, MP. 2009. Taxonomy was the foundation of Darwin's evolution. *Taxon* 58: 43–49.

Wyhe, J. van. 2016. The impact of A. R. Wallace's Sarawak Law paper reassessed. *Studies in History and Philosophy of Science Part C: Studies in History and Philosophy of Biological and Biomedical Sciences* 60: 56–66.

Further Reading

Archibald, JD. 2014. *Aristotle's Ladder, Darwin's Tree: The Evolution of Visual Metaphors for Biological Order.* Columbia University Press, New York.

Barsanti, G. 1992. *La scala, la mappa, l'albero: immagini e classificazioni della natura fra sei e ottocento.* Sansoni, Florence.

Craw, R. C. 1992. Margins of cladistics: Identity, differences and place in the emergence of phylogenetic systematics, 1864–1975. In: Griffiths, P. (ed.), *Trees of Life: Essays in the Philosophy of Biology.* Kluwer Academic, Dordrecht, pp. 65–107.

Kressling, F. 2016. Lateral and vertical transfer in biology, linguistics and anthropology: an account of widely neglected ideas in the formation of evolutionary theories. *Evolutionary Biology* 43: 474–480.

Minaka, N. 2018. *Phylogenetic Thinking. A History and Philosophy of Systematic Biology in the Twentieth Century.* Keiso Shobo Publishing Co., Tokyo.

Minaka, N. 2018. *Systematic Thinking. Diagrams in Taxonomy and Phylogeny.* Shunjusha Publishing Co., Tokyo.

Nelson, GJ. & Platnick, NI. 1981. *Systematics and Biogeography: Cladistics and Vicariance.* Columbia University Press, New York.

O'Malley, MA. (ed.) 2010. The tree of life. *Biology and Philosophy* 25: 441–736.

A special issue of *Biology and Philosophy* consisting of 14 contributions, most focusing on the apparent problems presented by the species rank taxa and the apparent problems presented by prokaryotes.

Pécaud, S. 2018. *Cladistique et Evolution: Une Fondation Problematique.* Histoire et Philosophie des Sciences, 18. Classiques Garnier, Paris.

Pietsch, TW. 2012. *Trees of Life. A Visual History of Evolution*, Johns Hopkins University Press, Baltimore.

Souza, RG. 2017. Críticas ao "Tree Thinking": elucidando o significado das relações filogenéticas. *Pesquisa e Ensino em*

Ciências Exatas e da Natureza 1(2): 115–130.

Tassy, P. 1991 [1998, 2nd ed]. *L'arbre à remonter le temps*. Christian Bourgois, Paris (2nd edition Diderot multimedia, Paris).

Voss, J. 2010. *Darwin's Pictures: Views of Evolutionary Theory, 1837–1874*. Yale University Press, New Haven, CT.

Wyhe, J. van. 2009. Darwin online and the evolution of the Darwin Industry. *History of Science* 47: 459–473.

4

Essentialism and Typology

Many non-taxonomist biologists are now frustrated by the 'old-fashioned' typological approach with which taxonomists still describe species.

(Dayrat 2005, pp. 407–8)

Now one thing I've learned from 30 years of watching biologists is that whenever you find one systematist calling another one an essentialist or a typologist, you can be 100% sure that the name-caller is purely, simply, and entirely, wrong, and is just creating a smoke-screen to cover his or her tracks.

(Platnick 2009, p. 6)

Who are these '"old-fashioned" typological' taxonomists Dayrat refers to, and what does 'old-fashioned' typologist mean anyway?

> Part of the problem of species delineation is the fact that morphology, as an approach for delimiting species, has some limits. Traditional morphology-based taxonomy only discriminates what Cain (1954) called 'morphospecies', i.e. species exclusively established on morphology ... Traditional morphology-based taxonomy is not the study of life's diversity *per se*, but rather the study of one of its multiple facets, morphological diversity, which I refer to as 'morphodiversity'. (Dayrat 2005, p. 408)

Arthur Cain's term *morphospecies* (first introduced in Cain 1953, p. 82[1]) gained a certain amount of currency (and sadly still in use today), with the meaning often, but not always, meant in a derogatory fashion, a term of abuse aimed at morphologists (as Dayrat does). Cain's morphospecies concept is similar to that described by Mayr as the '*static and strictly morphological* species concept' (Mayr, 1942, p. 109), 'based on the degree of morphological distinction ... the species

---

[1] Originally defined as: 'A group of specimens which is considered sufficiently different morphologically from the most closely related forms known, to be given a specific name' (Cain 1953, p. 83)

concept with which Linnaeus started the science of systematics ...' (Mayr, 1942, p. 115, our emphasis).

Consider these words from Theodosius Dobzhansky (1900–1975, population geneticist and an architect of the *Modern Synthesis*, see Ford 1977) in a review of Cain's book *Animal Species and Their Evolution* (1954) referred to by Dayrat above:

> The species of classical taxonomy are morphological species, or 'morphospecies' according to Cain. Morphospecies are defined entirely on morphological characters; some taxonomists still persist in refusing to consider even the possibility that species can be defined in any other way. Morphospecies are *static and unchangeable in space and in time*, because they are essentially names attached to a single 'type' specimen or to a small series of specimens. (Dobzhansky 1954, p. 298, our emphasis)

Never mind that Dobzhansky had a poor understanding of the type method, in certain minds, a connection was being made between *static and unchanging entities* (i.e., species), Linnaeus, morphology and the type method in *nomenclature* (as we said above, the story is complex, for further details of the 'type–typology story' see Witteveen (2015a, 2015b) as well as Winsor (2001, 2003, 2006a)).

Platnick, above in the second epigraph, was referring to the group of systematists who have been attempting to create a new code of biological nomenclature called the *PhyloCode* (e.g., de Queiroz & Gauthier 1990, for further details see Chapter 14), an attempt to 'Evolutionize Taxonomy' (de Queiroz 1997[2]).

One of their central critiques, like Dayrat above, is that traditional ('Linnean') taxonomists are *typologists* and *essentialists* (in his view, both bad), hampered and hindered by the non-evolutionary Linnean classifications of the past. As Farber put it:

> The terms 'typological,' 'typological thinking,' and 'types' are currently derogatory labels often used to distinguish pre-Darwinian from post-Darwinian biology ..., used to distinguish the new systematics from the old systematics ..., or used to cast aspersions on certain contemporary positions in science by associating them with an allegedly archaic, fuzzy-headed methaphysical [sic] position ... (Farber 1978, p. 91[3])

Even with just a brief reflection on how taxonomy has been undertaken and how it has progressed one might easily dismiss this whole series of

---

[2] 'Now, more than 130 years after the publication of Darwin's *Origin of Species*, taxonomists are finally freeing themselves from the bonds of ancient traditions and bringing about a reorganization of the very core of biological taxonomy (e.g., de Queiroz and Gauthier, 1990) (de Queiroz, 1992, p. 309)' (after Nixon & Carpenter 2000, p. 301). More on this in Chapter 14.

[3] The ellipses replace three references – all are authored by Ernst Mayr.

accusations, from Dobzhansky to de Queiroz, as abject nonsense as 'no modern taxonomists are really implicit essentialists … Believing in a stable system of typification makes you neither an essentialist nor a typologist' (Nixon et al. 2003, p. 112, see also Platnick & Rosen 1987 and Farris & Platnick 1989, among others). As Farber went on: 'This polemic usage does violence to the historical record and confuses contemporary debates rather than clarifies them' (Farber 1978, p. 91). How, then, did this situation arise? The story is complex. We will try to simplify it.

The Essentialism Story

The most extraordinary thing about the essentialism story is the contrast between the enormity of its reputation and the flimsiness of its basis in historical evidence.

(Winsor 2006a, p. 150)

Consider the following two quotations, the first from Richard Dawkins, the second from Polly Winsor – both are relatively lengthy, but worth citing here in full.

> Paleontologists will argue passionately about whether a particular fossil is, say, *Australopithecus* or *Homo*. But any evolutionist knows there must have existed individuals who were exactly intermediate. It's essentialist folly to insist on the necessity of shoehorning your fossil into one genus or the other. There never was an *Australopithecus* mother who gave birth to a *Homo* child, for every child ever born belonged to the same species as its mother. The whole system of labelling species with discontinuous names is geared to a time slice, the present, in which ancestors have been conveniently expunged from our awareness (and 'ring species' tactfully ignored). If by some miracle every ancestor were preserved as a fossil, discontinuous naming would be impossible. Creationists are misguidedly fond of citing 'gaps' as embarrassing for evolutionists, but gaps are a fortuitous boon for taxonomists who, with good reason, want to give species discrete names. Quarrelling about whether a fossil is 'really' *Australopithecus* or *Homo* is like quarrelling over whether George should be called 'tall'. He's five foot ten, doesn't that tell you what you need to know? (Dawkins 2015, pp. 84–85)

Dawkins argues that if all the 'products' of evolution, that is, if all the *stages of change* were available for study, then taxonomy would become (almost) superfluous, a possibly foolish venture, as every 'taxon' would grade imperceptibly into every other one.

> People who expected taxonomists to gradually fill in the gaps of the great chain of being, because nature makes no leaps, complained that the Linnaean hierarchy breaks apart what is continuous. But as hundreds of naturalists paid closer attention, lumps and clumps of taxa emerged from the fog, and the conviction grew that organisms were truly related, essentially similar, linked by something

called affinity. *Any naturalist who maintained that genera were nothing more than human inventions would never see in the living world any evidence for evolution.* Darwin's final theory said no such thing'. (Winsor 2013, p. 74)

Winsor's argument is the opposite of Dawkins: 'Taxonomy was the foundation of Darwin's evolution' (Winsor 2009). Why do these two articulate commentators differ so much from one another? Passing over the reasons for their disagreement at this moment, the pertinent question seems to be: Do taxonomists *discover* the taxonomic hierarchy (*sensu* Winsor), or do they *impose* it (*sensu* Dawkins)?

The Dawkin's (2015) quote comes from a recent book entitled *This Idea Must Die: Scientific Theories That Are Blocking Progress* (Brockman 2015). Dawkins's chapter was entitled *Essentialism*. The traditional view of progress in taxonomy, as practised before Darwin (Mayr 1982) – now known as the 'essentialism story' or the 'essentialism myth' (Winsor 2006a) – was that most, if not all, taxonomists were, prior to the liberating effect of Darwin's evolutionary insights, *fixists* and *essentialists*, searching to find the *essential* characters (*sensu* Plato) for static unchanging taxonomic groups. Recently summarised by Marc Ereshefsky thusly: 'Prior to the acceptance of evolutionary theory, essentialism was the standard mode of classification in biological taxonomy' (Ereshefsky 2001, p. 95, from Winsor 2003, p. 387; see O'Hara 1997, p. 324) and Dawkins suggested that, contrary to simple observation (Winsor 2009), 'Ernst Mayr blamed this [essentialism] for humanity's late discovery of evolution' (Dawkins 2015, p. 84). Of course, Ereshefsky and Dawkins (and O'Hara) happen to be (horribly) wrong.

Typological Thinking Versus Population Thinking

We noted earlier a paper of Mayr's in which he first discussed the 'typological versus population thinking' contrast (Mayr 1959). Although not yet fully formed in his mind at that time (that came later in Mayr 1982, see Chung 2003), the general idea promoted by Mayr was that early taxonomists were essentialists, searching for the *defining characters* of taxa forcing them to be entities that could not evolve with respect to these characters: that is, they were *typologists*. Mayr contrasted this with *population thinking*. Of course, Mayr did not use the word essentialist or essentialism. That came later.

What Darwin had done with his version of evolutionary theory, according to Mayr, was to liberate the taxonomist from this *fixist* methodology, allowing them to understand and document the *natural variation* found in *populations* of organisms, hence Mayr's term *population thinking*. Yet, this understanding of population thinking as applied to taxonomy was a creation entirely of Mayr's own making in his zeal to promote his particular view of the *origin of species*[4], as opposed to their discovery, rather than being based on anything Darwin said – or did, for that matter (see below).

Winsor (2003) recently examined the *practice* of many early (pre-Darwinian) taxonomists and, while there were a variety of ways used to find (*discover*) taxa, she came to the conclusion that most, if not all, pre-Darwinian taxonomists were not 'essentialists' at all with respect to the determination (characterisation) and discovery of higher taxa – many proposed what came to be called *polythetic* groups (see Chapter 5[5]): in fact, although many taxonomists *did* (and still do) find defining characters for taxa (e.g., angiosperms = flowers), practically all taxonomists were *empiricists* (Winsor 2003, 2006a, also Stevens 1994) – including Linnaeus (Winsor 2006b; Müller-Wille 2007). It was Arthur Cain (1921–1999) who tainted Linnaeus.

Along with dabbling in numerical taxonomy (Cain & Harrison 1958), experimental 'population-evolutionary' studies (Cain & Sheppard 1954) and critiques of phylogenetic systematics (Cain 1967), Cain examined the taxonomy of Linnaeus (Cain 1958). How did Linnaeus actually do his taxonomy? Again, it is a complex business untangling Linnaeus's approach as well as Cain's understanding of it, but Cain thought that Linnaeus had adopted an Aristotelian formal logic approach to defining species (taxa), which led to defining *static, unchanging* species (Cain 1958; Winsor 2001, 2006a). Yet, as one noted Linnean scholar remarked: '. . . Linnaeus's practice here was *empirical* rather than in accordance with *formal logic* (cf. Cain, 1958)' (Stearn 1959, p. 94, our emphasis, citation after Winsor 2006a, p. 144).

Essentialism has been mentioned a few times above, but that word had not found its way into taxonomic practice until 1965. In that year David Hull, still a student, published a two-part paper entitled 'The Effect of Essentialism on Taxonomy – Two Thousand Years of Stasis' (Hull 1965). The story of its publication via the hands of Karl Popper is told by Winsor (2006b, pp. 165–166).

Hull discussed some of Popper's recent thoughts, in particular his notion of *methodological essentialism*,[6] which Hull saw happening in current taxonomy as the philosophy of *typology*. Here, then, is the first connection between typology and Popper's version of essentialism (Winsor 2006b, p. 167, see Platnick & Rosen

---

[4] It's possible that the essence of this idea began with Mayr's thesis although we have not looked into the matter in detail (Mayr 1926).

[5] 'The essentialist method developed by followers of Plato and Aristotle required definitions to state properties that are always present. Polythetic groups do not obey that requirement, whatever may have been the ontological beliefs of the taxonomist recognizing such groups' (Winsor 2003, p. 387).

[6] 'I use the name *methodological essentialism* to characterize the view, held by Plato and many of his followers, that it is the task of pure knowledge or "science" to discover and to describe the true nature of things; i.e. their hidden reality or essence' (Popper 1950, p. 34, after Hull 1965, p. 317).

(1987) and Levit & Meister (2006)) – and Winsor goes on to suggest that Hull got his ideas on typology from Cain (1958) via Simpson (1961) (Winsor 2006b, p. 167).

Thus, the historical inaccuracies of 'Mayr–Cain–Hull (and de Queiroz–O'Hara, etc.)' on typology and essentialism are simply incorrect: 'Mayr's "essentialism story" – as it has been aptly termed by Mary P. Winsor – is the familiar story of the dominance of a false idea, or rather, ideology, which held European naturalists and philosophers in a 'strait jacket' for more than two thousand years, until they were freed from it through the heroic effort of a single nineteenth-century gentleman naturalist (Charles Darwin, of course)' (Müller-Wille 2011, p. 62).

We return to Darwin as the hapless hero below.

Population Thinking and Tree-Thinking

Mayr wanted to convert the retarded and recalcitrant taxonomists from being ignorant fixist typologists to becoming modern evolutionary population biologists – so, too, later, did Robert O'Hara want to effect a similar conversion with his *tree-thinking*:

> because more and more systematists have come to realize that in the evolutionary world the notion of classification as an object of systematics can be largely dispensed with ... and the analogy of systematics to classification is in fact a *relict* of the pre-evolutionary period, when living diversity was viewed ahistorically. (O'Hara 1994, p. 14, italics ours)

Significant here is his final sentence: 'the analogy of systematics to classification is in fact a *relict* of the pre-evolutionary period, when living diversity was viewed ahistorically'. A relict? O'Hara later rehearses the essentialism argument extending it to yet another branch of comparative biology:

> The term 'population thinking' was coined by Ernst Mayr in 1959. In coining the term Mayr did not claim to be describing something new; rather he intended to capture with the term a way of thinking that had swept through systematics and evolutionary biology generally in the first half of the twentieth century ... To understand the idea of population thinking it is necessary to contrast it with the mode of thought it replaced, which Mayr calls typology or essentialism. (O'Hara 1997, p. 324)

Again we return to one of Winsor's conclusions, that 'it is the relationship between the Linnaean hierarchy and branching evolution that is primary' (Winsor 2009). It is the Linnaean hierarchy that is *discovered* by searching for relevant evidence, an empirical approach (largely) untainted by theories of whatever kind. According to O'Hara, his contrasting term 'group-thinking' 'equates "systematics" with "classification"' (O'Hara 1997, p. 324). Perhaps that is the case, if one wishes to take an empirical stance (see Chapter 3). There may indeed be problems with 'group-thinking' (*sensu* O'Hara) but these can be, and are, better addressed empirically. Rather than 'tree-thinking', perhaps a better explanation for the taxonomic method is '*relationship-thinking*', a term that refers to scientific classification or any other cladistic research programme (Ebach et al. 2008; see Chapter 3 for more on tree-thinking).

Population Thinking and Homology-Thinking

We do not intend to tackle 'homology-thinking' in detail just to note that it is supposed to explain '... the properties of a homologue by citing the *history of a homologue*' (Ereshefsky 2012, p. 381, also Ereshefsky 2007). Thus, another version of what might be collectively called '*Mayr–thinking*' is introduced (see Wagner 2014, p. 424, 2015; Neisser 2015, etc.): '*Mayr–thinking*' because all these various versions cite Mayr (1959) as a significant influence and all require their subject matter to be based on the *explanation* (i.e., history) rather than the *discovery* (i.e., evidence).

Summary

The rhetoric in current debates sometimes implies that only an evolutionist can construct a meaningful reference system for living things … but … many European naturalists believed that organisms were all 'related' and that their classifications should express these relationships.

(Winsor 2009, p. 1)

If taxonomic groups have synapomorphies, those characters might (shudder) be viewed as essences (i.e. as both necessary and sufficient for group membership, once further character transformations are traced, and allowed for). But essences are evil, so taxonomic groups cannot have synapomorphies.

(Farris & Platnick 1989, p. 305)

The 'typology-essentialism-morphology' critique is currently being dismantled as it is nothing more than a longstanding series of misreadings of history and subsequently created myths concerning nineteenth and twentieth century biology designed to prop up the original *Modern Synthesis* – created and enhanced by, among others, Ernst Mayr, Arthur Cain, Michael Ghiselin and David Hull, and later, Kevin de Queiroz, Richard O'Hara and Marc Ereshefsky (for an earlier appraisal of Darwin see Loewenberg 1965, p. 36).

It may come as no surprise that many of these people were heavily involved with revitalising the reputation of Charles Darwin and placing him centre stage of evolutionary studies in the context of the *Modern Synthesis* – 'population thinking', 'tree-thinking' and 'homology-thinking' are all of a piece relative to systematics: 'wrong and harmful in their basic message'.[7]

Of course, Dawkins (and anyone else for that matter) could have simply peeked into Darwin's *Origin* to see what he actually thought (see quote in Chapter 3).

Although taxonomists knew it all along, it took a few decades and the persistence of a few historians to challenge the orthodoxy. The primary lesson learnt is that taxonomists should continue to be the empiricists they always were and always will be.

> The essentialism myth or story 'in its broad sweep across the history of systematics ... is not merely inaccurate in particulars, it is wrong and harmful in its basic message'. (Winsor 2006a, p. 3)

References

Agassiz, L. 1869. *De l'espèce et de la classification en zoologie*. Bailiere, Paris.

Brockman, J. 2015. *This Idea Must Die: Scientific Theories That Are Blocking Progress*. Harper Perennial, New York.

Cain, AJ. 1953. Geography, ecology and coexistence in relation to the biological definition of the species. *Evolution* 7: 76–83.

Cain, AJ. 1954. *Animal Species and Their Evolution*. Harper & Row, New York.

Cain, AJ. 1958. Logic and memory in Linnaeus's system of taxonomy. *Proceedings of the Linnean Society of London* 169: 144–163.

Cain, AJ. 1967. One phylogenetic system. *Nature* 216: 412–413.

Cain, AJ. & Harrison, GA. 1958. An analysis of the taxonomist's judgement of affinity. *Proceedings of the Zoological Society, London* 131: 85–98.

Cain, AJ. & Sheppard, PM. 1954. Natural selection in *Cepaea*. *Genetics* 39: 89–116.

Chung, C. 2003. On the origin of the typological/population distinction in Ernst Mayr's changing views of species, 1942–1959. *Studies in History and Philosophy of Biological and Biomedical Sciences* 34: 277–296.

Dawkins, R. 2015. Essentialism. In: Brockman, J. (ed.), *This Idea Must Die: Scientific Theories That Are Blocking Progress*. Harper Perennial, New York, pp. 84–87.

Dayrat, B. 2005. Towards integrative taxonomy. *Biological Journal of the Linnean Society* 85: 407–415.

de Queiroz, K. 1992. Phylogenetic definitions and taxonomic philosophy. *Biology and Philosophy* 7: 295–313.

de Queiroz, K. 1997. The Linnaean hierarchy and the evolutionization of taxonomy, with emphasis on the problem of nomenclature. *Aliso* 15: 125–144.

---

[7] 'It is not that I hold Darwin himself responsible for these troublesome consequences. In the different works of his pen, he never made allusion to the importance that his ideas could have for the point of view of classification. It is his henchmen who took hold of his theories in order to transform zoological taxonomy' (Agassiz 1869, p. 375).

de Queiroz, K. & Gauthier, J. 1990. Phylogeny as a central principle in taxonomy: phylogenetic definitions of taxon names. *Systematic Zoology* 39: 307–322.

Dobzhansky, T. 1954. [Review of: Cain, AJ. 1954. *Animal Species and Their Evolution*. Harper & Row, New York] Nature 173 (13 February): 286–287

Ebach, MC., Morrone, JJ. & Williams, DM. 2008. An outline of the foundations of systematics and biogeography. *History and Philosophy of the Life Sciences* 29: 21–24.

Ereshefsky, M. 2001. *The Poverty of the Linnaean Hierarchy: A Philosophical Study of Biological Taxonomy*. Cambridge University Press, Cambridge, UK.

Ereshefsky, M. 2007. Psychological categories as homologies: lessons from ethology. *Biology and Philosophy* 22: 659–674.

Ereshefsky, M. 2012. Homology thinking. *Biology and Philosophy* 27: 381–400.

Farber, PL. 1978. A historical perspective on the impact of the type concept on insect systematics. *Annual Review of Entomology* 23: 91–99.

Farris, JS. & Platnick, NI. 1989. Lord of the flies: the systematist as study animal. *Cladistics* 5: 295–310.

Ford, EB. 1977. Theodosius Grigorievich Dobzhansky. 25 January 1900–18 December 1975. *Biographical Memoirs of Fellows of the Royal Society* 23: 58–89.

Hull, DL. 1965. The effects of essentialism on taxonomy: two thousand years of stasis. *The British Journal for the Philosophy of Science* 15: 314–326, 16: 1–18 [reprinted in *Concepts of Species*, C. N. Slobodchikoff (ed.), Berkeley: University of California Press, 1976; in *The Units of Selection*, Marc Ereshefsky (ed.), Cambridge: MIT Press, 1992].

Levit, GS. & Meister, K. 2006. The history of essentialism vs. Ernst Mayr's "Essentialism Story": A case study of German idealistic morphology. *Theory in Biosciences* 124: 281–307.

Loewenberg, BJ. 1965. Darwin and Darwin studies 1959-1965. *History of Science* 4: 15–54.

Mayr, E. 1926. Die Ausbreitung des Girlitz (*Serinus canaria serinus* L.). Ein Beitrag zur Tiergeographie. *Journal für Ornithologie* 4: 571–671.

Mayr, E. 1942. *Systematics and the Origin of Species*. Columbia University Press, New York.

Mayr, E. 1959. Darwin and the evolutionary theory in biology. In: *Evolution and Anthropology: A Centennial Appraisal*. The Anthropological Society of Washington, Washington, DC, pp. 1–10 [reprinted in Mayr, E. 1976. *Evolution and the Diversity of Life: Selected Essays*. Harvard University Press, Cambridge, MA, pp. 26–29; reprinted as Mayr, E. 1976. Typological thinking versus population thinking. In: *Evolution and the Diversity of Life: Selected Essays*. Harvard University Press, Cambridge, MA, pp. 26–29 and Sober, E. (ed.). 1994. *Conceptual Issues in Evolutionary Biology*. The MIT Press. Bradford Books, Cambridge, MA, London, pp. 157–160].

Mayr, E. 1982. *The Growth of Biological Thought: Diversity, Evolution, and Inheritance*. Harvard University Press, Cambridge, MA.

Müller-Wille, S. 2007. Collection and collation: theory and practice of Linnaean botany. *Studies in History and Philosophy of the Biological and Biomedical Sciences* 38: 541–562.

Müller-Wille, S. 2011. Making sense of essentialism. *Critical Quarterly* 53: 61–67.

Neisser, J. 2015. *The Science of Subjectivity.* Palgrave Macmillan, Basingstoke, Hampshire.

Nixon, KC. & Carpenter, JM. 2000. On the other "Phylogenetic Systematics". *Cladistics* 16: 298–318.

Nixon, KC, Carpenter, JM. & Stevenson, DW. 2003. The PhyloCode is fatally flawed, and the "Linnean" system can easily be fixed. *Botanical Review* 69: 111–120.

O'Hara, RJ. 1994. Evolutionary history and the species problem. *American Zoologist* 34: 12–22.

O'Hara, RJ. 1997. Population thinking and tree thinking in systematics. *Zoologica Scripta* 26: 323–329.

Platnick, NI. 2009. [*Letters to Linnaeus*]. The Linnean Society of London, London.

Platnick, NI. & Rosen, DE. 1987. Popper and evolutionary novelties. *History and Philosophy of the Life Sciences* 9: 5–16.

Popper, K. 1950. *The Open Society and Its Enemies.* Princeton University Press, Princeton, NJ.

Simpson, GG. 1961. *Principles of Animal Taxonomy.* Columbia University Press, New York.

Stearn, WT. 1959. Four Supplementary Linnaean Publications: Methodus (1736), Demonstrationes Plantarum (1753), Généra Plantarum (1754), Ordines naturales (1764). In: Linnaeus C., *Species Plantarum: A Facsimile,* Vol. 2. Ray Society, London, pp. 73–102.

Stevens, PF. 1994. *The Development of Biological Systematics: Antoine-Laurent de Jussieu, Nature, and the Natural System.* Columbia University Press, New York.

Wagner, G. 2014. *Homology, Genes, and Evolutionary Innovation.*

Princeton University Press, Princeton & Oxford.

Wagner, G. 2015. What is "homology thinking" and what is it for? *Journal of Experimental Zoology Part B Molecular and Developmental Evolution* 326: 3–8.

Winsor, MP. 2001. Cain on Linnaeus: the scientist-historian as unanalysed entity. *Studies in History and Philosophy of Science Part C: Studies in History and Philosophy of Biological and Biomedical Sciences* 32: 239–254.

Winsor, MP. 2003. Non-essentialist methods in pre-Darwinian taxonomy. *Biology and Philosophy* 18: 387–400.

Winsor, MP. 2006a. Linnaeus's biology was not essentialist. *Annals of the Missouri Botanical Gardens* 93: 2–7.

Winsor, MP. 2006b. Creation of the essentialism story: an exercise in metahistory. *History and Philosophy of the Life Sciences* 28(2): 149–174.

Winsor, MP. 2009. Taxonomy was the foundation of Darwin's evolution. *Taxon* 58: 43–49.

Winsor, MP. 2013. Darwin and taxonomy. In: Ruse, M. (ed.), *The Cambridge Encyclopedia of Darwin and Evolutionary Thought.* Cambridge University Press, Cambridge, UK, pp. 72–79.

Witteveen, J. 2015a. 'A temporary oversimplification': Mayr, Simpson, Dobzhansky, and the origins of the typology/population dichotomy (Part 1 of 2). *Studies in the History and Philosophy of Biological and Biomedical Sciences* 54: 20–33.

Witteveen, J. 2015b. 'A temporary oversimplification': Mayr, Simpson, Dobzhansky, and the origins of the typology/population dichotomy (Part 2 of 2). *Studies in the History and Philosophy of Biological and Biomedical Sciences* 57: 96–105.

Further Reading

Liagouras, G. 2019. Population thinking vs. essentialism in biology and evolutionary economics. *OSF Preprints.* January 7. http://doi:10.31219/osf.io/r4g8x.

McOuat, GR. 2009. The origins of natural kinds: keeping "Essentialism" at bay in the age of reform. *Intellectual History Review* 19: 211–230.

Müller-Wille, S. 2013. Systems and how Linnaeus looked at them in retrospect. *Annals of Science* 70: 305–317.

Pratt, V. 1985. System-building in the eighteenth century. In: North, JD. & Roche, JJ. (eds), *The Light of Nature.* International Archives of the History of Ideas, Vol. 110. Springer Nature, Switzerland AG, pp. 421–431.

Richards, R. 2010. *The Species Problem: A Philosophical Analysis.* Cambridge University Press, Cambridge, UK.

Varma, CS. 2008. Threads that guide or ties that bind: William Kirby and the essentialism story. *Journal of the History of Biology* 42: 119–149.

Wilkins, J. 2009. *Defining Species: A Sourcebook from Antiquity to Today.* American University Studies, Peter Lang, New York.

Wilkins, JS. 2013. Biological essentialism and the tidal change of natural kinds. *Science and Education* 22: 221–240.

5

Monothetic and Polythetic Taxa

How are taxa defined? How should they be defined? Should they be defined at all? In this chapter we discuss two approaches to the formation of groups, captured by the notions of *polythetic* and *monothetic* classes.

Polythetic Classes, Monothetic Classes and Phenetics

The distinction between polythetic and monothetic was succinctly demonstrated by Stamos (2005). He contrasted what he referred to as *cluster classes* and *essentialistic classes* in classification, the two being roughly equivalent to polythetic and monothetic groups, respectively. The comparison is as follows:

For *essentialistic classes* a particular character or characters are required to define a taxon. These characters are seen as *essential* to the definition of any particular taxon. We pass over the meaning of essential here as we have pursued that subject in Chapter 4.

In the table below there are five individuals (specimens 1–5) and five characters (A–E):

Individuals	1	2	3	4	5
Characters	A	A	A	A	
	B	B	B	B	B
	C	C	C		C
	D	D	D	D	
	E	E	E	E	E
Class	Y	Y	Y	N	N

To group the five individuals together certain characters can be used to form taxa. For the five individuals, the first three have all of the characters A–E and so would be recognised as a class (taxon) (columns 1–3). If the characters are understood as *essential*, then the 4th and 5th specimens lack one or more of the five and so would not be included in the taxon (columns 4, 5). Groups 1–3 would be *essentialistic classes* or *monothetic groups*.

For cluster classes things differ. In the table below there are five individuals (specimens 6–10) and five characters (A–E):

Individuals	6	7	8	9	10
Characters	A	A	A		A
		B	B	B	
		C	C	C	
		D		D	D
	E	E		E	
Class	2/5	5/5	3/5	4/5	2/5

None of the individuals 6–10 have all five class characteristics. Rather, they all have some. Depending on how one understands the relation between the numbers of characters required to recognise any particular group, this would dictate which taxa are formally recognised. As Stamos explained it,

> With essentialistic classes, membership conditions are individually necessary and jointly sufficient, so that membership in the class is either yes or no. With cluster classes, on the other hand, membership conditions are neither individually necessary nor jointly sufficient. Instead, membership is either a matter of degree, determined by how many of the defining set of characteristics the individual has …. (Stamos 2005, p. 81)

These approaches to grouping are not new. Winsor has shown that '[t]here is already evidence suggesting that in the 18th and 19th centuries normal taxonomic practice permitted polythetic groups' (Winsor 2003, p. 391).

The word *polythetic* is relatively new. It was coined by Peter Sneath (1962, p. 290) and has since been associated with the pheneticist point of view. For example, Jensen wrote:

> Numerical taxonomists recognized the distinction between monothetic and polythetic groups and argued that natural taxa must be polythetic; they had to do this to avoid being described as essentialists. (Jensen 2009, p. 53)

Farris, commenting from a numerical cladistic perspective, referred to polythetic groups as 'compromise groups':

> Such groups would correspond closely to the distribution of most of a suite of closely associated features, but would not necessarily exactly reflect the distribution of any one feature. (Farris 1979, p. 500)

He continued:

> The general idea of compromise groups is no doubt reasonable as far as it goes, but it raises the further problem of assessing which of several conceivable compromises is most informative, most in accord with the concept of naturalness. (Farris 1979, p. 500)

This is a reasonable point: what amount of compromise could serve to identify any particular taxon, and how would one arrive at that assessment? Pheneticists associated polythetic classes with the general notion of 'overall similarity', a quantity that could be measured. But the association of polythetic groups with phenetics need not be the case as Winsor's survey of past taxonomic practice adequately demonstrates. Schuh and Farris noted that:

> Polythetic methods are simply those that construct groups by consideration of all relevant features, rather than by selecting some features as necessary and sufficient for group membership. Parsimony most certainly fits this description, and so the differences between phenetic and phylogenetic classification can have nothing to do with whether methods are polythetic. (Schuh & Farris 1981, p. 345)

For Stamos, if *clusterists* sought polythetic groups, then, *essentialists* sought monothetic groups.

Polythetic Classes, Monothetic Classes and Cladistics

With respect to what we understand as cladistics, differences rest primarily with how characters are formulated. For example, creating (or discovering) 'essentialistic' classes might seem similar to taxa found by using compatibility analysis (see Chapter 8), in as much as compatibility is 'a form of character weighting, seeking to give maximum weight to characters that evolve once and display no homoplasy' (Scotland & Steel 2015, p. 493), or as Le Quesne put it more directly many years ago, 'Sokal and Sneath (1963) point out that phenetic relationships derived from similarity coefficients lead to polythetic taxa. Methods derived from the uniquely derived character concept will, on the contrary, lead to monothetic taxa' (Le Quesne 1972, p. 288). Translating these ideas into taxonomic rather than phylogenetic terminology, maximum weight is given to the unique characters possessed by all the members of the set in question. That is, the compatibility approach is a method that sets out to find essentialistic classes, as Le Quesne suggests above.

With respect to characters, Wagner parsimony operates differently:

> The distinction is that compatibility methods recognize only perfect correlations—sets of fully congruent characters—whereas the Wagner [parsimony] method

> more realistically accepts some imperfect correlations, which makes possible a
> better fit to all available evidence. (Farris & Kluge 1979, p. 405)

Although the notion of 'better fit' is invoked, the meaning is that characters
might partially support one node and partially support another.

Both approaches – compatibility and parsimony – might be better understood by
recognising the underlying model: '[Wagner parsimony] implements a model of
character evolution that requires synapomorphy (evidence of relationship) to have
unique origin (optimized as 1 at a node with distal 0's as reversal(s), or vice versa)'
(Nelson 1993, p. 360), which is exactly that of compatibility, as noted by Scotland
and Steel (2015, with the exception of reversals, see Chapter 13 and Mooi & Gill
2016; Rineau et al. 2015 and Mavrodiev 2016). For the moment we are interested in
what 'more realistic' (sensu Farris & Kluge 1979, cited above) might mean.

Sankoff offered an explanation for differences between Wagner parsimony and
compatibility analyses:

> . . . maximizing parsimony is equivalent to minimizing the total number of extra
> steps over all characters. Maximizing compatibility is equivalent to minimizing the
> number of characters requiring at least one step. (Sankoff in Le Quesne 1975,
> p. 426)

One meaning that might be attached to Sankoff's statement is that the difference is
found in the way characters (and here we mean observable features, homologues)
are initially represented rather than any subsequent analysis. A review, again written
some time ago, on characters and character states noted 'problems in the term
"character" seem to have first surfaced in the field of numerical taxonomy' (Colless
1985, p. 229). Much of Colless' subsequent discussion is concerned with the desire
to successfully capture observations on organisms allowing complete, or near complete, description – that discussion continues today, under the topic of ontologies.
His viewpoint may be said to also fall into the phenetic approach to systematics, the
attempt to discover 'truth' in the guise of 'safe, sure knowledge' (Patterson's apt
summary of the phenetic enterprise, Patterson 1987, p. 5). But exploration of
characters and their representation might yield further insights. For example, the
conventional systematist's data matrix is composed of sets of binary characters, each
a string of states, while the implication behind compatibility analyses is that each
character is a mini-tree, specifying a particular relationship. The first is linear, the
second hierarchical. Alternatively, the first is a matrix of homologues, the second a
series of homology statements derived from the homologues. Other alternatives
were introduced some years ago but received relatively little attention.

Many taxonomists, pre-Darwinian or otherwise, simply did not work in that way,
and a more realistic view is that both monothetic and polythetic classes would be,
and probably are, recognised in any single classification.

References

Colless, DH. 1985. On "Character" and related terms. *Systematic Zoology* 34: 229–233.

Farris, JS. 1979. The information content of the phylogenetic system. *Systematic Zoology* 28: 483–519.

Farris, JS. & Kluge, AG. 1979. A botanical clique. *Systematic Zoology* 28: 400–411.

Jensen, RJ. 2009. Phenetics: revolution, reform or natural consequence? *Taxon* 58: 50–60.

Le Quesne, WJ. 1972. Further studies based on the uniquely derived character concept. *Systematic Zoology* 21: 281–288.

Le Quesne, WJ. 1975. Discussion of the preceding papers. In: Estabrook, GF. (ed.), *Proceedings of the 8th International Conference on Numerical Taxonomy.* W.H. Freeman, San Francisco, pp. 416–429.

Mavrodiev, EV. 2016. Dealing with propositions, not with the characters: the ability of three-taxon statement analysis to recognise groups based solely on 'reversals', under the maximum-likelihood criteria. *Australian Systematic Botany* 29: 119–125.

Mooi, R. & Gill, A. 2016. Hennig's auxiliary principle and reciprocal illumination revisited. In: Williams, DM., Schmitt, M. & Wheeler, QD. (eds), *The Future of Phylogenetic Systematics: The Legacy of Willi Hennig.* Cambridge University Press, Cambridge, UK, pp. 258–285.

Nelson, G. 1993. Reply to Harvey. *Cladistics* 8: 355–360.

Patterson, C. 1987. Introduction. In: Patterson, C. (ed.), *Molecules and Morphology in Evolution: Conflict or Compromise?* Cambridge University Press, Cambridge, UK, pp. 1–22.

Rineau, V., Grand, A., Zaragüeta, R. & Laurin, M. 2015. Experimental systematics: sensitivity of cladistic methods to polarization and character ordering schemes. *Contributions to Zoology* 84: 129–148.

Schuh, RT. & Farris, JS. 1981. Methods for investigating taxonomic congruence and their application to the Leptopodomorpha. *Systematic Zoology* 30: 331–351.

Scotland, RW. & Steel, M. 2015. Circumstances in which parsimony but not compatibility will be provably misleading. *Systematic Biology* 64: 492–504.

Sneath, PHA. 1962. The construction of taxonomic groups. In: Ainsworth, GC. & Sneath, PHA. (eds), *Microbial Classification.* Cambridge University Press, Cambridge, UK, pp. 289–332.

Sokal, RR. & Sneath, PHA. 1963. *Principles of Numerical Taxonomy.* W.H. Freeman, San Francisco.

Stamos, D. 2005. Pre-Darwinian taxonomy and essentialism – a reply to Mary Winsor. *Biology and Philosophy* 20: 79–96.

Winsor, MP. 2003. Non-essentialist methods in pre-Darwinian taxonomy. *Biology and Philosophy* 18: 387–400.

Further Reading

Mavrodiev, EV. 2016. Dealing with propositions, not with the characters: the ability of three-taxon statement analysis to recognise groups based solely on 'reversals', under the maximum-likelihood criteria. *Australian Systematic Botany* 29: 119–125.

Williams, DM. & Ebach, MC. 2017. What is intuitive taxonomic practice? *Systematic Biology* 66: 637–643.

6

Non-taxa or the Absence of –Phyly:
Paraphyly and Aphyly

... a somewhat tiresome point of terminology ...

(Dawkins 2004, p. 209)

We will treat the terms *monophyly* and *monophyletic group* in more detail in Chapter 7. Briefly, it refers to a *taxon* characterised by at least one synapomorphy (also further discussed in Chapter 7). Many recent definitions of monophyly have been based on ancestry. This book is focused on classification, so here monophyly is considered to be an empirical concept, matching evidence to conclusions. Monophyletic groups are taxa; but not all taxa are monophyletic – they are, for the most part, assumed to be so.

The term *paraphyly* and *paraphyletic group* are frequently used in connection with monophyly. Paraphyly has also acquired a definition based on ancestry. Here we discuss it, and some related concepts, in terms of evidence and conclusions.

Many named taxa have never been characterised by synapomorphies and hence are of unknown status relative to their '-phyly': there is as yet no definitive evidence to support their monophyly. In our view, all such groups should be referred to as *aphyletic* 'meaning that they require taxonomic revision' (Ebach & Williams 2010, p. 124; Wilkins & Ebach 2013, p. 144; Williams & Ebach 2017) and have no status with respect to any phylogenetic interpretation (Williams & Ebach 2017).

To start, consider a simple example: six species (A–F) belong to the genus Z, which is supported by several synapomorphies (homologies) (Figure 6.1a, the black box Z_1 represents the synapomorphies of genus Z). Further investigations reveal that three of the six species (A–C) have characters of their own, shared only among themselves relative to D–F. At this time, further investigations yield no more data relevant to the relationships of A–F; therefore, under the assumption that homology = taxon, as every discovery of a synapomorphy is equivalent to discovering a taxon (see Chapter 7), there are two groups: the larger group genus

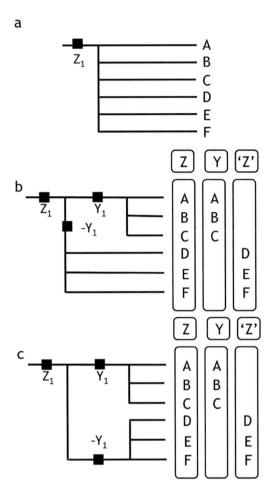

Figure 6.1 (a) Cladogram of six species (A–F) in the genus Z (its synapomorphies represented by the black box Z_1); (b) Cladogram of six species (A–F), with the discovery of new synapomorphies for A–C (represented by the black box Y_1) and three possible schemes for naming the resulting groups; (c) a further alternative with A–C defined by Y_1 and D–F defined by lacking Y_1.

Z (species A–F) and a smaller group with species A–C (Figure 6.1b, black box Y_1 represents the synapomorphies of genus Z). What of the classification of A–F? Excluding the names of the six individual species and the original genus Z, there is one additional taxon, the newly discovered group A–C.

Ignoring its rank (see Chapter 3), to add a new name for the discovery of A–C might seem unnecessary as the species A–F share the same generic name (genus Z), and in any case a simple diagram would illustrate the relationship DEF(ABC) (Figure 6.1b). Such a view might render classification in general superfluous as long as we have the diagram, any diagram, to illustrate the relationships.

Some hold this point of view: that classification is irrelevant, and the diagram will do:

> The focus of systematics has shifted massively away from classification: it is the phylogenies that are central, and it is nearly irrelevant how they are then used in taxonomy. (Felsenstein 2001, p. 467)

> Systematists get so worked up declaiming the centrality of classification in systematics that I have argued the opposite. (Felsenstein in Franz 2005, p. 495)

> Many phylogeneticists now see nomenclature and classification as largely irrelevant to phylogenetics … (Hillis 2007, p. 331)

For the most part, these views are wedded to the idea of 'tree-thinking', the literal interpretation of a branching diagram as an evolutionary 'event-o-gram', another myth we discuss elsewhere (Chapter 3). The creator of tree-thinking, Robert O'Hara, suggested the following:

> We are only now coming to realize that the Natural System is in fact the branching chronicle of events in evolutionary time, and that the analogy of systematics to *classification* is mistaken. (O'Hara 1991, p. 272)

This viewpoint seems mistaken, as classification is prior to any assumptions of interpreting a diagram as 'the branching chronicle of events in evolutionary time'. After all, without a classification (a scheme of relationships) what is there to explain or interpret (see Chapters 3 and 7)?

Below we will first discuss paraphyly, with a few examples; second we will discuss aphyly, with a few examples and finish with some general points about 'non-taxa'.

Paraphyly

Imagine that A–F in Figure 6.1 are genera rather than species, and some include many hundreds, perhaps thousands, of species. This is not an uncommon situation: the plant genus *Astragalus*, for example, has upwards of 3000 species; *Bulbophylum*, *Carex* and *Euphorbia* all have more than 2000 species (Frodin 2004, Carine pers. comm., 2019); *Agrilus* (jewel beetles) has more than 3000 species, and is possibly the largest animal genus; *Lasioglossum* (a genus of bees) has over 1700 species; others are *Anomala* (currently with more than a 1000 species), *Aphodius* (c. 700 species), *Onthophagus* (1200 species) and *Cryptocephalus* (840 species), all beetle groups; *Enicospilus* (wasps) currently has 700 species (Gavin Broad & Max Barclay, pers. comm., 2019); and many microscopic 'protists' (diatoms, ciliates, etc.) have genera that exceed even the 3000 species noted above; these are just a few examples.

Instead, then, let us imagine that A–F has in total 3000 species. Examination of just 3 of the 3000 shows them to be a definable group, equivalent to A–C of

Figure 6.1b. D–F now has only 2997 species, those without the synapomorphies of the 3 species A–C. Is D–F paraphyletic? Who knows? In terms of evidence, D–F could be defined by having the synapomorphies of the larger group A–F (Figure 6.1c, black box Z_1) and lacking the synapomorphies of A–C (Figure 6.1c, black box $-Y_1$).

This example is not too dissimilar to that offered by Brummitt (1997a, 1997b, 2002). It goes something like this: Brummitt provided a diagram (reproduced here as our Figure 6.2a after Brummitt 1997a, p. 725, fig. 1 and Brummitt 1997b, fig. 4, slightly modified in Brummitt 2002, p. 32, fig. 1, our Figure 6.2b; the latter was reproduced in Nelson et al. 2003, fig. 1, juxtaposed with an almost identical and earlier diagram from Rosa 1918, pp. 137–138, our Figure 6.2c – the tree has been inverted in our reproduction). Brummitt refers to his diagram as a 'very simplified representation of a hierarchy descent from a stem species E'. The point of his diagram is thus: imagine that a new character (or characters) has arisen at the point marked 1 thus enabling the recognition of genus A as monophyletic (Figure 6.2a). Brummitt goes on to note that 'we must immediately also create a paraphyletic taxon' because 'species B must always be referable to a genus, which by definition must be different from that including species A. The genus including B must be paraphyletic' (Brummitt 1997a, p. 726; similar examples can be found in Flegr 2013, Seifert et al. 2016, and many others). Note that species B is placed at a node (as are species E and D) with the view that the 'genus including B must be paraphyletic … If there has been negligible character change between B and C, a genus of present day species including C will be paraphyletic' (Brummitt 1997a, p. 726).

Let us deal with the diagram first. There are two ways:

One might assume that evidence has been found to support each branch point and the nodes are not species but characters (as in Rosa's tree, Figure 6.2c). That it is a cladogram, in fact. A classification is, thus, relatively straightforward.

The cladogram (in Figure 6.2c) has two branches (reading from top to bottom), each of those have two branches and so on. Thus, if the entire cladogram was considered to be a Class, the first two bifurcations would be Orders, the bifurcations within each order Families, and so. One might alternatively assume the diagram to be an actual genealogy with species at the nodes (which Brummitt has), in which case, the relationships of A—E (or rather E—A) would be a linear series suggested D evolved from E, etc.:

$$E \rightarrow D \rightarrow B \rightarrow AC$$

This can be represented as the following:

$$E(D(B(AC))) \text{ (as in Figure 6.2a)}$$

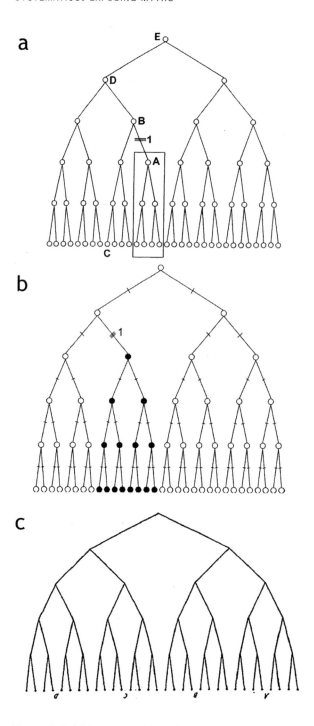

Figure 6.2 (a) Diagram of the relationships of some species (redrawn from Brummitt 1997a, p. 725, fig. 1) and (b) a modified version (redrawn from Brummitt 2002, p. 32, fig. 1). (c) Diagram of the relationships of some species from Rosa 1918, pp. 137–138 (redrawn from Rosa 1918 after Nelson et al. 2003, fig. 1 – our version has been inverted to match those of Brummitt).

This would yield:

Class E

    [Order E, Family E, Genus E, Species E]

Order D

[etc.]

Family B

[etc.]

Genus A

Genus C

A little cumbersome, to be sure, and an (unnecessary) excess of names and categories, but it does permit the classification of all the species in the diagram (see also Chapter 14) – and the two classifications are essentially the same.

Let us consider the latter view a little more. To begin, we could recast the argument in terms of an even simpler diagram with just two species (Species 1 and Species 2) in the same genus that share character 1 (Figure 6.3a), but only Species 2 has character 2 (Figure 6.3a). Given that the genus is defined by character 1, if Species 2 is moved to a new genus, because of character 2, the old genus is defined as those organisms with character 1 but lacking character 2, as in our first example. Brummitt assumes that the old genus is now paraphyletic and *ancestral* to the new one hence paraphyletic (ancestral) taxa are necessary in classification.

Monotypic taxa are usually considered to lack 'phyly' (see below), but suppose if the example is expanded to a genus of 50 species, from which 20 are removed to a new genus (Figure 6.3b), the old genus would be defined as those organisms with character 1 but lacking character 2 (found in species 31–50) and the old genus would be considered ancestral to the new one.

Brummitt's argument is based on the idea (i.e., a particular evolutionary model) that a widely distributed character will be older than a more narrowly distributed character and that the old taxon is really older than the new one – that is, in his example, character 1 is really older than character 2. This is a simple generational model suggesting that taxa with character 1 evolved from organisms/taxa without that character. Of interest, then, would be to investigate what the status is of the old genus after the new one is defined, which yields the same question: are Taxa D–F rendered paraphyletic? As we suggested above, there is no evidence for this: the old genus might very well be mono-, para- or even polyphyletic.

The general point here is that with respect to taxonomy and classification, regardless of the model invoked, *evidence of monophyly is not evidence of paraphyly.*

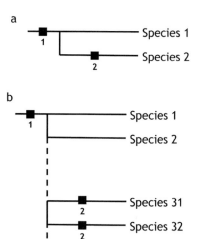

Figure 6.3 (a) Cladogram of two species; (b) cladogram of 50 species. In both, boxes 1 and 2 are characters.

Example 1: Gerald Mayr published a paper entitled 'Osteological evidence for paraphyly of the avian order Caprimulgiformes (nightjars and allies)' (Mayr 2002). His cladogram of relationships is reproduced as our Figure 6.4 (after Mayr 2002, p. 85, fig. 1). His cladogram has four nodes (1–4 on Figure 6.4). The text discusses the evidence available that supports each of the four nodes. The Caprimulgiformes do not group at any single node (Figure 6.4, the Caprimulgiformes are indicated by a line, added by us) and no characters support that group. Thus, his study provides evidence for the monophyly of several other groups (nodes 1–4) some of which were thought to compose the Caprimulgiformes.

Node 4 is the Apodiformes, the swifts, the treeswifts and the hummingbirds. While the temptation might be to suggest that Apodiformes evolved from Caprimulgiformes, the statement would rely on the non-taxon status of the Caprimulgiformes.

Brummitt argued that phylogenetic trees and cladograms are not the same thing:

> I think it is because so many people think only in terms of cladograms, and not phylogenetic trees, that they cannot see the absurdity of what they are saying in arguing for universal monophyly in Linnaean classification. I don't want to push this point, but I do feel that it is often too superficial to simply chop up a cladogram into supposed clades (can one really call them clades if there are only extant taxa and no ancestors included in the analysis?) and assume that this makes either good taxonomy or an accurate representation of what actually happened in the evolution of the group. (Brummitt 2002, p. 37)

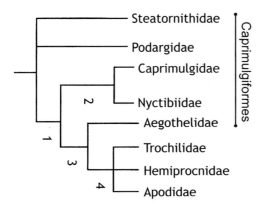

Figure 6.4 Cladogram from Mayr showing the 'paraphyly of the avian order Caprimulgiformes (nightjars and allies)' (redrawn from Mayr 2002, fig. 1).

Brummitt's argument, rather, proposes a generational model for taxa and the model has been found wanting by the results of taxonomic investigation – yet Brummitt appears to believe *his model* to be correct rather than *the data to hand and its resulting classification*; he embraces the model rather than the evidence:

> In *Taxon* earlier this year Rieseberg & Brouillet (1994) have pointed out that many plant species are likely to be paraphyletic taxa, *because they have given rise to other species*. (Brummitt 1996, p. 373, our italics)

> I would prefer to remove any stigma of unacceptability from paraphyletic taxa. Our task is to produce an optimally practical classification, and indicate *which genera have evolved from which other genera*, which families from which other families, and so on. (Brummitt 1997a, p. 731, our italics)

> ... so *one family cannot have evolved from within another family* because that would be an incomplete clade and so paraphyletic. (Brummitt 2014, p. 89, our italics)

This model is not new and has been utilised before to dispute findings from taxonomic (systematic) results. Ernst Mayr, for example, discussed a similar model a few decades ago (Mayr 1974, p. 112, fig. 4, illustrating the classic paraphyletic Reptilia, reproduced here as our Figure 6.5a), which he discussed again, in a different format, some 30 years later (Mayr 1995, p. 428, fig. 2, reproduced here as Figure 6.5b). Mayr and Bock wrote:

> In a Darwinian classification, whenever a taxon *gives rise (almost always by budding)* [cladogenesis] *to a derived new higher taxon*, this taxon is given the rank appropriate to its degree of difference. (Mayr & Bock 2002, p. 181, our emphasis)

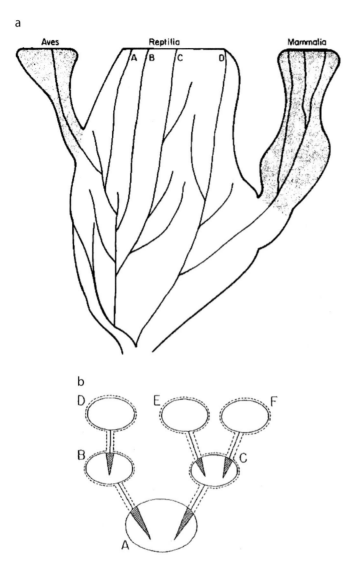

Figure 6.5 (a, b) A model of species origin as represented by Mayr (1974, p. 112, fig. 4, with permission), illustrating the classic paraphyletic Reptilia, discussed again in Mayr (1995, p. 428, fig. 2, with permission); the model is the same as that in Brummitt.

'. . . almost always by budding' is an empty statement masquerading as knowledge of an unobservable evolutionary process, such as cladogenesis.

There is one aspect of Brummitt's critique that is somewhat ironic:

> Taxa should be recognisable by characters, not by position in a putative phylogeny. (Brummitt 1996, p. 371)

... phylogenetic relationships are not observable and are mere hypotheses, whereas characters used for recognition of Linnaean taxa are directly observable. (Brummitt 1997a, p. 726)

We agree: relationships are discovered by characters rather than the interpretation of a diagram: *evidence of monophyly is not evidence of paraphyly.*

Aphyly

Let us assume, instead, that we can be free and easy with names, changing them as and when we see fit. For our example in Figure 6.1b, we might agree that there are two taxa to name: the original taxon A–F, which already has a name, genus Z, and the new group A–C. Our six species are:

ZA ZB ZC ZD ZE ZF

To represent the relationship discovered by our investigations, A–C cannot remain in the genus Z. A common solution to this problem would be to transfer the three to new genus Y:

YA YB YC or Y(A–C)

Leaving D–F in Z: ZD, ZE and ZF.

If taxonomic names mean anything, then the implication behind these two names is that there are two sets of three taxa (Z and Y) and they are related thus: (D–F) (A–C) rather than as the evidence has it: DEF(ABC).

In the former classification (D–F) is not paraphyletic but *aphyletic*. One might see the problem as what can be done to classify D, E and F to convey *their* relationships with respect to genus Y(A–C) or elsewhere. An alternative interpretation is to understand the group (D–F) as *paraphyletic* and then all sorts of evolutionary interpretations flow to explain that situation. The latter is, of course, merely fiction – even the diagram does not support that conclusion. Aphyly might be defined as indicting an *ambiguous taxon*. This means:

1. Members of one taxon share a closer relationship to members of another taxon than they do to their own group (e.g., fishes and dinosaurs);
2. There is no explicit statement about the relationship of members of a single taxon (e.g., an unresolved statement of relationship as suggested by a polytomy);
3. There are too few members of a single taxon to establish whether it is monophyletic or not (i.e., a monotypic genus).

Existing terms that identify these kinds of *ambiguous taxa* are: non-monophyly, paraphyly, polyphyly and monotypy. All can be synonymised under the single term *aphyly*. Unlike non-monophyly, paraphyly, polyphyly, etc., aphyly does not seek or require phylogenetic explanation. Rather, it is used to highlight the need for taxonomic revision, the need to find further data to determine the monophyly or otherwise of any inclusive taxon.

Some other terms have been used that are similar: *Restkörper* was proposed by Willi Hennig prior to his identification and characterisation of paraphyly (Hennig 1936); *merophyly* was proposed more than 30 years ago by the Brazilian entomologist Nelson Bernardi (1981); and *metaspecies* and *metaphyly* were proposed more recently (Donoghue 1985 and Archibald 1994). We will briefly discuss Hennig's 'Restkörper', Bernardi's 'merophyly' and *metaspecies* and *metaphyly* and then offer a few examples of how aphyly has been, and can be, used.

Restkörper: Hennig introduced the word *Restkörper* as a term for the heterogenous group(s) excluded when other smaller groups were characterised and deemed monophyletic, similar to the examples above (Hennig 1936, p. 172). He gave a clearer account in 1948, in the first volume of his *Die Larvenformen der Dipteren* where he included a figure that contrasted two ways to classify the relationships among nine species, A–H, J (he excluded a taxon I, Hennig 1948, our Figure 6.6a).

In tree A Hennig recognises five groups: A–C (Group I), E–F (Group II), H + J (Group III) and two more inclusive groups of Group I + D and Group II + Group III (Figure 6.6b), in effect recognising only the monophyletic groups in the tree. In contrast, in tree B (Figure 6.6c), whose relationships are identical to tree A, he drew attention to four groups: A–C (Group I), F + G (Group II), J (Group III) and D + E + H (Group IV), where Hennig wrote: 'In the following, the group that is illustrated in Fig. 1B as II can be described as "Restkörper"' (Hennig 1948, pp. 11–12, our translation; 'Im folgenden sind Gruppen die den in Fig. 1 B als II bezeichneten Gebilden entsprachen, als "restkörper" bezeichnet'; Hennig referred to 'Restkörperbildung' in the plate legend to fig. 1 B group II). Hennig thus used 'Restkörper' to mean all species left behind after a monophyletic group had been discovered. 'Restkörper' was translated by Schmitt as 'remaining body' (Schmitt 2013, p. 137) and by Meier as 'rest bodies' (Meier 2005, pp. 47, 57), but it might be better translated as 'the bits left over' or more simply the 'remainder', the 'flotsam and jetsam' (*sensu* 'the non-A group' of Schmitt 2001 and Rieppel 2005, and the 'paraphyly' of Meier 2005). For us, that is equivalent to aphyly.

Hennig gave another, perhaps clearer, version of the use of 'Restkörper' in 1953 (Hennig 1953, pp. 9–10). In this example, Hennig identified group A, indicated on the tree in his fig. 2, as those species left behind after the monophyletic group B had been identified; to group A Hennig attached the name 'Restkörper' (Figure 6.7, in the text Hennig writes: 'Die Arten (*a, b, c*) der Restkörpergruppe A (in Fig. 2)', Hennig 1953, p. 9).

Hennig may have been inching his way towards the concept of paraphyly (Schmitt 2013), a term he first used in 1962 in a footnote (Hennig 1962, p. 35) as

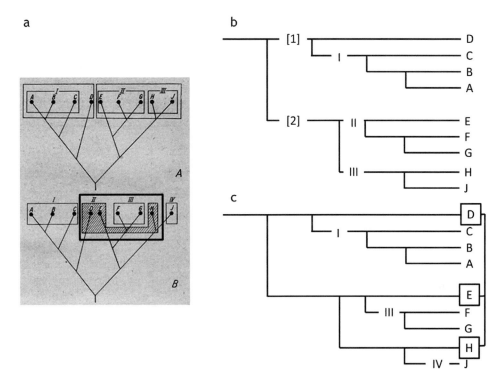

Figure 6.6 (a) Hennig's diagrams illustrating the meaning of *Restkörper* (Hennig 1948), both redrawn as (b) and (c).

commentary on the phylogenetic study of Odonata by Fraser (1954), utilising Fraser's diagram of relationships to equate the notion of symplesiomorphy (Fraser's 'archaic characters') with paraphyly.

While these developments are of some significance in the evolution of Hennig's thoughts on kinds of taxonomic groups, his trajectory from 'Restkörper' to paraphyly, it is the first term, 'Restkörper', we are interested in, as it is more general and of greater utility for practising taxonomists and seems related to, if identical with, our aphyly (Ebach & Williams 2010).

Merophyly: Bernardi coined the term *merophyly* to correspond to the identification of 'unnatural or non-monophyletic groups (both poly and paraphyletic)' (Bernardi 1981):

> A merophyletic (= non-monophyletic) group is a group resulting from the smallest monophyletic group of which it is a part upon the exclusion of at least 1 member species. Merophyletic groups may be either paraphyletic or polyphyletic. (Bernardi 1981, p. 323, from the English abstract)

Merophyly has been commonly used in Latin American taxonomic literature, and differs little from Bernardi's original usage. Interestingly, there have been several

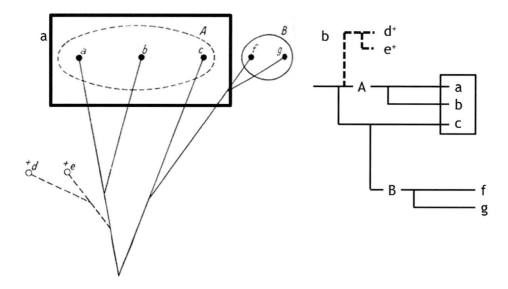

Figure 6.7 (a, b) Hennig gave another, perhaps clearer, version of the use of *Restkörper* (Hennig 1953, pp. 9–10); in the text Hennig writes: 'Die Arten (*a, b, c*) der Restkörpergruppe A (in Fig. 2)' (Hennig 1953, p. 9).

usages that make the term almost redundant, such as 'merophyletic paraphyletic group' (Wanderley-Nogueira et al. 2010, p. 74) and 'merophyletic polyphyletic groups' (Soares-Cavalcanti et al. 2009, p. 52). These tautologies beg the question as to why the authors used merophyly at all? The only alternate usage of merophyly was by Marques and Migotto (2001, p. 481) when referring to a polytomy.

Bernardi utilised ancestry as a criterion of demarcation between kinds of groups (poly- and paraphyletic groups), in the same fashion as Archibald did later when describing his metataxa concepts (metaspecies, mixotaxa and ambitaxa) (Archibald 1994).

Metaphyly: Donoghue introduced the term *metaspecies* to describe situations similar to those we discussed above (Donoghue 1985, p. 178). His primary example was illustrated in four diagrams (reproduced here as Figure 6.8), with six species A–F and three derived characters, a–c, at various places on each cladogram. In each cladogram, A–F and D–F are monophyletic, supported by characters b and a respectively. In their cladogram 1 (our Figure 6.8, cladogram 1), A–C is 'unresolved'; in their cladogram 2 (our Figure 6.8, cladogram 2), A–C is monophyletic supported by character c; in their cladogram 3 (our Figure 6.8, cladogram 3), in addition to the monophyletic D–F, C–F is also monophyletic, supported by character c, but A and B are 'unresolved' and group A–C is termed 'positively paraphyletic'; finally, in their cladogram 4 (our Figure 6.8, cladogram 4), which is identical to their cladogram 1, D–F is monophyletic and called a 'cladospecies'; A–C is 'unresolved' and called a 'metaspecies' (Donoghue 1985, p. 178, figs 1–4).

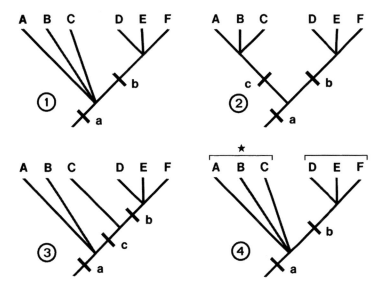

Figure 6.8 Donoghue used the term *metaspecies* as illustrated in these four cladograms with six species, A–F, and three derived characters, a–c, at various places on each of the four cladograms (after Donoghue 1985, p. 178, figs 1–4, with permission).

Reasons for using 'metaspecies' were that:

> Ackery and Vane-Wright (1984; and R. I. Vane-Wright, personal communication) reached a similar conclusion, and suggested that 'positively' monophyletic units be called 'cladospecies' and that the smallest unresolved units be called 'paraspecies.' The term 'paraspecies' might perpetuate the failure to distinguish clearly between 'positively paraphyletic' groups and unresolved groups, and for this reason de Queiroz et al. (MS in prep.) use 'metaspecies' for the smallest unresolved groups. (Donoghue 1985, p. 179)

It is simpler to conclude that species A–C of cladograms 1 and 4 are aphyletic, while species A and B in cladogram 3 are aphyletic and either to include species C in D–F or leave C as a monotypic genus (Figure 6.8). Either way, the concept of aphyly removes the necessity of describing something as 'positively paraphyletic' when there can be no such thing.

In a similar fashion, Archibald noted:

> I define a metataxon as a previously named portion of a polytomy for which positive evidence of monophyly or paraphyly is lacking or ambiguous. This definition follows the general intent of Gauthier et al. (1988), although slightly broadening their usage. (Archibald 1994, p. 28)

While we think Archibald's use of ancestry to determine the –phyly of any taxon is problematic, we find more troublesome the notion of 'positive evidence of ... paraphyly': *evidence of monophyly is not evidence of paraphyly.*

Example 1: For four species A–D, two synapomorphies are known, one that relates all four (A–D), the other that relates only two, C and D (Figure 6.9a).

Species in taxa A–D were already known and named as genus Y. With discovery of the synapomorphy that unites C and D, beyond the larger group (genus Y), the relationships of A and B remain unknown. In time, with further relevant data, taxa A and B may be found to be each other's closest relative, B may be more closely related to C+D than A, or vice versa, and A may be more closely related to C+D than B (as in Nelson and Platnick's Interpretation 1; there are further permutations that would correspond to Nelson and Platnick's Interpretation 2; Nelson & Platnick 1980). Regardless of any of those outcomes, we concluded that prior to the discovery of further relevant characters, A and B are aphyletic. But how should A and B be classified prior to the acquisition of further relevant data? We suggest four propositions.

Proposition 1: They could remain in genus Y with the information that C+D are closely related being conveyed only in the text accompanying the descriptions:

Genus   Y: A–D [informal identity: CD]

Proposition 2: C+D might be named a subgenus of Y:

Genus   Y: A–D

Y: A

    Y: B

Subgenus X: CD

Proposition 3: A–D might be raised to the level of family, C and D could be named as a new genus X with A and B remaining in genus Y but the generic name placed in inverted commas to indicate their aphyletic nature (this convention, using a non-monophyletic genus name inside inverted commas, was adopted by Parenti 1981).

Family   Y: A–D

Genus   'Y': A

    'Y': B

Genus   X: CD

Proposition 4: A–D might be raised to the level of family and species A and B could each be placed in a monotypic genus to indicate their unknown relationships beyond membership of Family Y.

Family   Y: A–D

Genus   Ya: A

Genus   Yb: B

Genus   X: CD

To some extent this proposition follows Hennig's alternatives, with the exception that A + B is not recognised as a group as it is aphyletic.

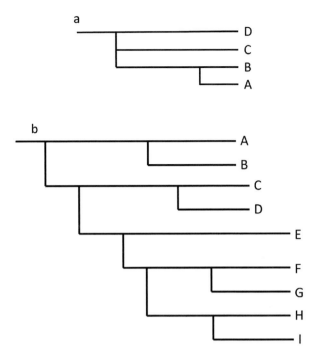

Figure 6.9 (a) An example of four species A–D with two known synapomorphies, one that relates all four (A–D), the other that relates only two, C and D. (b) An example from Pyle and Michel (2008, p. 42, fig. 1); their concern was to discuss 'the relationship of the trinity of nomenclature, taxonomy and type specimens' (Pyle & Michel 2008, p. 42).

Example 2: Pyle and Michel presented a figure to illustrate 'different nomenclatural interpretations for species A through I, mapped to a hypothesized phylogeny' (Pyle & Michel 2008, p. 42, fig. 1, redrawn here as Figure 6.9b). Their overall concern was to discuss 'the relationship of the trinity of nomenclature, taxonomy and type specimens' (Pyle & Michel 2008, p. 42), but it also demonstrated the relationship between (potential) groups and their names, offering three naming schemes for the same diagram and a fourth for a dissenting view on the diagram's underlying phylogeny. Their diagram is interesting in its relevance to the concept of aphyly.

From the perspective of characters (synapomorphies), the assumption is that there is information (evidence) relevant to each branch point. Figure 6.10 is labelled with respect to the seven nodes that represent synapomorphies uniting various taxa: Node 1: A+B, Node 2: C–I, Node 3: C+D, Node 4: E–I, Node 5: F–I, Node 6: F+G, Node 7: H+I. Here the node number is a proxy for the

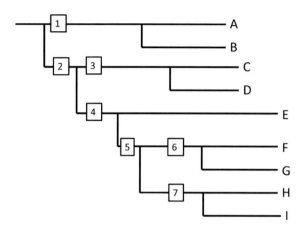

Figure 6.10 Pyle and Michel (2008, p. 42, fig. 1) with components added and numbered.

synapomorphy(ies) that unite the included taxa. If we add hypothetical characters, the situation might be so:

Species	Node # T = terminal	Characters/synapomorphies * = autapomorphy	Included nodes
E–I	4	1–5	4
E	T	1–5, 11*	4
F–I	5	1–5, 6–8	4 + 5
F+G	6	1–5, 6–8, 9	4 + 5 + 6
F	T	1–5, 6–8, 9, 12*	4 + 5 + 6
G	T	1–5, 6–8, 9, 13*	4 + 5 + 6
H+I	7	1–5, 6–8, 10	4 + 5 + 7
H	T	1–5, 6–8, 10, 14*	4 + 5 + 7
I	T	1–5, 6–8, 10, 15*	4 + 5 + 7

All species E–I have Characters 1–5. Species F–I have additional characters (6–8) that place them in a subgroup (node 5). Species E has no further resolution other than being a member of the larger group E–I (node 4). In this case, although its relationships are thereby determined (basal to group F–I), it is, in our sense, aphyletic.

The nodes are also proxies for names that might be applied and a conventional approach would be to make species E a monotypic genus. Pyle and Michel's Taxonomist 1 suggests that E–I be named a single genus (*Eus*), an action we have no disagreement with. Within that genus, however, there are two subgroups, F+G and H+I, and species E. Should these be recognised as well, and if so, how? There is no logical reason for that to be undertaken, but if it was, one might be tempted to modify Taxonomist 1's original scheme and recognise F+G and H+I as subgenera:

Genus	Subgenus	Species
A		A
		B
C		C
		D
E		E
	E1	F
		G
	E2	H
		I

What, then, of species E? Its relationships, beyond being a member of genus E, are unknown. One might make this species a monotypic subgenus, E3 (as noted below), which would make it aphyletic:

Genus	Subgenus	Species
A		A
		B
C		C
		D
E	E3	E
	E1	F
		G
	E2	H
		I

Why is this of any significance? Species E is clearly in need of further investigation. It has synapomorphies uniting it with the larger group F–I, autapomorphies of its own but no characters that relate it more closely to any other taxon. Aphyly highlights these species (or other taxa) as (potentially) problematic and offers a specific indication for further study.

An earlier discussion between Platnick and Wiley on monotypic taxa focused on issues of origin and ancestors and the relative status of a basal monotypic taxon, similar to species E in our example (Platnick 1976, 1977a; Wiley 1977). Rather than argue with either Platnick or Wiley, they both recognised the situation as problematic and, in our view, aphyly solves this by drawing attention to a taxonomic problem that only further relevant data can address.

Where aphyly becomes more useful is when the residual group has more than one species. In the example above, if the branch leading to species E actually represents not one but 10 species, then the outstanding question is how to deal with those 10 in terms of names. One might name each species as a monotypic genus. While this would satisfy the dictates of the data, it leads to the somewhat cumbersome introduction of many new names that one might predict will disappear relatively quickly. It would be far easier to retain the group as a genus but acknowledge that it is aphyletic. The situation is more complex when the residual group is, say, 1000 species for conventional naming procedures but not for the concept of aphyly.

Example 3: Triceratium The diatom genus *Triceratium* has been in existence since the mid-1800s and accumulated no less than 1300 names (not all are accepted, of course, but it is problematic to determine which names are), many of which are known only from fossils. Piecemeal revisions of small groups of species is being undertaken but it is impossible to determine the current '-phyly' of *Triceratium* as these groups are removed. To that extent, *Triceratium* is obviously aphyletic. In a recent effort to disentangle some of the relationships of the genus, a group of highly diverse fossil representatives were removed from *Triceratium* and placed in two new genera, *Entogoniopsis* and *Trilamina*, and their relationships determined relative to existing taxa, the previously known genera *Medlinia*, *Entogonia* and *Sheshukovia* (Witkowski et al., 2015). *Entogoniopsis*, *Trilamina*, *Medlinia* and *Entogonia* are considered to be more closely related to each other than to any other remaining member of *Triceratium*; they form a monophyletic group. Some species of *Sheshukovia* are better placed in *Entogoniopsis*, which left the remaining species of *Sheshukovia* of indeterminate '-phyly'. Thus, conclusions for this study are: *Entogoniopsis*, *Trilamina*, *Medlinia* and *Entogonia* are each a monophyletic group; *Entogoniopsis* + *Trilamina* + *Medlinia* + *Entogonia* together form a monophyletic group; the pairs *Entogoniopsis* + *Trilamina* and *Medlinia* + *Entogonia* are a monophyletic group and the pair sister to each other; *Triceratium* and *Sheshukovia* are both aphyletic (Witkowski et al. 2015). While it might be an obvious conclusion that *Triceratium* needs attention, it is less so with *Sheshukovia*, which has a relatively modest 50+ species.

Note that our original suggestion of aphyly was made with reference to characters, the evidence supporting specific relationships; Bernardi's merophyly was created with reference to specific instances of ancestry (i.e., paraphyly and polyphyly); and our examples above (one of our own and one taken from Pyle & Michel 2008) were made with reference to a diagram of relationships, a diagram that need not have any specific implications of ancestry. Bernardi's merophyly was limited, inasmuch as it did not deal with polytomies or monotypic taxa. Thus our aphyly can be generalised to situations that require a classification to be derived from a diagram of relationships, a specific diagram of ancestry or a study of taxon characters (synapomorphy).

Example 4: Dryandra The confusion between a classification derived from a set of relationships, as opposed to a notion of ancestry, appeared in the recent debate over the transfer of *Dryandra* to *Banksia* (Mast and Thiele 2007; George 2014; Thiele et al. 2015). The debate began with George (2014) disputing the transfer of *Dryandra* into *Banksia* by Mast and Thiele (2007), based on an earlier molecular analysis (Mast et al. 2005). The analysis recovered a relationship in which the genus *Dryandra* was nested within *Banksia* (Figure 6.11). As a genus, *Dyrandra* was monophyletic. To complicate matters, the aphyletic *Banksia* Linnaeus 1781 has priority over *Dryandra* Brown 1810. The question is how to deal with an aphyletic *Banksia*. Mast and Thiele (2007) proposed three options: '(1) reducing *Dryandra* to a synonym of *Banksia*; (2) maintaining *Dryandra* at generic rank and splitting *Banksia* into several new genera, and (3) establishing two genera equivalent to the clades formed at the basal split of *Banksia* (Mast & Thiele 2007, p. 64)'. The authors chose the first option, meaning that the more populous *Dryandra* was synonymised into one of two *Banksia* subgenera (*B. Banksia*).

The taxonomic revision of Mast and Thiele (2007) was disputed by George (2014) who stated that the taxonomic study 'was based on the use of holophyly (monophyly s. str.) as an essential criterion for recognition of taxa' and that 'paraphyly should be accepted in biological classification' (George 2014, p. 32). The confusion between a set of relationships determining a classification versus notions of ancestry had clouded matters. Arguing for artificial taxa names to be retained based on quasi-evolutionary models like the one George (2014) proposes is not a taxonomic solution, nor does it resolve the aphyletic *Banksia*. Monophyly defined in the same way it is identified (i.e., as a taxon that its 'members' are more closely related among themselves than to any other taxon) means there is less focus on what causes monophyly (i.e., common ancestor) and more on taxonomic revision. In arguing for a particular evolutionary model, George (2014) has effectively turned a discussion about taxonomy and natural classification into one about evolution, in which aphyletic taxa (i.e., paraphyly) should be recognised within a taxonomic classification.

Thiele et al. (2015), responding to George (2014), believe, 'paraphyly is preferred in order to allow the retention of a morphologically convenient and homogeneous group in preference to a monophyletic but inconveniently heterogeneous one' (Thiele et al. 2015, p. 201). We think not. The problem, rather, is the history of debate surrounding the definition of paraphyly, namely that it represents a taxon that includes an ancestor but not all its descendants (e.g., Ashlock 1971; Nelson 1972). The definition gives paraphyly a veneer of validity, seemingly representing an incomplete taxon, such as the traditional Dinosauria (see Chapter 14). It is this phylogenetic definition of paraphyly that taxonomists, such as George, use to

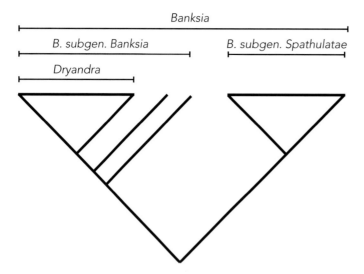

Figure 6.11 The relationships of *Dryandra* and *Banksia* (redrawn from Mast et al. 2005). The analysis recovered a relationship in which the genus *Dryandra* was nested within *Banksia*.

justify their own 'incomplete' groups. What if paraphyly were to be defined as an artificial taxon, without any evolutionary explanation? Unfortunately, it is too late to revise the term; however, the adoption of aphyly is timely. Using a neutral term with no definition other than the '–phyly' is unknown, aphyly is ideal for describing ambiguous taxon names. With such a definition, taxonomists can get on with describing and revising taxa in light of new evidence, without the need for untestable evolutionary scenarios. Discussions about the meaning of paraphyly and its role in classification has obscured that task, something which aphyly, had it been adopted sooner, would have avoided.

Summary

Aphyly allows the recognition of areas of further study in cases where some persons might be tempted to offer evolutionary explanations to cover the *lack of data* rather than its *presence* (e.g., Stepanek & Kociolek 2019). It should be noted that paraphyly has long been a source of confusion, even by its creator – Hennig Platnick illustrated some of the various and conflicting versions Hennig used (Platnick 1977b, p. 196, figs 1–4) – this is not an issue we discuss further here (see Vanderlaan et al. 2013). Perhaps, like Patterson (1982, as but one example from many), paraphyly and polyphyly (Nelson's *non-monophyly*, Nelson 1971, p. 472) might best be dropped from use altogether, and, as we have demonstrated, only the terms monophyly and aphyly used as they are both of practical value to the taxonomist: the first term for proposing natural taxa, the second term for those taxa that need further investigation.

We hope to have demonstrated that rather than a '[...] tiresome point[s] of terminology [...]', the relationship between evidence and conclusions seems paramount. *And we repeat: evidence of monophyly is not evidence of paraphyly.*

References

Ackery, PR. & Vane-Wright, RI. 1984. *Milkweed Butterflies: Their Cladistics and Biology.* Cornell Univ. Press, Ithaca, New York.

Archibald, JD. 1994. Metataxon concepts and assessing possible ancestry using phylogenetic systematics. *Systematic Biology* 43: 27–40.

Ashlock, PD. 1971. Monophyly and associated terms. *Systematic Zoology* 20: 63–69.

Bernardi, N. 1981. Parentesco filogenético, grupo monofilético e conceitos correlatos: novas definições. *Revista Brasileira de Entomologia* 25: 323–326.

Brummitt, RK. 1996. In defence of paraphyletic taxa. In: LJG. van der Maesen (ed.), *The Biodiversity of African Plants* (Proceedings of the 14th Plenary Meeting of the Association for the Taxonomic Study of the Flora of Tropical Africa [AETFAT] Congress, Wageningen). Kluwer Academic Press, Dordrecht, pp. 371–384.

Brummitt, RK. 1997a. Taxonomy versus cladonomy, a fundamental controversy in biological systematics. *Taxon* 46: 723–734 [an enhanced version was distributed to some fellow botanists as a reprint "with added summary"].

Brummitt, RK. 1997b. *Proposing the motion 'that this house believes that Linnean classification without paraphyletic taxa is nonsensical': the theoretical case.* Royal Botanic Garden, Kew.[1]

Brummitt, RK. 2002. How to chop up a tree. *Taxon* 51: 31–41.

Brummitt, RK. 2014. Taxonomy versus cladonomy in the Dicot families. *Annals of the Missouri Botanical Garden* 100:89–99.

[1] This publication was privately printed and circulated to interested parties by Brummitt. The title page has the following statement: "This paper is cited as for publication in '*Linnean* 13 (in press)' among the references in a paper by the same author in *Taxon* 46: 734 (1997). After the latter paper was accepted for publication, the Linnean Society took a decision not to publish any of the papers from the debate in *The Linnean.* The present paper is effectively published in its present form by distribution to more than 25 libraries by the author in December 1997. All papers presented at the debate are available on request from the Linnaean Society of London". Enquiries to the Linnean Society revealed no further documentation for this meeting and, so it seems, no further papers were submitted (Dorothy Fouracre, pers. comm., 2018).

The motion was noted in *The Linnean* 13 (3): "A debate has been arranged for the evening of 6th March, starting at 4.30 (tea at 4pm). The motion is 'that this house believes that Linnaean classification without paraphyletic taxa is nonsensical'. For the motion will be Dr. R.K. Brummitt and Dr A.J. Paton, both of the Royal Botanic Gardens, Kew, against Professor C.J. Humphries, Botanical Secretary and Dr. P.L. Forey FLS, both of the Natural History Museum". The motion was accepted.

Dawkins, R. 2004. *The Ancestor's Tail.* Weidenfeld & Nicolson, London.

Donoghue, MJ. 1985. A critique of the biological species concept and recommendations for a phylogenetic alternative. *The Bryologist* 88: 172–181.

Ebach, MC. & Williams, DM. 2010. Aphyly: A systematic designation for a taxonomic problem. *Evolutionary Biology* 37:123–127.

Felsenstein, J., 2001. The troubled growth of statistical phylogenetics. *Systematic Zoology* 50: 465–467.

Flegr, J. 2013. Why *Drosophila* is not *Drosophila* any more, why it will be worse and what can be done about it? *Zootaxa* 3741 (2): 295–300.

Franz, NM. 2005. On the lack of good scientific reasons for the growing phylogeny/classification gap. *Cladistics* 21: 495–500.

Fraser, FC. 1954. The origin and descent of the order Odonata based on the evidence of persistent archaic characters. *Proceedings of the Royal Entomological Society London*, Series B, 23: 89–94.

Frodin, DG. 2004. History and concepts of big plant genera. *Taxon* 53: 753–776.

Gauthier, J., Estes, R. & de Queiroz, K. 1988. A phylogenetic analysis of Lepidosauromorpha. In: Estes, R. and Pregill, G. (eds.), *Phylogenetic Relationships of the Lizard Families.* Stanford University Press, Stanford, California, pp. 15–98.

George, AS. 2014. The case against the transfer of *Dryandra* to *Banksia* (Proteaceae) 1. *Annals of the Missouri Botanical Gardens* 100: 32–49.

Hennig, W. 1936. Beziehungen zwischen geographischer Verbreitung und systematischer Gliederung bei einigen Dipterenfamilien ein Beitrag zum Problem der Gliederung systematischer Kategorien höherer Ordnung. *Zoologischer Anzeiger* 116: 161–175.

Hennig, W. 1948. *Die Larvenformen der Dipteren.* Akademie-Verlag, Berlin, Teil 1.

Hennig, W. 1953. Kritische Bemerkungen zum phylogenetischen System der Insekten. *Beiträge zur Entomologie* 3 (Sonderh.): 1–85.

Hennig, W. 1962. Veränderungen am phylogenetischen System der Insekten seit 1953. *Deutsche Akad. Landwirtschaftswiss. Berlin: Tagungsber.* 45: 29–42.

Hillis, DM. 2007. Constraints in naming parts of the Tree of Life. *Molecular Phylogenetics and Evolution* 42: 331–338.

Marques, AC. & Migotto, AE. 2001. Cladistic analysis and new classification of the family Turulariidae (Hydrozoa, Anthomedusae). *Papéis Avulsos de Zoologia* 41: 465–488.

Mast, AR. & Thiele, KR. 2007. The transfer of *Dryandra* R.Br. to *Banksia* L.f. (Proteaceae). *Australian Systematic Botany* 20, 63–71.

Mast, AR., Jones, EH. & Havery, SP. 2005. An assessment of old and new DNA sequence evidence for the paraphyly of *Banksia* with respect to *Dryandra* (Protaeceae). *Australian Systematic Botany* 18: 75–88.

Mayr, E., 1974. Cladistic analysis or cladistic classification? *Zeitschrift für zoologische Systematik und Evolutionsforschung* 12:94–128.

Mayr, E. 1995. Systems of ordering data. *Biology and Philosophy* 10: 419–434.

Mayr, G. 2002. Osteological evidence for paraphyly of the avian order Caprimulgiformes (nightjars and allies). *Journal für Ornithologie* 143: 82–97.

Mayr, E. & Bock, WJ. 2002. Classifications and other ordering systems. *Journal of*

Zoological Systematics and Evolutionary Research 40: 169–194.

Meier, R. 2005. Role of dipterology in phylogenetic systematics: the insight of Willi Hennig. In: Wiegmann, BM. & Yeates, DK. (eds), *The Evolutionary Biology of Flies.* Columbia University Press, New York, pp. 45–62.

Nelson, GJ. 1971. Paraphyly and polyphyly: Redefinitions. *Systematic Zoology* 20: 471–472.

Nelson, GJ. 1972. Phylogenetic relationship and classification. *Systematic Zoology* 21:227–231.

Nelson, GJ. & Platnick, NI. 1980. Multiple branching in cladograms: Two interpretations. *Systematic Zoology* 29:86–91.

Nelson, GJ., Murphy, DJ. & Ladiges, PY. 2003. Brummitt on paraphyly: A response. *Taxon* 52: 295–298.

O'Hara, RJ. 1991. Representations of the natural system in the nineteenth century. *Biology and Philosophy* 6: 255–274.

Parenti, L. 1981. A phylogenetic and biogeographic analysis of cyprinodontiform fishes (Teleostei, Atherinomorpha). *Bulletin of the American Museum of Natural History.* 168 (4): 335–557.

Patterson, C. 1982. Morphological characters and homology. In: Joysey, KA. & Friday, AE. (eds), *Problems of Phylogenetic Reconstruction.* Academic Press, London, pp. 21–74.

Platnick, NI. 1976. Are monotypic genera possible? *Systematic Zoology* 25: 198–199.

Platnick, NI. 1977a. Monotypy and the origin of higher taxa: A reply to E. O. Wiley. *Systematic Zoology* 26: 355–357.

Platnick, NI. 1977b. Paraphyletic and polyphyletic groups. *Systematic Zoology* 26: 195–200.

Pyle, RL. & Michel, E. 2008. Zoobank: Developing a nomenclatural tool for unifying 250 years of biological information. *Zootaxa* 1950: 39–50.

Rieppel, O. 2005. Proper names in twin worlds: Monophyly, paraphyly, and the world around us. *Organisms Diversity & Evolution* 5: 89–100.

Rieseberg, LH. & Brouillet, L. 1994. Are many plant species paraphyletic? *Taxon* 43: 21–32.

Rosa, D. 1918. *Ologenesi.* R. Bemporad, Firenze. [Reprint: Vergata, A. La (ed.). 2001. *Daniele Rosa, Ologenesi.* Biblioteca della Scienze Italiana, no. 32. Giunti, Firenze.]

Schmitt, M. 2001. Willi Hennig (1913–1976). In: Jahn, I. & Schmitt, M. (eds), *Darwin & Co. Eine Geschichte der Biologie in Portraits.* C.H. Beck, Munich, pp. 317–343.

Schmitt, M. 2013. *From Taxonomy to Phylogenetics–Life and Work of Willi Hennig.* Leiden: Brill.

Seifert, B., Buschinger, A., Aldawood, A., Antonova, V., Bharti, H., Borowiec, L., Dekoninck, W., Dubovikoff, D., Espadaler, X., Flegr, J., Georgiadis, C., Heinze, J., Neumeyer, R., Ødegaard, F., Oettler, J., Radchenko, A., Schultz, R., Sharaf, M., Trager, J., Vesnić, A., Wiezik, M. & Zettel, H. 2016. Banning paraphylies and executing Linnaean taxonomy is discordant and reduces the evolutionary and semantic information content of biological nomenclature. *Insectes Sociaux* 63: 237–242.

Soares-Cavalcanti, NM., Wanderley-Nogueira, AC., Belarmino, LC., dos Santos Barros, P. & Benko-Iseppon, AM. 2009. Comparative in silico evaluation of MYB transcription factors in eucalyptus, sugarcane and rice transcriptomes. In: *Computational Intelligence Methods for Bioinformatics and Biostatistics.* Springer: Berlin, Heidelberg, pp. 44–55.

Stepanek, JG. & Kociolek, PJ. 2019. Molecular phylogeny of the diatom genera *Amphora* and *Halamphora* (Bacillariophyta) with a focus on morphological and ecological evolution. *Journal of Phycology,* doi:10.1111/jpy.12836

Thiele KR., Weston, PH. & Mast, AR. 2015. Paraphyly, modern systematics and the transfer of *Dryandra* into *Banksia* (Proteaceae): a response to George. *Australian Systematic Botany* 28, 194–202.

Vanderlaan, TA., Ebach, MC., Williams, DM. & Wilkins, JS. 2013. Defining and redefining monophyly: Haeckel, Hennig, Ashlock, Nelson and the proliferation of definitions. *Australian Systematic Botany.* 26: 347–355.

Wanderley-Nogueira, AC., da Mota Soares-Cavalcanti, N., Belarmino, LC., Barbosa-

Silva, A., Kido, EA., Do Monte, SJH., Pandolfi, V., Calsa-Junior, T. & Benko-Iseppon, AM. 2010. In silico screening for pathogenesis related-2 gene candidates in *Vigna unguiculata* transcriptome. In: *Computational Intelligence Methods for Bioinformatics and Biostatistics.* Berlin Heidelberg: Springer, pp. 70–81.

Wiley, EO. 1977. Are monotypic genera paraphyletic? - A response to Norman Platnick. *Systematic Zoology* 26: 352–355.

Wilkins, JS. & Ebach, MC. 2013. *The Nature of Classification: Relationships and Kinds in the Natural Sciences.* Palgrave Macmillan.

Williams, DM. & Ebach, MC. 2017. Aphyly: identifying the flotsam and jetsam of systematics. *Cladistics* 34:459–466.

Witkowski, J., Sims, PA., Strelnikova, NI. & Williams, DM. 2015. *Entogoniopsis* gen. nov. and *Trilamina* gen. nov. (Bacillariophyta): a survey of multipolar pseudocellate diatoms with internal costae, including comments on the genus *Sheshukovia* Gleser. *Phytotaxa* 209: 1–89.

PART III

The Cladistic Programme

Every scientific discipline has an over-abundance of technical terms. Taxonomy is no exception. Problems arise when terms are used with one meaning and interpreted with another, a common enough occurrence with language in general. The word 'Tree' is a good example. *Sequoia sempervirens*, the Californian Redwood, is the largest and oldest known tree – tree here means a large, woody plant with many branches (Figure P.1a); Ernst Haeckel's well-known diagrams depicting the phylogenetic relationships of organisms are also called trees, some of his earlier examples being interpreted as similar to an oak tree (*Quercus*) (Figure P.1b); Arthur Cayley (1821–1895), a founding father of graph theory (Bondy & Murty 2008), a branch of mathematics using *graphs* with branching structures he called trees (Otter 1948, Figure P.1c). All of these different kinds of trees crop up in taxonomic studies – they have one thing in common: they branch. One might exploit the branching aspect and force a direct equation, such that some might understand phylogenetic trees to be a 'kind' of mathematical tree, and phylogenetic research might then be understood to be a branch of graph theory. Be that as it may – and we return to some of those equations below – words are used with very different meanings.

It would exhaust the patience of our readers to document all the varying definitions of the very many terms that have been used in taxonomy and systematics – and are still being introduced. In this book we endeavour to define as precisely as possible how we will use certain terms and naturally hope others will follow our usage. We do not offer extended discussion on alternative uses except in cases where it is necessary, as discussions of definitions rarely produce significant light, most often doing the opposite: generate copious amounts of heat.

In Chapter 1 we provided very brief definitions of a few terms commonly encountered in taxonomic or systematic studies, such as cladistics, cladogram, monophyly, paraphyly, homologue, homology, synapomorphy, heterobathmy, taxon, phylogeny and so on. We hope readers did not find our definitions too idiosyncratic. Here we expand on some of the more important terms so that our usage is explained in more detail and the methodology of taxonomy may be seen clearer.

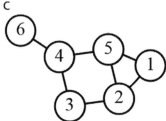

Figure P.1 (a) *Sequoia sempervirens* (the Coast Redwood, after Wikipedia, 'Redwood Highway at Jedediah Smith State Park near Crescent City, California USA, the Simpson-Reed Grove of redwoods', photograph by Acroterion. https://commons.wikimedia.org/wiki/User:Acroterion, 1st October, 2009). (b) One of Haeckel's phylogenetic trees (Haeckel 1866, taf. VII). (c) Graph from graph theory.

References

Bondy, JA. & Murty, USR. 2008. *Graph Theory*. Springer.

Haeckel, E. 1866. *Generelle Morphologie der Organismen: Allgemeine Grundzüge der organischen Formen-Wissenschaft, mechanisch begründet durch die von C. Darwin reformirte Decendenz-Theorie*. Berlin.

Otter, R. 1948. The number of trees. *Annals of Mathematics*, 2nd ser. 49(3): 583–599.

7

Parameters of Classification:
Ordo Ab Chao

Cladogram

A cladogram is simply a branching diagram (the word is derived from the Greek *klados* meaning branch); it is non-reticulate; it summarises current knowledge about organisms (Nelson 1979). A cladogram relates all taxa, fossil and Recent, based on evidence derived from organisms and their parts (homologues) and, ultimately, their interrelationships (monophyly, homology). The branching aspect (the specific relationship) is referred to as its *cladistic parameter* (Nelson 1979, p. 12; Williams & Ebach 2008).

Cladograms relate to *phylogenetic (phyletic) trees* but are distinguished from them because a 'cladogram is an atemporal concept; a phyletic tree is a cladogram to which the temporal aspect has been added' and, further, '*a cladogram is a synapomorphy scheme*; a phyletic tree, a phylogeny' (Nelson 1979, p. 8).[1] In its most general sense, *synapomorphies* are defining characters,[2] defining particular groups of organisms (taxa). We return to the concept of 'trees' later, but at this point it should be clear that cladograms are simply diagrams that depict the topographical relationships between organisms (taxa) as supported by evidence, the evidence being their defining characters (homologues).

Nelson described three distinct forms of cladogram, which we follow here:

1. 'Fundamental cladograms ... summarize data about the interrelationships between terminal taxa' (Nelson 1979, p. 5).

---

[1] By atemporal we mean that if a 400-million-year-old Devonian trilobite is included, it will be found at the terminal branches along with all other Recent taxa. Regardless of the taxon's age, it is treated equally to all other taxa.

[2] The definition of synapomorphy is problematic and discussed further below.

Fundamental cladograms are straightforward as they represent information about taxon interrelationships based on some direct assessment of the data to hand (Nelson 1979, p. 5). It is important to note that at this point no prescription is needed to determine how any particular fundamental cladogram might be found – that is, *no method need be stated.*

2. General cladograms: 'the single most important principle of cladistics is that diverse fundamental cladograms may be combined to form a single general cladogram' (Nelson 1979, p. 7).

   General cladograms summarise relational information from several sources relative to the taxa at hand. They relate to a set of methods that have been referred to as *consensus or supertree methods* (dealt with in more detail in Chapters 9 and 11) – the equation is not entirely exact.

   General cladograms are not necessarily the product of *any particular consensus method.*

3. 'Derivative cladograms (classifications) ... specify the cladistic aspect of hierarchical classifications ... A derivative cladogram is a graphic representation of a hierarchical classification ... The word 'derivative' implies derivation from a fundamental cladogram' (Nelson 1979, p. 5).

More simply put, *derivative cladograms* are classifications (Nelson 1979, p. 5). They are, in many cases, the classification used or adopted by whoever undertakes an analysis or study and are usually, but need not be, derived from a particular fundamental cladogram or a particular general cladogram. The expectation is that a derivative cladogram (the classification) represents exactly, at least in as much as it can, the results found in the fundamental cladogram(s). It is unlikely that a derivative cladogram (classification) would depart in any significant way from the results of any particular study, but it is not unknown and more common than one might expect (see Example 2 below).

Derivative cladograms extend to all classifications, regardless of how that classification was produced. This was clearly illustrated in the many diagrams of organism relationships, from Theophrastus (371–c. 287 BC) to Willi Hennig (1913–1976), in Nelson and Platnick (1981, figs 2.1–2.77). Each relationship diagram can be represented by a simpler one that highlights only its *cladistic parameter* (Figure 7.1). One might object to comparing the diagrams found in Theophrastus with those found in Hennig, as each would probably have had different ideas in mind as to what constitutes a classification and, perhaps more significantly, what constitutes evidence for it. Yet certain aspects of taxonomy (systematics), whether intended or not, are reflected in the branching structure found in the usual kinds of hierarchical classifications, which capture the cladistic

a

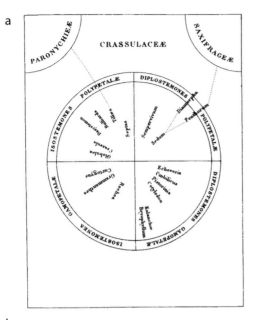

b

I. Crassulaceae Legitimae
    Isostemoneae
        Polypetalae
            Genera: *Tillaea, Bulliarda, Dasystemon, Septas, Crassula,*
                Globulea
        Gamopetalae
            Genera: *Curtogyne, Grammanthes, Rochea*

    Diplostemoneae
        Gamopetalae
            Genera: *Kalanchoe, Bryophyllum, Cotyledon, Pistorinia,*
                Umbilicus, Echeveria
        Polypetalae
            Genera: *Sedum, Sempervivum*
II. Crassulaceae Anomale
            Genera: *Diamorpha, Penthorum*

c

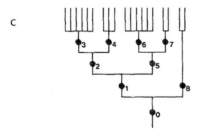

Figure 7.1 (a) Candolle's diagram of the relationships of plants in the family Crassulaceae (Candolle 1828, pl. II); (b) Candolle's written classification; (c) a *derivative cladogram* from the written classification (after Nelson & Platnick 1981, p. 109, fig. 2.47, with permission of the authors).

parameter regardless of how the diagram was conceived (for more detail on classifications and diagrams see Chapter 3).

Example 1 (Figure 7.1): Candolle (1828, pl. 2, reproduced in Nelson & Platnick 1981, p. 109, fig. 2.47, and Pietsch 2012, p. 48): Candolle's diagram (Figure 7.1a) depicts in a circular fashion the relationships of plants in the family Crassulaceae, a not uncommon mode of representation at that time (e.g., some examples in Pietsch 2012[3]). Nevertheless, it is reasonably easy to see how Candolle's written classification (Figure 7.1b) relates to the diagram (Figure 7.1a). The circle in the diagram is divided into four quadrants, with the pair of quadrants at the top and bottom of each half of the circle being a named group: Isostemoneae and Diplostemoneae. Each quadrant has a name, the upper left and right being Polypetalae and consisting of six and four genera, respectively; the lower left and right being Gamopetalae, and consisting of three and six genera, respectively. This arrangement is easily turned into a *derivative cladogram* by reconstructing the hierarchical (branching) aspect of the classification (Figure 7.1c).

Example 2 (Figure 7.2): Simonsen (1979, fig. 3), relationships of diatoms (for a more detailed account see Williams 2007): In 1979 Reimer Simonsen (1931–2012, Crawford et al. 2013) published a critical review of diatom morphology, phylogeny and evolution (Simonsen 1979). He primarily focused on the evolution of a group of diatoms colloquially known as 'centric' diatoms (due to the radial symmetry of their silica encasing), following upon his earlier more speculative essay (Simonsen 1972). In 1979 he presented a new classification alongside what he called a full 'pedigree' of diatoms – a phylogenetic tree (Figure 7.2a, b). The black box in Figure 7.2a encloses what is usually considered to be the groups that encompass the 'centric' diatoms, which in Simonsen's classification are named as the Order Centrales and in his 'pedigree' are aphyletic (or, to some, paraphyletic[4]); the black box in Figure 7.2b encloses what is usually considered to be the pennate diatoms (so called because of the bilateral symmetry of their silica encasing), which in Simonsen's classification are named the Order Pennales and in his 'pedigree' are monophyletic. Within the Pennales two of the families, Diatomaceae and Protoraphidaceae, are together usually considered to be the 'araphid' diatoms, which in Simonsen's classification comprise the Suborder Araphineae (= 'araphid' diatoms) and in his 'pedigree'

---

[3] It's not entirely unpopular now: https://blog.education.nationalgeographic.org/2015/09/22/circle-of-life/

[4] See Chapter 6 for a more detailed account of this term. It is discussed in passing later in this chapter.

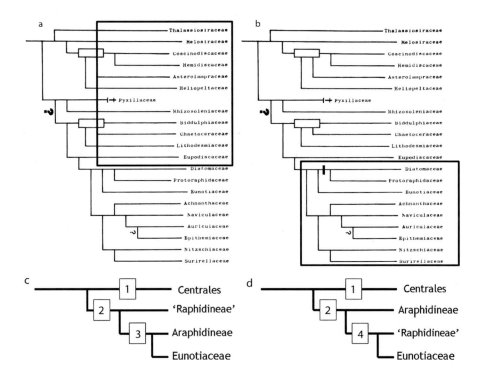

Figure 7.2 Simonsen's diagram of the relationships of diatoms (redrawn and modified from Simonsen 1979, fig. 3, with permission from Schweizerbart science publishers, www.schweizerbart.de). (a) The black box encloses what is usually considered to be the groups that encompass the 'centric' diatoms; (b) the black box encloses what is usually considered to be the pennate diatoms; (c and d) alternative derivative cladograms (see text and Williams 2007 for a fuller account).

are monophyletic (node marked with a vertical black bar in Figure 7.2b). All the remaining pennate diatom families, from Eunotiaceae to Surirellaceae, grouped together as Suborder Raphineae, in his 'pedigree' are paraphyletic.[5]

On the original 'pedigree' diagram, Simonsen added rectangular boxes to two branches of the tree, placing them at two nodes; one grouping some 'centric' diatoms (four families), the other grouping some of the other 'centric' diatoms (three families). Each group is monophyletic on the diagram. These were referred to as 'uncertain - not common - ancestors' (Simonsen 1979, fig. 3) of all the groups distal to the box in question. The exact meaning of Simonsen's words is hard to tell, but it is evident that his intention was to add a note of doubt as to the

---

[5] It has since been shown that neither the 'centric' nor the 'araphid' diatoms are monophyletic (e.g., Theriot et al. 2009)

possibility of the distal groups sharing direct 'common' ancestry – but they are monophyletic on the diagram. Whatever was intended, the relationships expressed in his 'pedigree' can be summarised in a simpler diagram (Figure 7.2c). In this diagram the paraphyletic 'centric' diatoms are grouped together (Figure 7.2c, branch 1), the pennate diatoms are grouped together (Figure 7.2c, node 2) and the paraphyletic 'araphid' diatoms are more closely related to one of the pennate groups, the Eunotiaceae (Figure 7.2c, node 3).

Comparison with Simonsen's classification (Figure 7.2d), suggests that the 'centric' diatoms do form a group (Figure 7.2d, branch 1), the pennate diatoms are grouped together (Figure 7.2d, node 2) and Eunotiaceae is more closely related to the other raphid diatom groups than to the 'araphid' diatoms, the more conventional view (Figure 7.2d, lower diagram, node 4).

It might be, of course, that Simonsen's classification was never intended to represent his phylogenetic relationships at all. The basic details of Simonsen's classification are as follows (the nodes and monophyly refer to Figure 7.2a–d):

Order Centrales (branch/node 1, paraphyletic)

Suborder I: Six Families
Suborder II: Two Families
Suborder III: Four Families

Order Pennales (node 2, monophyletic)

Suborder VI: Two Families (= 'araphid' diatoms, monophyletic)
Suborder V: Seven Families (including Eunotiaceae (node 4, monophyletic)

There is, then, conflict between evidence for the 'pedigree' (the phylogenetic tree) and evidence for the classification (the characters): the tree and classification are not the same. If the classification is considered to be a *derivative cladogram*, then of the groups (and nodes) mentioned, only one is paraphyletic, the 'centric' diatoms. The phylogeny has more paraphyletic groups, which is maybe not surprising as the structure was largely derived from an ancestry narrative rather than the evidence.

Example 3 (Figure 7.3a): Hughes (2007), relationships of trilobites: In cases when the interrelationships between taxa are *aphyletic* (disputed), classifications tend to lean toward evolutionary explanations and results in basic derivative cladograms. Hughes (2007), for example, took a more 'evo-devo' (evolutionary developmental biology) approach to classifying the orders of trilobites (†Trilobita), namely through cataloguing their morphological and ontogenetic development through time. The evolutionary history of the orders of trilobites plotted against time shows

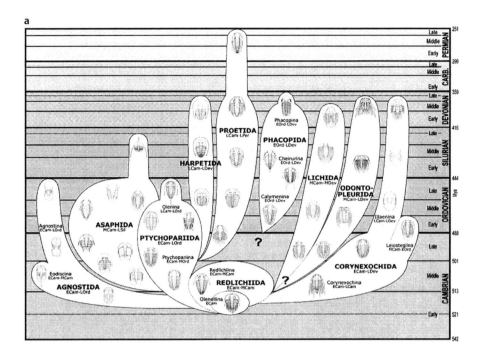

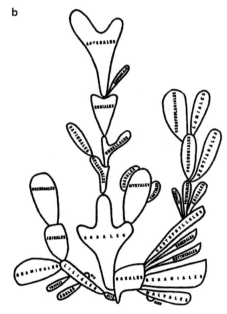

Figure 7.3 (a) evolutionary 'bubbleweed' diagram (modified from 'Trilobite Systematic Relationships Given that trilobites are Arthropods, what is their place among the known arthropod groups?', www.trilobites.info/triloclass.htm#orders, with permission from Sam M. Gon III); (b) a similar style of diagrams was used by Bessey ('Bessey's cactuses') approximately one hundred years ago (after Bessey 1915, fig 1).

lineages emerging as 'thought' bubbles from older bubbles, resulting in the evolutionary 'bubbleweed' of Hughes (2007, fig. 6, reproduced as our Figure 7.3a[6]). Even though the 'bubbleweed' is supposed to show the evolution of the trilobite orders, it does, however, also show an unresolved classification:

Redlichiida
    Ptychopariida
        Asaphida
        Harpetida
        Proetida
    Corynexochida
Lichida
Phacopida
Agnostida

Note the relationships of Phacopida and Lichida are unknown, partly because there is no temporal overlap between Phacopida and Redlichiida, and in the case of the Lichida, no known synapomorphies. Moreover, Redlichiida and Corynexochida are known to be aphyletic, and the Harpetida, Ptychopariida, Asaphida and Proetida are thought to belong to the subclass Libristomata (Fortey 1990, 1997; Ebach & McNamara 2002), on 'the basis of the natant hypostomal condition in which the hypostome was detached from the rest of the calcified ventral cephalic shield' (Hughes 2007, p. 411). Given this fundamental cladogram (the relationship of the Libristomata to everything else), the classification (in which aphyletic taxa are marked with an asterisk) would appear as:

    Libristomata
        Ptychopariida
        Asaphida
        Harpetida
        Proetida
Agnostida
Redlichiida*
Corynexochida*
Lichida
Phacopida

Higher-level trilobite classification is still largely unknown, and the use of cladistics in trilobitology is fairly recent (see Lieberman & Karim 2010). Regardless,

---

[6] These kinds of diagrams are somewhat similar to those published by Bessey for plants approximately one hundred years ago (Bessey 1915; we have reproduced Bessey's fig. 1 as our Figure 7.4b).

the revised diagram by Hughes (2007) does present a classification of sorts, one in which the interrelationships of trilobite orders is unknown. A newer classification, like the one above, based on known synapomorphies, is a step ahead toward a potential monophyletic classification.

Every derivative cladogram, every classification, is based on evidence of some kind, even if that evidence is not explicitly stated. Thus all classifications, past and present, are valid representations of knowledge at that time, inasmuch as the data supports (or is assumed to support) the conclusions (the taxa). Research into the classification of any group of organisms might be thought of as exploring the connection between the relationships determined and the evidence that supposedly supports it – converting aphyletic taxa into their constituent monophyletic taxa.

For the most part, previous discussions concerning cladograms and their meaning has neglected to explore the relationship between fundamental and derivative cladograms – in fact, as far as we can tell, there has never been any explicit mention of derivative cladograms in the literature since its description in 1979 (Nelson 1979), yet many classifications, if not all classifications, are derived directly from an original fundamental cladogram, or part of one, or from available data interpreted or analysed in one way or another.

Trees

As we will argue, *cladograms* are not the same as *trees*. In the following we use the term 'tree' interchangeably with *phylogeny*, *phylogram* and *phylogenetic (phyletic) tree* or any other word or combination of words that imply a phylogenetic interpretation of results. We understand the term *phylogeny* and *phylogenetic tree*, and so on, to apply to a particular kind, or particular kinds, of cladogram, of which more below.

First we offer a cautionary note on the common understanding (an understanding that is assumed in most cases) of the many published diagrams now called *phylogenies* and *phylogenetic trees*. In the current systematics literature numerous studies are now entitled 'The phylogeny of . . .', 'The phylogenetic relationships of . . .', '. . . multilocus phylogeny of . . .', and so on. In our view, this is a misuse of the terms *phylogeny* and *phylogenetic tree*. If a *phylogeny* or *phylogenetic tree* is a kind of cladogram, then it is the particular kind that is said to depict specific types of evolutionary relationships. Instead, these many studies on 'phylogeny' (usually, but not always) refer to the analysis of some quantity of DNA sequence data using one or another piece of phylogenetic reconstruction software to find organism relationships. This begs another question: What is phylogeny reconstruction? The usual, and somewhat glib, answer would be that it is a method of analysis (software) 'that produces *phylogenetic trees*'. It is usually thought to be the results from using a program or programs that are part of a statistical tree-building package specifically designed for the analysis of aligned DNA sequences to yield a tree diagram of some sort or other. For example, 'MrBayes 3 is a program for Bayesian inference and model choice across a large space of phylogenetic and evolutionary models' (Ronquist et al. 2011); 'RAxML (Randomized Axelerated Maximum Likelihood) is a program for sequential and parallel Maximum Likelihood based inference of large phylogenetic trees' (Stamatakis 2016); and the 'objective of the *MEGA* software has been to provide tools for exploring, discovering, and analyzing DNA and protein sequences from an evolutionary perspective' (Kumar et al. 2015), the latter being a large package of programs not limited to statistical approaches but including a constellation of methodologies that are currently available.

Regardless of intent, these methodologies find relationships (usually represented in a tree-like form) based on data of some kind and thus these results are better understood as *fundamental cladograms* rather than *phylogenetic trees* in spite of the objectives of the study (see also Dayrat 2005 below).

What, then, are phylogenetic trees? As far as we can determine, there are at least two different kinds (and suspect there are more):

1. The first kind is derived directly from a *fundamental cladogram*, found by whatever means, with its structure interpreted as a phylogeny. That is, the fundamental cladogram is seen as a direct representation of history with the branch points equivalent to real speciation events, the nodes equivalent to ancestors (real or hypothetical) and synapomorphies equivalent to real evolutionary novelties. This is probably the most common interpretation. In Figure 7.4a, branch points 1a, 2a and 3a are interpreted as real speciation events: branch point 1a giving rise to the common ancestor of A+B (2a) and C +D (3a); branch points 2a and 3a giving rise to A and B, and C and D, respectively. One might even interpret 1a, 2a and 3 as ancestors, albeit hypothetical. The lengths of the branches represent 'anagenetic'[7] change, and any synapomorphies found supporting nodes 1–3 are considered true evolutionary novelties. To some this is an inescapably obvious interpretation: 'Like most biologists, I accept that phylogeny reflects the reality of evolution by natural selection, that branching points are the result of real splitting events, and that clades are based on common ancestry' (Benton 2000, p. 634). To others less so: 'A phylogenetic theory is more than just systematics of the pre-Darwinian kind with an historical explanation imposed upon it' (Ghiselin 1972, p. 132). And as Patterson later commented, take out the 'more than' and it is 'an accurate description of work in the century following

---

[7] Commonly understood to be the amount of evolutionary change along any single branch prior to further splitting.

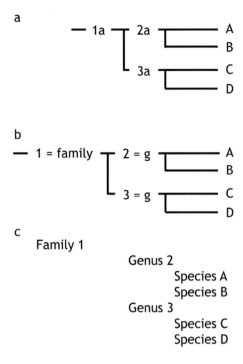

a

b

c

Family 1
 Genus 2
  Species A
  Species B
 Genus 3
  Species C
  Species D

Figure 7.4 (a) One kind of *fundamental cladogram* understood as a direct representation of history with the branch points equivalent to real speciation events, nodes equivalent to ancestors. (b) A *fundamental cladogram* as classification, with node 1 a family, nodes 2 and 3 genera, and A–D species; (c) is a written version.

Darwin' (Patterson 1977, p. 632). In this case the fundamental cladogram is assumed to be isomorphic with the *phylogenetic tree*. But this is an *interpretation*, not an aspect of the data.

2. The second kind is also derived from a *fundamental cladogram*, found by whatever means, but in this case its structure includes a *set of logically possible phylogenetic trees* rather than just one. Some theoretical examples are: Figure 7.5a is taken from Platnick (1977, fig. 2) and shows the array of trees possible from a set of three-taxon cladograms; Figure 7.5b is taken from Harper and Platnick (1978, fig. 1), the latter being a more complex set of phylogenetic trees derived from the cladogram A(BC). Neither of these arrays exhausts the range of possibilities (see Nelson & Platnick 1981, figs 3.2–3.9). The point being that many such phylogenetic trees are possible from one fundamental cladogram.

a

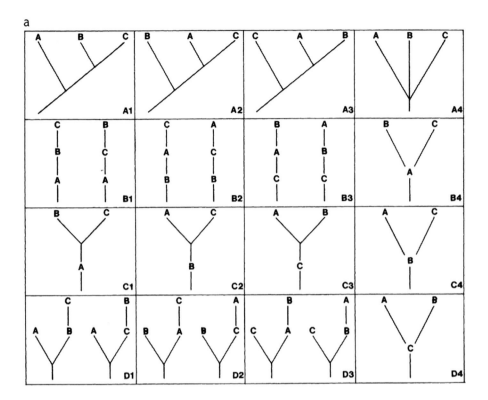

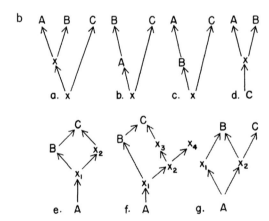

Figure 7.5 A second kind of *fundamental cladogram*, which includes a *set of logically possible phylogenetic trees* rather than one. (a) The diagram is after Platnick (1977, fig. 2, with permission) and shows the array of trees possible from a set of three taxon cladograms; (b) the diagram is taken from Harper & Platnick (1978, fig. 1, with permission) and is a more complex set of phylogenetic trees derived from the cladogram A(BC).

The first kind of phylogenetic tree shown in Figure 7.5a has become synonymous with the idea of 'tree-thinking' (for more on this see Chapters 3 and 14); the second type of phylogenetic tree shown in Figure 7.5b is more complex and relates to the derivation of different kinds of trees from any particular fundamental cladogram rather than assuming an isomorphic interpretation. The first assumes the fundamental cladogram to be a direct (or near enough) record of history; while the second also takes the tree(s) to have captured aspects of the historical record but offers a more complex way to deal with different permutations of possible ancestors all compatible with that cladogram. Even if the first kind is taken to be a *phylogeny* or *phylogram* and the second *a set of phylogenetic trees* or *set of phylograms*, both *interpretations* are framed with reference to a fundamental cladogram (a synapomorphy scheme) and thus regardless of what any particular researcher requires (or claims to require), cladograms are always found prior to any phylogenetic interpretation. Put another way, phylogenies, however conceived, are always based on a fundamental cladogram, which is effectively a classification, and not the other way around as is so often assumed. Classification always comes first.[8]

Example 4: Figure 7.6 reproduces a recently published phylogeny of birds (Prum et al. 2015). The authors note that it is a 'Time-calibrated phylogeny of 198 species of birds inferred from a concatenated, Bayesian analysis of 259 anchored phylogenomic loci using ExaBayes' (Prum et al. 2015). Regardless of the many assumptions behind the analysis, the diagram is, in effect, simply a fundamental cladogram consisting of 190 nodes representing various sets of relationships, as the authors clearly describe in their introductory paragraphs: 'Five major clades form successive sister groups to the rest of Neoaves: (1) a clade including nightjars, other caprimulgiforms, swifts, and hummingbirds; (2) a clade uniting cuckoos, bustards, and turacos with pigeons, mesites, and sandgrouse; (3) cranes and their relatives; (4) a comprehensive waterbird clade, including all diving, wading, and shorebirds; and (5) a comprehensive landbird clade with the enigmatic hoatzin (*Opisthocomus hoazin*) as the sister group to the rest' (Prum et al. 2015, p. 569). To assume it is a phylogeny (the authors mention the word 'ancestor' only three times) does not make it so – but their description of the cladogram (the specific relationships) is empirically grounded and has no need of appeals to ancestors, etc. for it to be of value.

---

[8] Just to highlight the issue, Dayrat noted the following: 'The problem is that most "phylogenetic trees" or 'phylogenies' currently published *simply are cladograms* [..., i.e. classifications]' (Dayrat 2005, p. 349, our emphasis). He then goes on to explain what has previously been examined in detail and given the term *aboristics*, as in the study of trees, (Nelson & Platnick 1981, p. 265). Dayrat goes on to write that 'Completing the actual lines of descent of the Tree of Life [e.g., *aboristics*] *is an unattainable goal*, but we ought to attempt to approach it' (Dayrat 2005, p. 352, our emphasis).

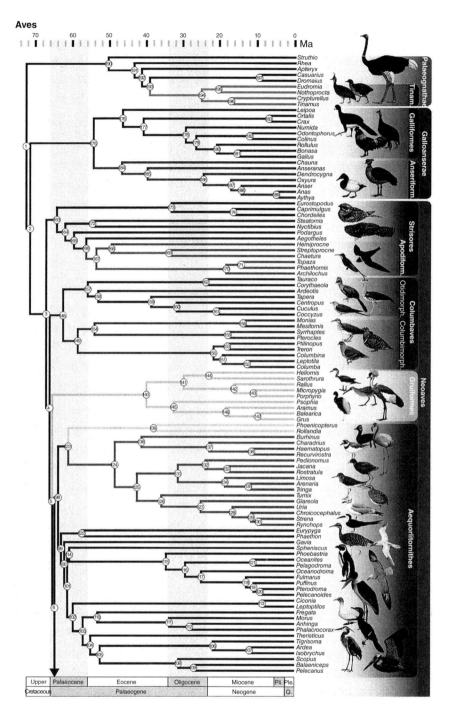

Figure 7.6 A 'Time-calibrated phylogeny of 198 species of birds inferred from a concatenated, Bayesian analysis of 259 anchored phylogenomic loci using ExaBayes' (after Prum et al. 2015, with permission). This is a fundamental cladogram of 190 nodes.

Example 5: The diatom genus Rhoicosphenia *(Figure 7.7)*: In a study of the 'enigmatic'[9] diatom genus *Rhoicosphenia*, a suite of morphological and cytological characters, and sequences from three genes, SSU, LSU and rbcL, were examined (Thomas et al. 2016). The results proposed that '... members of the Achnanthidiaceae are basal to *Rhoicosphenia*, and *Rhoicosphenia* is basal to the Cymbellales, or a basal member of the Cymbellales, which includes the Gomphonemataceae' (Thomas et al. 2016, p. 1, abstract). The authors offered a diagram of relationships ('Maximum Likelihood phylogram from three-marker concatenated alignment', Thomas et al. 2016, p. 10; for us, this is a cladogram, see Figure 7.7a and our simplified version Figure 7.7b) and a classification:

SUPERORDER: Cymbellidae (Achnanthidiaceae + Cocconeidaceae + *Rhoicosphenia* + Cymbellales)
    Unnamed Clade (Cocconeidaceae + *Rhoicosphenia* + Cymbellales)
    Unnamed Clade (*Rhoicosphenia* + Cymbellales)
        ORDER: Cymbellales (*Rhoicosphenia*)
            Suborder: Cymbellineae, Suborder nov.
                Family: Cymbellaceae (*Adlafia, Anomoeoneis, Cymbella, Cymbopleura, Didymosphenia, Encyonema, Encyonopsis, Geissleria, Gomphoneis, Gomphonema, Placoneis, Reimeria*)

Their classification can be simplified thus:

ORDER: Cymbellales
    Rhiocosphenia
    Suborder: Cymbellineae
        Family: Cymbellaceae
            Adlafia, Anomoeoneis, Cymbella, Cymbopleura, Didymosphenia, Encyonema, Encyonopsis, Geissleria, Gomphoneis, Gomphonema, Placoneis, Reimeria

Their order Cymbellales corresponds to node 1 in Figure 7.7b and their suborder Cymbellineae and family Cymbellaceae correspond to node 2 in Figure 7.7b. Regardless of the resolution in the diagram, regardless of ranks adopted, and regardless of their interpretation of the diagram as a 'phylogram', the cladogram and classification are one and the same and both represent the empirical component of the study.

Phylogenies are, then, special kinds of cladogram. The examples above raise a further interesting, if not central, question to the method or methods of taxonomy. We have written above for the bird example: the '... cladogram is empirically grounded ...'. Well, is it? Certainly the data collected are observations of some kind. But the use of ExaBayes (in the bird example) involves many in-built or

---

[9] Enigmatic because its relationships have been ambiguously interpreted.

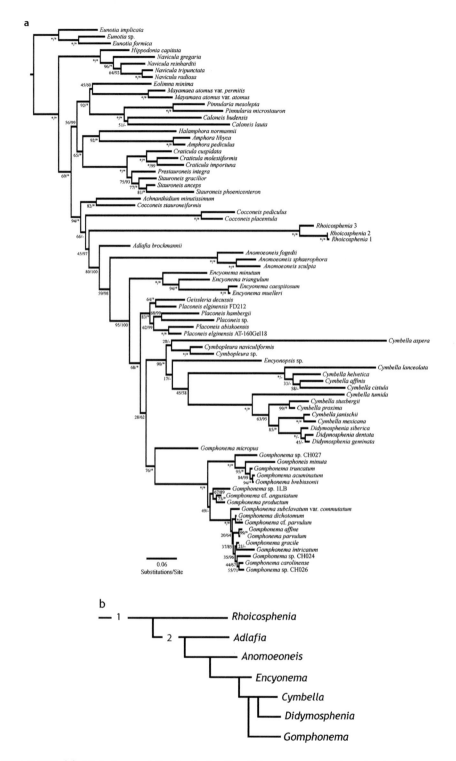

Figure 7.7 (a) A 'phylogram' that includes the diatom genus *Rhoicosphenia* (Thomas et al. 2016, p. 10, part of fig. 2, with permission); (b) a simplified version of part (a). This is a fundamental cladogram.

user-inspired assumptions. How far do those assumptions affect the final result? How much of the final result, the cladogram and its nodes, is real? By real we mean having meaning beyond merely being found by this or that set of programs, using this or that set of assumptions: are the results method-independent, or method-independent enough to reflect an accurate summary of the natural world? The short answer is: *there is no way of knowing.* Beyond making the grand claim that the results really do represent phylogeny, what is their value in scientific terms? We answer this more fully later, but briefly its value rests with the nodes recovered and the relationships expressed, its *cladistic parameter.* We take this up in more detail in Chapter 8 (for further discussion of trees in systematics see Chapter 3).

Before we proceed, four relatively minor digressions are required. The first relates to the term *dendrogram*; the second to *Nelson Trees (or Nelson cladograms)* sensu Wiley (1981) and Wiley & Liebermann (2011); the third deals with a set of ideas that arose with the first molecular cladograms and their relationship to trees; and finally, the fourth deals with some concepts rarely, if ever, encountered nowadays: the relationship between cladograms, trees and scenarios (see Boxes 7.1–7.4).

Box 7.1 Dendrograms

In response to Nelson's (1979) definition of a cladogram (which we use and follow in this book), entomologist Charles Michener (1918–2015) suggested that *dendrogram* was a better general term for branching diagrams and that Nelson's analysis of branching diagrams might better be called *Dendronomics* (McGinley & Michener 1980, p. 92). We disagree.

Ernst Mayr claimed the term dendrogram as his own: 'I had long used this term, in lectures and in manuscript, for any branching, tree-like diagram serving to indicate degrees of relationship. The term was first published in Mayr et al. (1953, p. 58, 175)' (Mayr 1978, p. 85). Indeed, the term dendrogram was first used by Mayr et al. in connection with a diagram depicting the relationships of 37 species of river duck in the genus *Anas* (Mayr et al. 1953, pp. 57–58, reproduced as Figure B7.1.1b; the diagram was based on an earlier version published by Delacour & Mayr 1945, p. 7, fig. 1, reproduced as Figure B7.1.1a). Mayr noted of his diagram:

> This is not a phylogenetic tree, because it is not based on any information on fossil forms which might be ancestral connections of the various branches. Such a diagrammatic illustration of degree of relationships *based on degree of similarity* (morphological or otherwise) may be called a *dendrogram.* (Mayr et al. 1953, pp. 57–58, italics ours)

Box 7.1 *(cont.)*

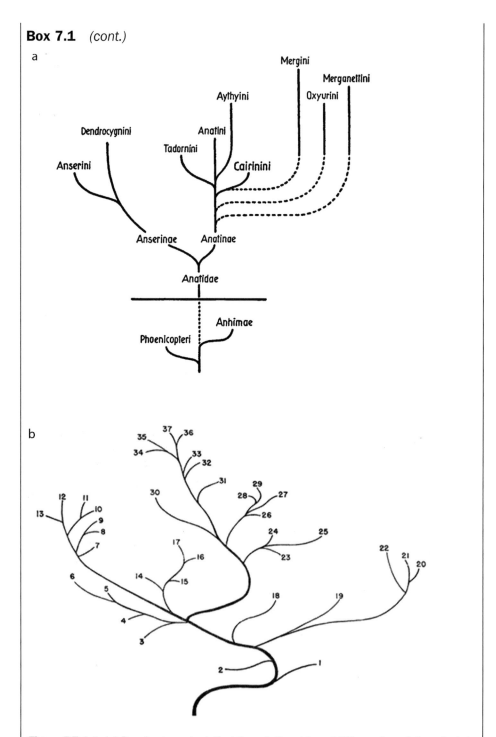

Figure B7.1.1 (a) Dendrogram depicting the relationships of 37 species of river duck in the genus *Anas* (after Mayr et al. 1953, pp. 57–58, with permission); (b) diagram of the relationships of river duck in the genus *Anas* (from Delacour & Mayr 1945, p. 7, fig. 1).

Box 7.1 *(cont.)*

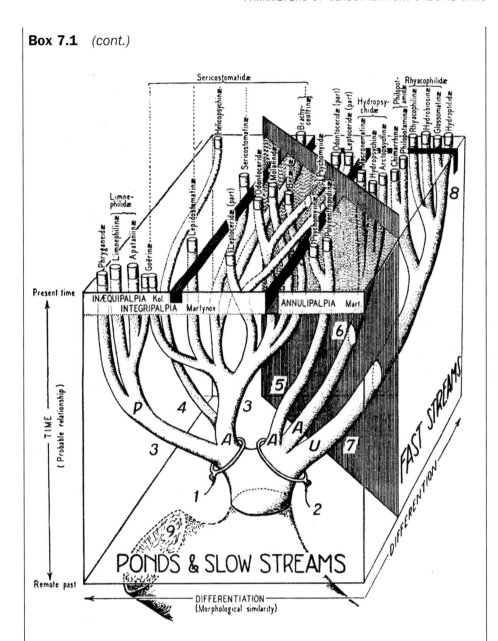

Figure B7.1.2 '[T]hree-dimensional dendrogram representing the evolution of caddisworm case construction' (after Milne & Milne 1939, p. fig. 1, with permission).

In that book's glossary, Mayr gave the following definition:

Dendrogram. A diagrammatic drawing in the form of a tree designed to indicate degrees of relationships as suggested by *degrees of similarity* (cf. Phylogenetic tree). (Mayr et al. 1953, p. 304, italics ours)

Box 7.1   *(cont.)*

Interestingly, 'Phylogenetic tree' was defined as:

> A diagrammatic presentation of assumed lines of descent, based on paleonto-logical, morphological, or other evidence. (Mayr et al. 1953, p. 304)

A more detailed treatment of the dendrogram concept was given in the book's section on 'The Presentation of Phylogeny', where the passage dealing explicitly with dendrograms appears to contradict the pair of definitions above by suggesting that dendrograms might very well be representations of phylogeny, illustrated with the wonderfully complex 'three-dimensional dendrogram representing the evolution of caddisworm case construction' (Figure B7.1.2, also reproduced in Mayr et al. 1953, p. 176, fig. 40 and in Pietsch 2012, fig. 151, after the original in Milne & Milne 1939, p. 540, fig. 1).

In Mayr's 'Origin and History of Some Terms in Systematic and Evolutionary Biology', he noted that 'It became apparent soon after 1953 that there were many different kinds of dendrograms' (Mayr 1978, p. 84) and gave separate definitions for cladogram and phenogram (Mayr 1978, pp. 84–85). With respect to cladogram, the word originally came into being as a description of a kind of dendrogram, with Camin and Sokal suggesting that '... the term cladogram [can be used] to distinguish a cladistic dendrogram from a phenetic one, which might be called a phenogram' (Camin & Sokal 1965, p. 312) – in this case the 'cladistic dendrogram' was constructed using an early form of parsimony (Camin–Sokal parsimony). At almost the same time Mayr suggested that in 'a cladogram ... the ordinate gives estimated time, the abscissa degree of difference' (Mayr 1965, p. 81). Later, Mayr wrote that a cladogram is 'a dendrogram based on the principles of cladism; a strictly genealogical dendrogram in which rates of evolutionary divergence are ignored' (Mayr 1969, p. 255), a definition similar to that given in his 1978 'Origin and History of Some Terms ...' paper: 'A cladogram is a dendrogram depicting the branching of the phylogenetic tree without respect to rates of divergence' (Mayr 1978, p. 82). Mayr had now removed from his definition of dendrogram the *'degrees of similarity'* criterion, yet both definitions of cladogram and phenogram were given in the context of types of dendrograms (Mayr 1978, pp. 84–85). By 2002, Mayr had dropped the term dendrogram altogether from his definition of cladogram: 'Cladogram – A diagram of a tree of clades – basically a branching phylogeny. Such a diagram depicts a sequence in the origins of uniquely derived traits that are found in all the members of the clade and not in any others. A cladogram is a diagram of the origin of characters' (Mayr & Bock 2002).

Mayr and his associates, naturally, made definitions to suit their own agenda and that applied to their definition of dendrogram, cladogram and other –grams proposed. But this approach was offered from those 'who argued that Hennig's philosophy was hope-lessly narrow-minded and, therefore, deserved to receive a special name' (Nelson 1979, p. 13). Therefore we reject the many definitions of Mayr and his associates for the same reason but also because it is clear that Mayr's original intention was to link dendrograms to diagrams reflecting *degree of relationships* with *degrees of similarity*, which meant excluding phylogenetic trees, as these are 'diagrammatic presentation[s] of assumed lines of descent ...' (Mayr et al. 1953, p. 311[1]). Thus, originally a dendrogram was a

Box 7.1 *(cont.)*

narrower concept than simply meaning branching diagrams. We now understand Mayr's term dendrogram and his definition of phylogenetic tree to both be examples of cladograms.[2]

Cladogram, as we use the concept, is broader, more useful and captures the one aspect of systematics that can be studied empirically, the *cladistic parameter* (see below).

References

Brower, AVZ. 2016. What is a cladogram and what is not? *Cladistics* 32: 573–576.

Camin, JH. & Sokal, RR. 1965. A method for deducing branching sequences in phylogeny. *Evolution* 19: 311–326.

Delacour, J. & Mayr, E. 1945. The family Anatidae. *Wilson Bulletin* 57: 3–55.

Mayr, E. 1955. Comments on some recent studies of song bird phylogeny. *Wilson Bulletin* 67: 33–44.

Mayr, E. 1965. Numerical phenetics and taxonomic theory. *Systematic Zoology* 14: 73–97.

Mayr, E. 1969. *Principles of Systematic Zoology*. McGraw-Hill, New York.

Mayr, E. 1978. Origin and history of some terms in systematic and evolutionary biology. *Systematic Zoology* 27: 83–88.

Mayr, E. & Bock, WJ. 2002. Classifications and other ordering systems. *Journal of Zoological Systematics and Evolutionary Research* 40: 169–194.

Mayr, E. Linsley, G. & Usinger, RL. 1953. *Methods and Principles of Systematic Zoology*. McGraw-Hill, New York/London.

McGinley, RJ. & Michener, CD. 1980. Dr. Nelson on dendronomics. *Systematic Zoology* 29(1): 91–93.

Milne, MJ. & Milne, LJ. 1939. Evolutionary trends in caddisworm case construction. *Annals of the Entomological Society of America* 32: 533–542.

Nelson, GJ. 1979. Cladistic analysis and synthesis: principles and definitions, with a historical note on Adanson's *Familles des Plantes*. *Systematic Zoology* 28: 1–21.

Pietsch, TW. 2012. *Trees of Life. A Visual History of Evolution*. Johns Hopkins University Press, Baltimore, MD.

[1] Later Mayr wrote 'Phylogenies can be established only by unequivocal evidence from comparative anatomy or by fossil finds. It is much safer to use a neutral term, such as "dendrogram" (Mayr et al. 1953, p. 58), for a diagram of hypothetical descent based exclusively on the comparison of living forms. In view of the mounting evidence for frequent evolutionary reversal and convergences in adaptive characters, it is quite inadmissible to apply the term "phylogenetic tree" to a diagram that portrays merely morphological sequences' (Mayr 1955, p. 41).

[2] As usual, Andrew Brower has written a thoughtful note on the subject of cladograms and what they mean. We have some sympathy for his views but see the term cladogram as having a much greater generality than he has allowed for. Understood in our terms, the central significance of cladistics might be better appreciated (Brower 2016).

Box 7.2   Wiley and Nelson Trees (or Nelson Cladograms)

Martin and Wiley offered a lengthy discussion of various kinds of 'trees' and various kinds of 'cladograms' (Martin & Wiley 2008; Martin et al. 2010, summarised in Wiley & Lieberman 2011, p. 92, further commentary in de Queiroz 2013). Their account, which differs from ours, attempts to exploit the mathematics behind graph theory (Martin & Wiley 2008). For them, there are three kinds of trees:

(1) *Phylogenetic trees:* 'By the term phylogenetic tree, we mean a tree that models (hypothesized) phylogenetic relationship among taxa by depicting taxa by edges [= branches], and speciation events by vertices [= nodes]'.

(2) *Hennig tree:* 'In a Hennig tree, taxa are represented by vertices [= nodes], not by edges [= branches]. An edge of a Hennig tree does not represent a lineage or anything else occurring in nature: rather, it represents a relationship among two vertices, or more empirically, the hypothesis of a relationship'.

(3) *Nelson cladograms*[1]: Martin & Wiley cite an unpublished manscript as their source (Nelson 1979): 'A cladogram may, therefore, be defined as a dendritic structure illustrating an unspecified relationship between certain specified terms that in the context of systematics represent taxa,' while 'a tree may be defined as a dendritic structure having one or more general as well as unique components (or combination of components).'[2]

The terminology they use – 'edges', 'vertices' and so on – comes directly from graph theory (as in Wiley & Lieberman 2011, pp. 103–104, discussed in their section 'Other Kinds of Tree Terminology'). As a consequence, Martin and Wiley conclude that Nelson's dendritic structure 'must mean "tree" in the mathematical sense (that is, an acyclic connected graph)' (Martin & Wiley 2008). Here, they follow Hendy and Penny: 'This widely accepted definition is: A tree is an acyclic connected graph' (Hendy & Penny 1984, p. 245, see also Semple & Steel 2003 and Morrison 2014, p. 4). Hendy and Penny '. . . end with a plea to those using these terms in biology to note that cladograms and evolutionary trees are but two "species" in the genus "tree"' (Hendy & Penny 1984, p. 247).

This usage, however, merely reflects the *mathematical structure of trees* derived from graph theory, which may have value for those interested in developing methods of 'building' phylogenetic trees (see Box 3.4: Unrooted 'Trees' and Networks) but is of limited interest for exploration of *relationships* undertaken here and, as we see it, the primary task of taxonomy. Our intentions are different – specifically, we are not interested in the many varied ways of approaching 'tree construction' (as they are all flawed in a general sense, see Chapters 8–12), nor are we simply interested in asserting that a tree diagram, when found, is a phylogeny. The general definition of cladogram relates to a diagram that expresses relationships; a cladogram has a very specific set of relationships and a tree, as discussed in the body of this chapter, has a much vaguer set of relationships that is often, but not always, based on a cladogram. Wiley's various trees are easily reinterpreted:

Phylogenetic trees = a fundamental cladogram found by application of weighted data analysis (see Chapter 8);

Hennig tree = a fundamental cladogram;

Nelson cladogram = a fundamental cladogram.

Thus, once again, all such trees can be reduced to the empirical reality of *any particular fundamental cladogram* – or, as we prefer, *any particular classification*.

Box 7.2 *(cont.)*

We began this chapter by contrasting some uses of the word 'tree': as a large, woody plant, a Haeckelian phylogeny diagram and part of the mathematical study called graph theory. It should be clear that attempts to see direct equivalence between these various uses are, in some cases, illusory, beyond merely sharing the branching aspect.

Wiley and Lieberman attempt to develop the connection between the mathematics of graph theory and phylogenetic methodology, as before them have Foulds et al. (1979), Hendy & Penny (1984), Semple & Steel (2003) and, more recently, Morrison (2014, p. 4); a detailed account of these approaches to trees in the mathematical-graph theory sense is given in Page & Holmes (1998), their chapter 2, and Semple & Steel (2003, their chapter 1). Such is not our purpose, nor has it ever been the purpose of generations of taxonomists. Here we are attempting to develop the idea of *cladistic analysis* as referring to 'the analysis of branching ... the analysis is of cladograms ... or classifications that may be represented by cladograms' (Nelson 1979, p. 3). That is, cladistics is the analysis of relationships not the calculation of, or manipulation of, or interpretation of, trees (see Chapter 12).

References

de Queiroz, K. 2013. Nodes, branches, and phylogenetic definitions. *Systematic Biology* 62: 625–632.

Foulds, LR., Hendy, MD. & Penny, D. 1979. A graph theoretic approach to the development of minimal phylogenetic trees. *Journal of Molecular Evolution* 13: 127–149.

Hendy, MD. & Penny, D. 1984. Cladograms should be called trees. *Systematic Zoology* 33: 245–247.

Martin, JL. & Wiley, EO. 2008. Mathematical models and biological meaning: taking trees seriously. Cornell University. https://arxiv.org/abs/0808.0287

Martin, J., Blackburn, D. & Wiley, EO. 2010. Are node-based and stem-based clades equivalent? Insights from graph theory. *PLoS Currents* 2: RRN1196. http://doi:10.1371/currents .RRN1196

Morrison, DA. 2014. Phylogenetic networks: a review of methods to display evolutionary history. *Annual Research & Review in Biology* 10: 1518–1543.

Nelson, GJ. 1979. Cladistic analysis and synthesis: principles and definitions, with a historical note on Adanson's *Familles des Plantes*. *Systematic Zoology* 28: 1–21.

Nelson, GJ. & Platnick, NI. 1981. *Systematics and Biogeography: Cladistics and Vicariance*. Columbia University Press, New York.

Nixon, KC. & Carpenter, JM. 1996. On consensus, collapsibility, and clade concordance. *Cladistics* 12: 305–321.

Page, RDM. 1989. Comments on component-compatibility in historical biogeography. *Cladistics* 5: 167–182.

Page, RDM. & Holmes, EC. 1998. *Molecular Evolution: A Phylogenetic Approach*. Wiley-Blackwell, Oxford.

Schuh, RT. & Farris, SJ. 1981. Methods for investigating taxonomic congruence and their application to the Leptopodomorpha. *Systematic Zoology* 30: 331–351.

Semple, C. & Steel, M. 2003. *Phylogenetics*. Oxford University Press, Oxford.

Wiley, EO. & Lieberman, BS. 2011. *Phylogenetics: Theory and Practice of Phylogenetics Systematics*. Wiley-Blackwell, Hoboken, NJ.

Box 7.2   *(cont.)*

[1] These are called *'Nelson Trees'* in Wiley & Lieberman (2011, p. 92). The term *Nelson Tree* was already in use for a type of consensus tree (Schuh & Farris 1981, p. 339). There has been much discussion on this topic (see, for example, Page 1989; Nixon & Carpenter 1996), which we will briefly refer to in Chapter 8.

[2] This passage was eventually published as part of Nelson & Platnick with a few changes: "A cladogram, therefore, may be defined as a branching, or dendritic, structure, or dendrogram, illustrating an unspecified relation (general synapomorphy) between certain specified terms that in the context of systematics represent taxa" (Nelson & Platnick 1981, p. 171). Here, Nelson and Platnick muddy the waters by using the term dendrogram but the addition of 'general synapomorphy' connects the notion of cladogram to evidence, even if the relation is still left unspecified.

Box 7.3   Molecular Cladograms and Trees

Various kinds of 'trees' were discussed in the context of analytical methodologies for molecular systematics. For example, Page and Holmes (1998) understood a cladogram to be a diagram of *relative recency of common ancestry*, such that if A, B and C are related as A(BC), then B and C share a more recent common ancestor than either does with A (Figure B7.3.1a). This is not simply a cladogram; it is a phylogram. That is, it is a cladogram with an added interpretation or explanation.

There are more complex mathematical trees called 'Additive Trees', which include branch lengths. The numbers on the branches in Figure B7.3.1b are said to add a notion of the 'amount' of evolutionary change. For example, taxon B has acquired two 'changes' since it shared a common ancestor with C (Figure B7.3.1b). This too is not just a cladogram; it is a phylogram, admittedly a more complex phylogram than that in Figure B7.3.1a.

'Ultrametric Trees' are a special kind of additive tree in which all the terminal taxa are equidistant from the root. This kind of tree is often associated with estimates of time along branches and is associated with molecular clock divergence estimates (Figure B7.3.1c). But this too is not just a cladogram; it is a phylogram but another even more complex phylogram.

If the three diagrams in Figure B7.3.1a–c are compared, then it can be easily appreciated they are identical with respect to the cladistic parameter: B and C are more closely related to each other than they are to A. So all three diagrams, if based on evidence to support that relationship, are *fundamental cladograms* and only become phylograms when interpreted as such. This kind of interpretation belongs to tree-thinking.

We discussed tree-thinking further in Chapter 3, but for now the significant point is regardless of the interpretation of any kind of 'tree', or the method of finding any particular fundamental cladogram; all the diagrams in Figure B7.3.1 are cladograms and all are accurately summarised by their cladistic parameter alone – which is effectively the classification.

As we note above, phylograms and phylogenetic trees, however complex, are always based on a fundamental cladogram (a classification) and not, as commonly thought, the other way around.

Box 7.3 *(cont.)*

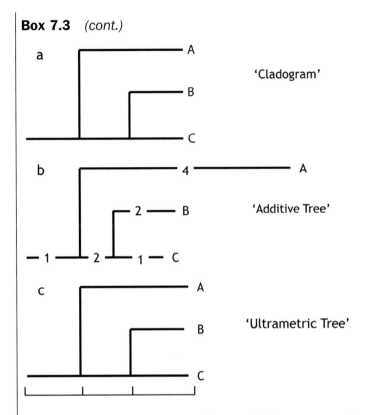

Figure B7.3.1 (a) Cladogram as a diagram of *relative recency of common ancestry* for A, B and C when related as A(BC), with B and C sharing a more recent common ancestor than either does with A; (b) 'Additive Tree', for A, B and C when related as A(BC) including branch lengths; (c) 'Ultrametric Trees' for A, B and C when related as A(BC) in which all the terminal taxa are equidistant from the root.

Reference

Page, RDM. & Holmes, EC. 1998. *Molecular Evolution: A Phylogenetic Approach*. Wiley-Blackwell, Oxford.

Box 7.4 Cladograms, Trees and Scenarios

An idea that emerged in the early 1980s concerning phylogenies and their relationship to cladograms was the increasing complexity of moving from cladogram to tree(s) and finally to scenario, or 'evolutionary scenario'. First documented by Tattersall and Eldredge (1977), the 'Cladogram – Tree – Scenario' sequence appears novel to Eldredge (Eldredge

Box 7.4 *(cont.)*

1979; although Hull, 1979, p. 421, footnote 1, implies Gareth Nelson originally outlined this sequence of increasing complexity). For Eldredge, a cladogram is 'a branching diagram illustrating the pattern of related organisms of derived characters, which one might describe as evolutionary novelties' (Tattersall & Eldredge 1977, p. 204; Eldredge 1979, pp. 167, 169), whereas a phylogenetic tree 'adds to the information contained in the cladogram by specifying the nature of the evolutionary relationship postulated' (Tattersall & Eldredge 1977, pp. 204; Eldredge 1979, pp. 168, 182) – Eldredge then suggested that while a phylogenetic tree is a diagram, it is 'not necessarily branching!' (Eldredge 1979, pp. 168). Eldredge noted that evolutionary relationships can be of two kinds: 'linear evolution' (direct ancestry) or 'speciation' (common ancestry). Finally, he suggested that a scenario is 'a phylogenetic tree fleshed out with further types of information, most commonly to do with adaptation and ecology' (Tattersall & Eldredge 1977, p. 204; Eldredge 1979, pp. 168, 192).

In the context of the above discussion, Eldredge's description of a cladogram is, however, a phylogram, and his 'phylogenetic tree' and 'scenario' relate in a more complex and unspecified way, both going extensively beyond the available data. Although, Eldredge's 'Cladogram – Tree – Scenario' sequence was the focus of some discussion (see, for example, Hull 1979, p. 420, Hull 1988 p. 245, Scott-Ram 1990, p. 135, Panchen 1992, p. 244, Arthur 1997, p. 57, Lee & Doughty 1997, among others), it is entirely superfluous to any empirical progress in classification and merely mimics the rather older narrative version of phylogenetic speculation.

References

Arthur, W. 1997. *The Origin of Animal Body Plans: A Study in Evolutionary Developmental Biology*. Cambridge University Press, Cambridge, UK.

Eldredge, N. 1979. Cladism and common sense. In: Cracraft, J. & Eldredge, N. (eds), *Phylogenetic Analysis and Paleontology*. Columbia University Press, New York, pp. 165–197.

Hull, DL. 1979. The limits of cladism. *Systematic Zoology* 28: 416–440.

Hull, DL. 1988. *Science as Process: An Evolutionary Account of the Social and Conceptual Development of Science*. University of Chicago Press, Chicago.

Lee, MSY. & Doughty, P. 1997. The relationship between evolutionary theory and phylogenetic analysis. *Biological Reviews* 72: 471–495.

Panchen, AL. 1992. *Classification, Evolution and the Nature of Biology*. Cambridge University Press, Cambridge, UK.

Scott-Ram, NR. 1990. *Transformed Cladistics, Taxonomy, and Evolution*. Cambridge University Press, New York.

Tattersall, I. & Eldredge, N. 1977. Fact, theory and fantasy in human paleontology. *American Scientist* 65: 204–211.

Taxon, Monophyly, Paraphyly, Aphyly

Taxon (pleural: *taxa*) is usually understood as a group[10] of named organisms of whatever rank. It is shorthand for *taxonomic group(s)*, derived from *taxonomy*.[11] If all nodes on any fundamental cladogram are named, these will be its constituent taxa, effectively its classification. All taxa are assumed to be monophyletic. Returning to Figure 7.4, node 1 could be a family, nodes 2 and 3, genera and A–D species (as in Figure 7.4b, written as Figure 7.4c). A similar process applies to the relationships of Anatidae (depicted in Figure B7.1.1, after Mayr et al. 1953, pp. 57–58 based on Delacour & Mayr 1945, p. 7, fig. 1). Here nine tribes are recognized, but none of their interrelationships are *necessarily* captured (noted as nodes 2–8 on Figure 7.8; node 1 is the family Anatidae) by just listing the tribe names.

Species are *usually* (conventionally) considered in terms of genealogical hypotheses of interbreeding populations rather than in terms of cladistic relationships; species are *usually* (conventionally) considered to be *the* units of evolution. Because of this, and unlike taxa in general, species are *usually* (conventionally) considered to be different from other taxa, special even, inasmuch as every species has an ancestor of some kind, a progenitor, a stem lineage and an explicit genealogy created by successive waves of interbreeding populations. Hennig provided one of his classic diagrams, which has been reproduced, discussed and dissected many times – perhaps more so than any of his other diagrams – detailing what he referred to as *tokogenetic relationships* (Figure 7.9a): 'Reflection of relationship – from organism to species or from species to organism – is its direction meaningful except as metaphysics? Relevant, perhaps, is what functions as the source of evidence. Does knowledge of synapomorphy arise from relationships of parts, or from relationships of species? If the latter, how then can it be evidence of the same?' (Nelson 2016, p. 205, see Figure 7.9b).

Theories behind the mechanics of species generation is a highly contentious field (interested readers could start with Wilkins 2018, the revised version of his

---

[10] We use the word *group* but that is not the exact meaning we wish to convey. Elsewhere we have explicitly suggested that *taxon = relationship*.

[11] Lam discusses some of the history of the term *taxon* redefining it for a more practical application: '. . . what exactly is, or should be, a taxon? It is undoubtedly meant to cover a natural group or natural groups, but this does of course not mean that every natural group deserves the name taxon'. (Lam 1957, p. 214)

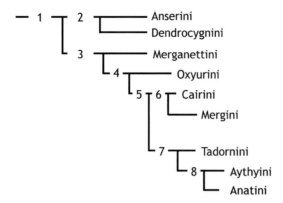

Figure 7.8 A classification of the relationships of Anatidae (after Mayr et al. 1953, pp. 57–58 based on Delacour & Mayr 1945, p. 7, fig. 1, with permission). Nine tribes are recognised but none of their interrelationships are *necessarily* captured (noted as nodes 2–8; node 1 is the family Anatidae) by just listing the tribe names.

earlier book (Wilkins 2009), and then digest the contents of de Pinna 2014, the latter may very well be, or perhaps should be, the last word on the subject of species – in our view, of course). At present there are upwards of 27 different species concept hypotheses (Wilkins 2011, p. 59, 'There are *seven* distinct definitions of "species", and *27* current variations and mixtures. And there are $n+1$ definitions of "species" in a room of n biologists'). Most of these views understand species as non-taxa. As such, they are of limited value in any classification as they are dependent on one of a number of generation hypotheses, something that by necessity (usually) invokes ancestors, stem lineages, genetic cascades, or whatever (Pavlinov 2013; Plutynski 2019). These additional hypotheses effectively relate to *phylograms* rather than *cladograms*. Therefore, it seems to us, discovery of species is required prior to any notions of generation – and species, in fact, turn out to be just another taxon.

We understand species, like other taxa, as a *cladistic relationship* (rather than a group) and therefore deal with them in our practical work as we would any other taxon (e.g., genus, family, order, etc.) as units to be discovered, which seems to be the general approach of most working taxonomists.[12]

The truncated terms for species will be used in this book: 'sp.' (singular) and 'spp.' (pleural).

---

[12] With respect to species as cladistic relationships: 'Reflection of relationship – from organism to species or from species to organism – is its direction meaningful except as metaphysics? Relevant, perhaps, is what functions as the source of evidence. Does knowledge of synapomorphy arise from relationships of parts, or from relationships of species? If the latter, how then can it be evidence of the same?' (Nelson 2016, p. 205).

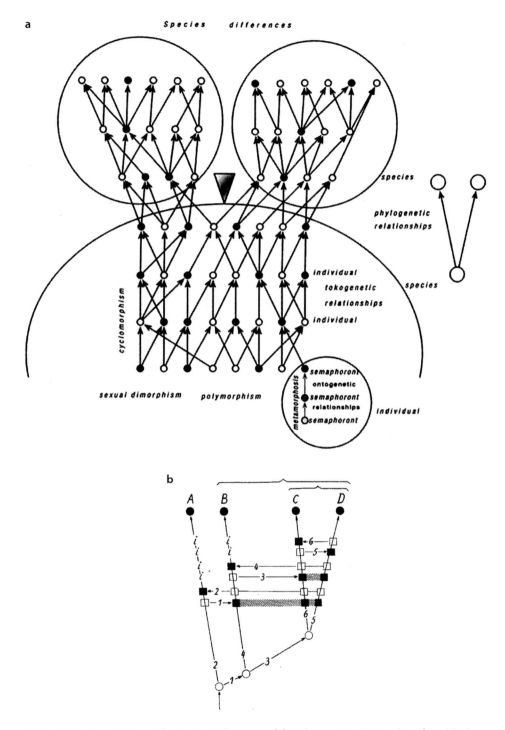

Figure 7.9 Two of Hennig's classical diagrams. (a) *Tokogenetic relationships* (modified from Hennig 1966, fig. 6); (b) an argumentation plan (modified from Hennig 1965, fig. 3).

Monophyly, monophyletic and *monophyletic 'group'* refers to a taxon that is (usually) characterised by at least one synapomorphy. We include the word 'usually' as in some cases the synapomorphy is not obvious or unique to all the component parts of the taxon in question (see Williams & Ebach 2017; Mavrodiev 2019 and Chapter 5) and in other cases (e.g., many precladistic classifications) synapomorphies are not always explicitly stated. While many definitions of monophyly relate to inclusive ancestry, this would apply only to trees. We are focused on cladograms and classification so in our sense monophyly is the result of an investigation (see Chapter 4 for further details). It should be clear that *taxon* and *monophyletic 'group'* are also equivalent to the cladistic parameter (and *group* is not quite the most appropriate word), thus:

Relationship: Homology: Monophyly: Taxon: Natural Classification

The term *group* may refer to any collection of objects or ideas, whereas a taxon refers to a specific relationship to the exclusion of all other individual species, genera etc.

Aphyly, paraphyly: As we stated above, a vast number of taxon names have yet to be demonstrated as representing monophyletic groups characterised by synapomorphies. These uncharacterised groups can be referred to as *aphyletic,* in need of revision. Application of aphyly avoids making the assumption that when a monophyletic group is discovered from within an already known and named taxon, the species left behind are apparently rendered paraphyletic. *Paraphyly* and indeed *polyphyly* are redundant terms. These topics are treated in more detail in Chapters 6 and 14.

Homologue

Homologues: An Introduction

To gain scientific credibility, statements of relationship are, or should be, based on evidence. Every statement of relationship has, or should have, empirical content. Where does that evidence come from? It is primarily derived from the parts of organisms, including some parts invisible to the naked eye but nevertheless accessible to study (cell structure, DNA and so on). In this sense, organisms can be conceived of as assemblages of parts, the parts abstracted from the whole, such that in certain animals it is possible to differentiate between legs and arms, and heads and tails, and in certain plants it is possible to differentiate between flowers and stems, and roots and shoots. Taxonomists often refer to these parts as *characters* or *taxonomic characters.* In comparative anatomy the parts are called *homologues.*

As do most discussions of homologues, we begin with Richard Owen and his *Lectures on Comparative Anatomy* where he defines homologue (Owen 1843, p. 379):

HOMOLOGUE: (Gr. *homos*; *logos*, speech) The same organ in different animals under every variety of form and function. (Owen 1843, p. 379)

Every biologist would, at some point, encounter the parts of organisms in their day to day activities.

Example 1. Box jelly fish: In Chapter 2 we mentioned a few details of the parts of the box jellyfish's structure: mastigophores; nematocysts; pedalia; rhopalial horns; perradial lappets and so on. These are box jellyfish homologues.

Example 2. The spider genus *Scaphiella*: Platnick and Dupérré's (2010) study of the spider genus *Scaphiella* has 842 illustrations, mostly of homologues: 'The goblin spider genus *Scaphiella* Simon is characterized by several unique characters, including peculiarly shaped macrosetae found medially on the chelicerae and basally on the palpal tarsi of both sexes' (Platnick & Dupérré 2010, p. 3).

There is no reason to believe that every part identified as a homologue is actually a real part – our studies can be mistaken by assuming that a part in one organism is the same as that in another. A major part of taxonomists' efforts are guided towards finding and characterising the parts of organisms. Such studies allow a greater understanding of how organisms are constructed and how they function.

Example 3. Forelimbs of Mammals: In Mammals the distal region of the skeleton of the upper limb includes two bones called the *radius* and *ulna*. These appear in many different guises – in the *fin* of a whale, the *wing* of a bird and the *forelimb* of a dog and a human (Figure 7.10). The radius and ulna are the names of homologues.

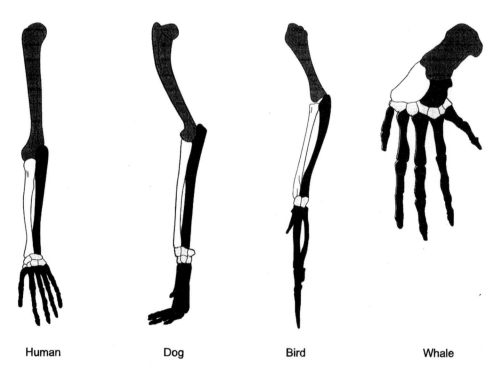

Human   Dog   Bird   Whale

Figure 7.10 Illustrations of: the *forelimb* of a human and a dog, the *wing* of a bird and the *fin* of a whale.

In his book *On the Nature of Limbs* (Owen 1849, recently republished with a series of commentaries, Owen 2007), Owen expanded on his meaning of homologue:

> When a bone in the skeleton of a dog, a horse, a mole, and a platypus, is proved to be the same or answerable bone to the one called 'scapula' in man, and is called by *the same name*; and when the answerable bone can be traced through birds, reptiles and fishes, where it is similarly recognised and indicated, –determined, in short, to be *the namesake* or *'homologue'* of the human scapula; it follows, that, whatever other or higher proposition respecting the nature of that bone and its relations to the fundamental pattern of the vertebrate skeleton can be demonstrated by the sum of its characters in all applies individually to every form of the bone ... (Owen 1849, p. 71, italics ours[13])

For various reasons some parts of organisms that have the same name are not actually the same. For example, we mentioned in Chapter 1 that birds have 'beaks' and 'wings', and often fly. But as *characters* 'beak' and 'wing' are open to interpretation as turtles also have 'beaks', and beetles also have 'wings'; the latter 'fly', the former do not – and neither turtles nor beetles are kinds of birds.

It could be that the terms 'beak' and 'wing' are insufficient to describe or characterize the part(s) they refer to. Consider for a moment the classic example of 'wings'. If a rudimentary comparison is made of the wings of beetles and the wings of birds, one immediately notices that bird wings have feathers but beetles wings do not. A more detailed examination would show that the wings of birds are composed of a series of bones, whereas the wings of beetles are membranous. In fact, beetles have two sets of wings, forewings – called elytra – that are hardened and not used for flying. It is the hind wings that are membranous and vibrate, allowing the animal to fly. The term 'wing', when used as a taxonomic character, might be considered insufficient, even incorrect, as it is a description of a collection of parts that allow a function to be performed, the act of flying; therefore a wing is not strictly speaking a structure but a collection of structures. Of course, shorthand allows that birds and beetles do actually have wings, but that merely indicates both animals have the potential to fly. As homologues, the term 'wing' is not useful. Rather, the elytra, feathers and bones (and their parts) are the homologues – and very different from one another.

Thus a major, if not *the* major, task of all taxonomists is moving from a general description of an organism's parts (such as 'wings') to a more accurate assessment of the parts (such as 'elytra'), identifying or discovering homologues, the 'parts that merit the same name' (Patterson 1988, p. 605). Much progress, if not all progress, is made in taxonomy by clarifying homologues.

---

[13] 'A "homologue" is a part or organ in one organism so answering to that in another as to require the same name ...' (Owen 1866, p. xii).

It is important to note that homologue is often confused with, or used interchangeably with, *homology*. This is a mistake. They are different. Owen discussed homology separately as, for him, it merited a much longer discussion, where he focused on three different 'kinds' of homology: general, special and serial (of which more later).

To close this section, we cannot stress enough how important the study of homologues is to comparative biology (taxonomy). This is the arena in which expertise is developed relative to any particular group of organisms; this is the empirical part of comparative biology (taxonomy).

False Homologues

Establishing homologues is often equated with some criterion of similarity, as if similarity can be a measure of how two structures are alike. Having a 'measure' of something implies the use of a quantifiable approach and often taxonomic characters are divided into *quantitative* and *qualitative*. We have so far been addressing what would be termed *qualitative taxonomic characters*. With respect to *quantitative taxonomic characters*, numerical taxonomy (as taximetrics, see Chapter 8) sought a variety of mathematical measures to discriminate characters one from another in the hope that such measurements would provide some exactitude.

Statements of similarity can, of course, deal with many kinds of parameters and many kinds of quantities, usually expressed as some numerical value of size or calculation of shape. The study of *morphometrics* (from the Greek *morph* meaning shape or form, and *metron* meaning measurement) addresses these concerns.

Example 4. Measurements: The leaves of any plant can be measured at various points and may fall into various size groups corresponding to variables such as 'length of leaf blade'. These lengths may form groups such as: length 8–12 cm, length 13–18 cm, length 18 > cm, and so on. Such variables are often referred to as *continuous* and while there have been numerous attempts to quantify the 'breaks' and 'gap size' in any continuous series ('Should that be two groups of 8–12 cm and 13–18 cm, or should that be three groups of 8–10 cm, 10–15 cm and 16–18 cm?'), the divisions are always somewhat arbitrary: all leaves change size during their development. Measurements need to be taken at comparable stages of growth, at a particular time of year, and so on – but even then, they still remain more or less a continuum.

These kinds of measurements are for the most part associated with the concept of *phenetic similarity*, an attempt at a purely quantitative assessment. Size does not necessarily constitute variation in form as the structures that make up a leaf (blade, petiole, stipules, vein, etc.) remain identical. If quantitative taxonomic characters are referred to as homologues, then they should be considered *false*, inasmuch as they are not structural and may indeed simply reflect the method applied rather than some actual feature. Should these characters be added to an existing dataset, any subsequent analysis may find unambiguous relationships – but the results may be based on artefactual resolution, or even be entirely wrong if based on false homologues.

This does not mean that measurements of this kind are useless. Size may be, and often is, a perfectly good way to distinguish specimens of different species, one from another – it is an excellent parameter for artificial classifications (Chapter 7) designed to *identify* particular organisms rather than *characterise* them.

Example 5. Craticula Grunow: *Craticula* is a genus of diatoms with c. 50 recognised species. Although the genus was first described in 1867, most species have been described since 1990. Many were previously included in the genus *Navicula* but have since been transferred on account of their craticula, the internal valve that forms inside the frustules (see Levkov et al. 2016). The key below is a guide to some common species:

1a. Lineolae often aligned to form distinct longitudinal lines (longitudinal ribs)	... 2
1b. Longitudinal lines absent	... *C. accomoda*
2a. Longitudinal lines distinctly recognisable, 8–20 in 10 µm	... *C. cuspidata*
2b. Longitudinal lines still recognisable, but >28 in 10 µm	... *C. halophila*
	etc.

In this key these species are distinguished by the number of longitudinal lines distinctly recognisable and by their numbers. These criteria help *identify* the species but not necessarily to *characterise* them. It would be helpful in the future if these kinds of characters were referred to as *false homologues*.

Analogues

Alongside his first definition of homologue, Owen defined 'analogue' (Owen 1843, p. 374):

> ANALOGUE. A part or organ in one animal which has the same function as another part or organ in a different animal. See homologue.

To demonstrate its meaning, Owen referred to *Draco volans*, the flying dragon:

> Its forelimbs being composed of essentially the same parts as the wings of a bird are homologous with them; but the parachute being composed of different parts, yet performing the same function as the wings of a bird, is analogous to them ... But homologous parts may be, and often are, also analogous parts in a fuller sense, viz. As performing the same functions ... the pectoral fin of the flying fish is analogous to the wing of the Bird, but, unlike the wing of the Dragon, it is also homologous with it. (Owen 1847, p. 175)

Analogy was thus 'defined' by Owen solely in terms of function: legs are for walking, wings are for flying.

Homology

Homology is the relation implied by *homologues*.

Relating Homologues

Consider a simple, hypothetical example of a single part that has been studied and identified as homologue A:

A

Further study reveals various manifestations of homologue A in other organisms so that alternatives can be recognised and named, such as homologue B:

A   B

Homologues A and B together might be considered to represent a single element, representing a single structure, which could be written:

A – B or **{A B}**

The hyphen in the first example above merely indicates there is a 'connection' of some sort between the two, as does enclosing them within curly brackets. One might even use both:

{A – B}

There is nothing real about the two symbols (the hyphen and the curly bracket). Real in the sense that one might be able to observe or discover what it is or means. They are not visible parts of anything – not accessible to investigation, unlike the homologues themselves.

If more homologues are found as more organisms are studied, the series grows:

A – B – C

{A B C}

{A – B – C}

One is then faced with the idea that homologues A, B and C are indeed the same but different. We will address this shortly. For the moment it is worth noting that to determine exactitude of comparison between A, B and C so that one can be sure they really are the 'same' has resulted in extensive discussions on the quality of auxiliary data that may or may not assist such studies. But to be sure requires a measure of certainty and comparisons cannot be fool-proof in any exact way; there is no such thing as 'safe, sure knowledge'.

To relate homologues, one to another, one might need to find meaning in the hypothetical hyphen(s) that separates the series, such as A – B – C above. To some

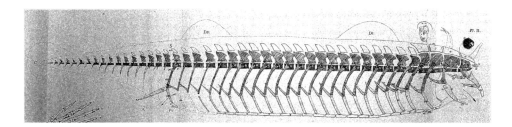

Figure 7.11 Owen's archetype (from Owen 1848, pl. II).

it may be sufficient to note that two or more homologues simply relate and the hyphen is merely representational, indicating that A, B and C are similar enough to be considered truly the same regardless of the fact that they do indeed differ. It is quite possible that most taxonomists work under this assumption and that the pursuit of exactitude in similarity considerations of this kind can only be found through further investigation.

For Owen, if homologues were to be considered 'truly the same', it would mean them having a common 'origin' in the archetype, an invented animal representing the common source of the array of homologues in the real organisms (Owen 1848, Figure 7.11). For most biologists, after Darwin published his *Origin of Species*, the meaning of that hyphen between homologues turned Owen's archetype into an ancestor. As Darwin wrote in his copy of Owen's *On the Origin of Limbs*:

> I look at Owen's archetypus as more than ideal, as a real representation as far as the most consummate skill & loftiest generalizations can represent the present forms of Vertebrata. – I follow him that there is a created archetype, the parent of its class. (after Nelson 2011, p. 138)

The implication is that any series of homologues, such as A – B – C, are transformations of one to another:

> Naturalists frequently speak of the skull as formed from metamorphosed verte-brae; the jaws of crabs as metamorphosed legs, the stamens and pistils of flowers as metamorphosed leaves ... Naturalists, however, use such language only in a metaphorical sense; they are far from meaning that during a long course of descent, primordial organs of any kind ... have been converted into skulls and jaws. Yet so strong is the appearance of a modification of this nature having occurred, that naturalists can hardly avoid employing language having this plain signification. *On my view these terms may be used literally* ... (Darwin 1859, pp. 438–439, our italics)

With this revised perception, or understanding, Darwin ushered in a new research programme called *evolutionary morphology*, designed to update and modernise Owen's *comparative anatomy*: 'The survey of such a series [of homologues] thus reveals a *process*, which involves the transformation of one and the same organ in different organisms'. (Gegenbaur 1870, p. 6, emphasis added, after Rieppel 2011, p. 179)

The homologues A, B and C thus became a dynamically related series of successive changes, the hyphen becoming an arrow, the arrow a representation of a process:

$$A \rightarrow B \rightarrow C$$

The systematic literature is replete with examples of such series of transformational similarities discussed as if this was homology, such as Gegenbaur's own 'homology of hand to forelimb' (Gegenbaur 1870). But such interpretations are purely hypothetical. Some homologues might be observed via developmental stages: here one might see a process that can be studied, rather than inferred – but that does not relate to establishing actual historical transformation series but of determining its polarity with reference to the direction of change through development. If we start with A – B – C, what evidence can be brought to bear on the issue of whether the homologues are related thus:

$$A \rightarrow B \rightarrow C$$

$$B \rightarrow A \rightarrow C$$

$$C \rightarrow B \rightarrow A$$

$$A \rightarrow C \rightarrow B$$

etc.

Developmental biology (ontogeny, embryology), the most significant source of transformational information even prior to Haeckel's Biogenetic Law stretched its meaning, is an empirical endeavour – one can study the development of any organism's parts and discover if one homologue is 'the same' as another. By that, we mean one might observe a transformation, 'And as a *transformation*, each character is related through ontogeny to its homologs. Hence its *relations* ("polarity") are phenomena open directly to investigation'. (Nelson 1985, p. 42)

Here we need to reiterate that series of homologues do not make homology, even when it is clear beyond doubt that one homologue really is a transformation of another, as in some established embryological series. Thus, when a series such as A $\rightarrow$ B $\rightarrow$ C is equated with what has been called *transformation homology*, the terminology is somewhat inaccurate: 'Homology is also asserted as a relationship between characters with different identities, as in fins and legs. This is transformational homology' (Wiley 2008). 'Characters with different identities' is the basis for discovering homologues, the conundrum of some things being the same but different. A better explanatory phrase for 'Characters with different identities' would be *transformational homologues* or better still *transformational similarities*.

Many systematists understand *transformational homologues* as the basis for the entries in a modern taxon $\times$ character data matrix. That is, each column of a data matrix represents *transformational homology* [=homologue]: 'All the states of a topographically identical character are considered transformationally homologous with one another ...' (Brower & Schawaroch 1996, p. 269); 'We argue that the recognition of structures which are alternate forms is a vital stage of primary homology assessment and is equivalent to the conceptualization of a transformational homology' (Hawkins et al. 1997, p. 275, abstract). We disagree with this use of the word homology: 'structures which are alternate forms' of each other are simply alternative homologues. Further discussion of this matter, with respect to data matrices, is in Chapter 8; further discussion concerning *relating homologues* is in the following section.

Homology As Relationship

In Chapter 1 it was noted that the central problem of biological classification is to discover the properties of organisms to reveal its relationships. These relationships are represented in *classifications* expressed in nested sets of taxa. A *taxon* is one relationship, with the organisms considered to be more closely related amongst themselves than they are to any other. Well-known examples are: vertebrates, mammals, angiosperms and *Homo sapiens*.

In Chapter 2 we discussed a particular species of box jellyfish, *Malo kingii*, and noted its complete classification in the animal kingdom:

Regnum: Animalia
    Phylum: Cnidaria
        Classis: Cubozoa
            Ordo: Carybdeida
                Familiae: Tamoyidae
                    Genus: *Malo*
                        Species: *Malo kingii*

Each proper name in the sequence above represents a particular taxon: Animalia, Cnidaria, Cubozoa, Carybdeida, Tamoyidae, *Malo* and *Malo kingii*. Each taxon is included in the preceding one such that the levels of generality extend from species (*Malo kingii*) to kingdom (Animalia). That is, the properties of the kingdom (animals) apply to the species (*Malo kingii*). If one was to fully describe *Malo kingii* in all its detail, it would be a list of the properties from each successive taxon as specified above.

Discoveries of taxa are based on evidence that reveals such relationships. For example, angiosperms can be recognised by several properties:

- Ovules enclosed within a carpel (a structure made up of an ovary, which encloses the ovules, and the stigma; a structure where pollen germination takes place);
- Double fertilisation leading to the formation of an endosperm (a nutritive tissue within the seed that feeds the developing plant embryo);
- Stamens with two pairs of pollen sacs;

- Features of gametophyte structure and development;
- Phloem tissue composed of sieve tubes and companion cells.[14]

Each of the five items above represents evidence for the relationship 'angiosperm'. So:

1. Evidence = Ovules enclosed within a carpel   = angiosperms
2. Evidence = Double fertilisation   = angiosperms
3. Evidence = Stamens with two pairs of pollen sacs   = angiosperms
4. Evidence = Gametophyte structure and development   = angiosperms
5. Evidence = Phloem ... sieve tubes and companion cells   = angiosperms

Thus taxon and evidence are one and the same. *Relationship* means that any two organisms (or their parts) are understood to be more closely related to each other than they are to any other organism (or their parts). It should be noted here that while an organism's parts provide material evidence, it is homology that is the relationship. In the example above, 'angiosperm' is shorthand for 'All organisms having "Stamens with two pairs of pollen sacs" and are more closely related to each than they are to any other organism'. The evidence is the homologues.

All organisms are related at some taxonomic level, even if that level is Life itself. The red roughy, robins and box jellyfish of Chapter 2 are all related as animals (Animalia, 'having breath'). The red roughy and robins are more closely related to each than they are to box jellyfish, as vertebrates. An individual bird from *Taeniopygia guttata* (the Zebra Finch) shares a greater relationship with any other individual bird from *Taeniopygia guttata* than it will with any individual from, for example, *Taeniopygia bichenovii* (the Double-Barred Finch): All finches are related amongst themselves as the family Estrildidae; all finches are related among themselves as birds, chordates and animals.

One often reads statements like the following: 'a mouse (*Mus musculus*) is related to a giraffe (*Giraffa camelopardalis*)'. While the statement is undoubtedly true, it makes no real sense. These kinds of statements crop up frequently in the taxonomic literature: 'Species A is *closely* related to Species B', or 'Species A is *very closely* related to Species B'. All life, extinct and extant, is related *in some way*. A two-taxon statement is thus empty of meaning:

Mouse: Giraffe

All useful comparisons require a third item to be meaningful. *A statement of relationship* is thus: 'A mouse (*Mus musculus*) and a rat (*Rattus rattus*) are more closely related to each other than they are to a giraffe (*Giraffa camelopardalis*)'. This is the simplest possible statement of relationship one can make. It can be represented parenthetically:

(Mouse: Rat) Giraffe

---

[14] Data from after the Tree of Life Project – www.tolweb.org/Angiosperms/20646

The meaning of the notation above is that the items inside the brackets are more closely related to each other than they are to items outside the brackets. It should be clear from this, then, that taxon = relationship. Thus it could be written:

(Mouse: Rat) 'Any Other Organism'

The (Mouse: Rat) 'Any Other Organism' relationship is captured by the taxon name Rodentia (rodents). But rodents include many more animals than just *Mus musculus* and *Rattus rattus*. So evidence that links the mouse and rat applies to further organisms and Rodentia informs us of their collective relationship.

If mice and giraffe are more closely related to each other than they are to the Zebrafish (*Danio rerio*), then the relationship is expressed so:

(Mouse: Giraffe) Zebrafish

Specifically, these can be referred to as *Cladistic Statements of Relationship*. They have a generality beyond biological systematics, a generality that will not be discussed further here.

Of course, not all statements of relationship turn out to be true. Some simply have no evidence to support them. For example, if it was thought that the giraffe was more closely related to jellyfish than to a mouse, the relationship could be stated thus:

Mouse (Jellyfish: Giraffe)

This is still a precise statement of relationship. But what evidence is there to support it? None, as far as we are aware. The example is contrived as no one has ever attempted to relate jellyfishes and giraffes most closely to each other relative to all other animals, but some relationships that are accepted do indeed turn out to be false.

Insectivores (Order Insectivora), for example, are a group of mammals based on their shared feeding habits of insects. Insectivores include moles, shrews and hedgehogs. While it is true that the group of mammals included are all insect eaters, that observation relates many other insect-eating organisms such as bats, possums, lizards, spiders and some humans. But even if these were added, the group would exclude insectivorous plants. While it might be useful to know that a hedgehog and shrew are both insectivorous, it does not provide evidence for their relationships. 'Insect-eating' describes a general mode of life rather than a particular structure of any organism – because of that one might initially doubt its use as evidence of relationships. It does not describe a part of an organism and therefore cannot be a homologue.

Above we noted that finding and studying homologues of organisms was the primary task of taxonomists. The second task is discovering the relationships between them. The first deals with homologues, the second with homology.

Owen's Homology and Analogy

As we noted above, Owen defined analogues in terms of function ('A part or organ in one animal which has the same function as another part or organ in a different animal'), hence the *relation* is one of function. Homology and analogy, as relations, can be combined in four permutations to show that while both homologues and analogues may be based on the same criteria (similarity in structure, relative position, etc.), the difference between analogous and non-analogous relations are ultimately governed by function, while the differences between homologous and non-homologous relations are ultimately governed by form.

Homology and analogy are different kinds of relations not opposites, 'antonyms' or 'antitheses'. Nevertheless, there has been a tendency to use the two concepts as if they were 'opposites' rather than different relations (e.g., Ghiselin 1976, 1984, p. 109; cf. Ghiselin 2016).

To consider homology and analogy as opposites would be to imply that one is true, the other false, under the general understanding that homology is the indicator of 'true' relationship (however conceived) among organisms based on hypotheses concerning their parts, their homologues. That dichotomy was a later invention, one that Lankester began (but did not intend) by dissecting the term homology and substituting two new terms of his own invention: *homogeny* for similarities that 'truly' indicate common ancestry (the true relationship), and *homoplasy* for similarities otherwise explained (Lankester 1870). There was no need for Lankester's intervention. Homology and analogy *complement* one another; homoplasy is a redundant term.

If homology and analogy are not opposites, how, then, do we use these terms when discussing function and relationship? Consider the wings of a bat, a bird and a grasshopper. The wings in these three animals perform the same function, namely to assist flying. However, the structure of the wings of bats and birds are the same: they are modifications of the mammalian forelimb. In this sense the wing of a bird and a bat are both homologous and analogous at the same time, when compared to a grasshopper. Further, the wings of a grasshopper are not homologous although they are analogous to the bat and bird, when compared to say a diatom or a jellyfish. In this sense there are four possible combinations of analogues and homologues:

1. A homologous and analogous statement of relationship, namely, ((Bat, Bird) Grasshopper): Synapomorphy = mammalian forelimb; Analogy = wings.

2. A homologous and non-analogous statement of relationship, namely, ((Bat, Jellyfish) Diatom): Synapomorphy = cell wall; Analogy = none.

3. A Non-homologous and analogous statement of relationship, namely, ((Bat, Grasshopper), Bird); Synapomorphy = 'wings'; Analogy = flight.

4. A Non-homologous and non-analogous statement of relationship, namely, ((Diatom, jellyfish), Bird); Synapomorphy = 'waterborne'; Analogy = none.

The difference between an analogy and a synapomorphy are false homologues. A false analogue is based on function, namely an observable process (e.g., tool use) or condition (e.g., adapted to cool temperate waters) without reference to the threefold parallelism (see Chapter 13). A false analogy would be a semantic issue or based on a conjecture (e.g., a tortoise and a sloth are both considered slow), or on an assumed but unobserved process (e.g., all dinosaurs are cold blooded). A false analogue is by definition a false homologue and is therefore a redundant term.

Box 7.5   Molecular Homology

The search to explore 'Molecules as Documents of Evolutionary History' (Zuckerkandl & Pauling 1965), and as a consequence develop the subject of molecular systematics and molecular homology, began as soon as it was possible to gain enough useful comparative molecular data (e.g., Pauling & Zuckerkandl 1963). The first serious effort was made using comparisons of myoglobin and the α, β, γ and δ human haemoglobin chains (Itano 1957; Ingram 1961). We note these early commentaries simply to draw attention to two aspects of these studies: When considering the evolution of haemoglobins, Ingram (1961) included a scheme of relationships, a tree of relationships (Figure B7.5.1, after Ingram 1961), while Itano (1957) relied almost entirely on a narrative (see Morgan 1988 for a history). What flowed from the steady accumulation of molecular data, from the 1950s to the present, and its relationship to phylogeny ('Evolutionary History'), was a plethora of methods to analyse the data of which all – or nearly all – assumed that the resulting phylogeny would be tree-like and hence that mode of representation was built into the various programs (the historical studies of Edna Suárez-Díaz are relevant: Suárez-Díaz & Anaya-Muñoz 2008; Suárez-Díaz 2009, 2010; see also Hagen 2009 and Marks 2009). Thus, the narrative approach gave way to the analytical approach, with results presented in the familiar tree-like fashion.

 With respect to the analytical, one might be forgiven for thinking of current molecular systematics as being linked to part of, or even all of, *bioinformatics* – there is obviously a connection, but bioinformatics had a beginning prior to and outside of molecular systematics (Hagen 2000; Gauthier et al. 2018[1]). Nevertheless, in spite of many decades in the development of methods of analysis, the term homology continues to be misused or misunderstood. For example, in their bioinformatics review, Gauthier et al. write that:

> In the first sequence-based phylogenetic studies, the proteins that were investi-gated (mostly homologs from different mammal species) were so closely related that visual comparison was sufficient to assess homology between sites. (Gauthier et al. 2018, p. 4)

This is not an unusual viewpoint: homology taken to mean similarity.

Box 7.5 *(cont.)*

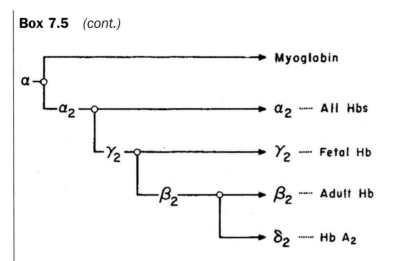

Figure B7.5.1 Evolution of haemoglobins (after Ingram 1961, with permission).

In a recent contribution to the subject of molecular homology, Inkpen and Doolittle summarise much of the previous literature observing that

> The initial switch to a molecular underpinning for homology may have been based on the dream that historical relationships are easier and cleaner to read from molecular rather than morphological data. But molecular data are much more equivocal than we thought in the mid 1960s. In molecular phylogenetics, just as in others areas of biology, and at other levels of the organizational hierarchy, we must contextualize homology if we are to talk about it at all (Inkpen & Doolittle 2016, p. 190)

And

> In 1987, David [sic[2]] Patterson mused: 'I intended to float the idea that molecular homologies are no more secure, and are possibly more precarious, than morphological ones' (1987, p. 18). We are not sure that molecular data are more precarious than other data, nor that their uncertainties are always similar in form to morphology, but faith that molecular phylogenetics can solve the homology problem once and for all is unfounded. (Inkpen & Doolittle 2016, p. 190)

In the midst of the development of cladistics, it did indeed seem reasonable to suggest that 'molecular homologies are no more secure, and are possibly more precarious, than morphological ones' (Patterson 1987, p. 18). Patterson explored molecular homology in terms of the three tests he first applied to morphological homology: similarity, conjunction and congruence (Patterson 1982, 1988). That exploration yielded some differences in molecular and morphological homology, inasmuch as a few new 'categories' of relations were proposed and some new explanations of those relations were offered, but overall Patterson understood morphological and molecular homology to be more or less the same.[3]

Box 7.5 *(cont.)*

Table B7.5.1 Patterson's eight 'categories' of homology for morphology and molecules

Morphology	Molecules
Homology	Orthology[1]
Homonomy	Paralogy[2]
Complement	Complement[3]
Two homologies	Two orthologies[4]
Parallelism	Xenology[5]
Homeosis [multiparallelism]	Paraxenology, plerology[6]
Convergence	[Convergence]
Endoparasitism [multiconvergence]	. . .

[1] 'Orthology is homology reflecting the descent of species, and paralogy is homology reflecting the descent of genes' (Patterson 1988, p. 610).

[2] '. . . paralogy is homology reflecting the descent of genes' (Patterson 1988, p. 610).

[3] 'The complement relation (presence of an orthology vs. its absence) and 'two orthologies' (presence of two orthologies with the inclusion relationship, such that the group circumscribed by one includes or equals that circumscribed by the other – e.g., cytochrome c and myoglobin; cf. notochord and femur) are the same in molecular as in morphological comparisons' (Patterson 1988, p. 612).

[4] '. . . presence of two orthologies with the inclusion relationship, such that the group circumscribed by one includes or equals that circumscribed by the other', see also footnote 27 (Patterson 1988, p. 612).

[5] Xenology 'is a form of homology (inferred common ancestry) in which the sequence (gene) homology is incongruent with that of the organisms carrying the gene, and horizontal gene transfer or transfection is the assumed cause. Xenology implies "foreign genes" (Gray and Fitch 1983, p. 64) . . .'(Patterson 1988, p. 612).

[6] 'Since molecular homology may reflect either species phylogeny or gene phylogeny, there are more kinds of homologous relation between molecular sequences than in morphology. The terms *paraxenology* and *plerology* are proposed for two of these kinds – respectively, the consequence of multiple xenology and of gene conversion' (Patterson 1988, p. 602); 'Gene conversion is not the name of a relation but of a process inferred to explain that similarity between paralogous sequences which implies a relationship between paralogues that is incongruent with that based on other paralogies, because the similarity is too great. The relation has no name, but *plerology* [from the Greek *pleres* (full of, complete)] might serve to emphasize the extra similarity observed' (Patterson 1988, p. 614).

Setting the tests of homology to one side, Patterson outlined eight 'categories' or 'kinds' of relation (Table B7.5.1). Brief explanations of some of these 'categories' are given in the footnotes to our Table B7.5.1, but our focus of interest here is orthology, paralogy and xenology, the molecular equivalents of homology, homonomy[4] and parallelism, respectively (Patterson 1988). Patterson offered concise definitions of orthology and paralogy:

> Orthology is homology reflecting the *descent of species*, and paralogy is homology reflecting *the descent of genes*. (Patterson 1988, p. 610, italics ours)

Box 7.5 *(cont.)*

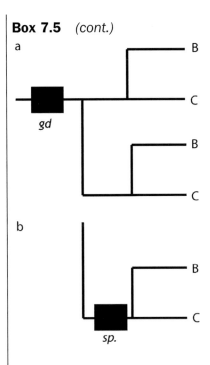

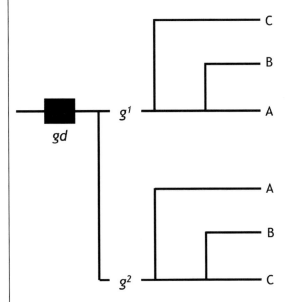

Figure B7.5.2 (a) *Gene duplication* (gd) producing an extra copy of BC; (b) the black box is the position of the inferred speciation event (sp.).

Figure B7.5.3 The *relationships of the characters [homologues]* (paralogy) do not reflect the *relationships of the taxa* (orthology) as the two parts of the paralogous tree differ: gene g^1 = C(AB), gene g^2 = A(BC).

Box 7.5   *(cont.)*

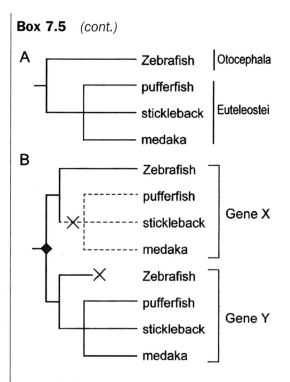

Figure B7.5.4 (A) A 'gene tree, containing four selected teleost species, is consistent with their relationships in the species tree ... (B) The hypothetical tree showing the possibility that gene loss occurred in an asymmetric pattern between Otocephala and Euteleostei, and that the zebrafish gene is paralogous to the genes from the other three species in Euteleostei' (Kuraku 2013, p. 120).

One might modify this to make the statements more general:

> Orthology is homology reflecting the *descent of taxa*, and paralogy is homology reflecting *the descent of characters.*

Nevertheless, orthology and paralogy are said to define particular relationships:

> Orthology is homology reflecting the *relationships of taxa,* and paralogy is homology reflecting the *relationships of characters [homologues].*

Do characters and taxa differ? Presumably (most) taxa are defined by *sets* of congruent characters, rather than just one, but essentially taxa and characters define the same relationship.

The standard (conventional) explanation for paralogy is *gene duplication:* producing an extra copy or copies of a gene, of which each may (or may not) have a separate history (e.g., Zhang 2003 provides a review). That is, in the conventional understanding, a gene duplication event may precede a speciation event (Figure B7.5.2a: the black box is the position of the gene duplication (gd); Figure B7.5.2b: the black box is the position of the inferred speciation event (sp.)). In Figure B7.5.2 the relations are obvious as there are only two terminals (taxa): it might be said that the *relationships of the characters [homologues]* (paralogy) reflect the *relationships of the taxa* (orthology). The relationship

Box 7.5 *(cont.)*

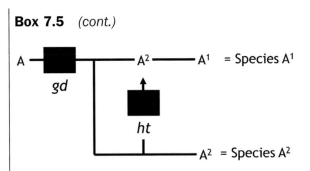

Figure B7.5.5 *Xenology* explained by the Horizontal (Lateral) Transfer of a particular gene: gene A is duplicated (black box, gd) into A^1 and A^2. Gene A^1 defines, or is part of the definition of, species A^1, and gene A^2 defines, or is part of the definition of species A^2; at some point A^2 transfers to species A^1 (black box, ht) thus becoming part of its definition.

in Figure B7.5.3 is made just a little more complex by adding a third taxon. Here the *relationships of the characters [homologues]* (paralogy) do not directly reflect the *relationships of the taxa* (orthology) as the two parts of the paralogous tree differ: for gene g^1 the relationship is C(AB); for gene g^2 the relationship is A(BC) (Figure B7.5.3). For these data it is impossible to know which of the two genes is the orthologue and which the paralogue. The terms seem superfluous with respect to the more general term homology. More complex problems obviously occur (e.g., Kuraku 2013, our Figure B7.5.4a after Kuraku 2013, fig. 2).

Xenology is the relation explained by the horizontal (lateral) transfer of a particular gene. In Figure B7.5.5 gene A is duplicated (Figure B7.5.5, black box, gd) into A^1 and A^2. Gene A^1 defines, or is part of the definition of, species A^1, and gene A^2 defines, or is part of the definition of, species A^2. At some point A^2 transfers to species A^1 (Figure B7.5.5, black box, ht) thus becoming part of its definition. Although our example is almost overly simple, what it shows is that xenology is another form of incongruence rather than another form of homology.

In terms of analysis – genes as characters – these are simply sets of *fundamental cladograms*. Thus, events (explanations, such as speciation, gene duplication, horizontal transfer, etc.) appear irrelevant: it is the congruence of individual fundamental cladograms that yield results. Molecular systematics, then, is yet another version of cladistic analysis when understood as *the analysis of relationships*.[5]

Nevertheless, there is much left unexplored: what are the homologues? Are they the bases, the genes, both? If the base pairs are the homologues, are they inherently phenetic? That is, do they have no obvious notion of synapomorphy? Or in an analytical sense, are they really the same as morphological characters? These problems – and we see them as interesting problems awaiting solution – require addressing (some valuable and insightful contributions have been made by Springer & Gatesy 2016, 2018; see also Gatesy & Springer 2017). It would appear to be of limited use to simply keep adding more and more data, going from 1-gene trees, to 3-gene trees, to 5-gene trees, to genomics

Box 7.5 *(cont.)*

('phylogenomics'), etc., without addressing and exploring these issues – otherwise these approaches really are just further excursions into phenetics.

The time is right to examine molecular homology from the point of view of relationships.

References

Doyle, JJ. 1992. Gene trees and species trees: molecular systematics as one-character taxonomy. *Systematic Botany* 17: 144–163.

Doyle, JJ. & Davis, JI. 1998. Homology in molecular phylogenetics: a parsimony perspective. In: Soltis, DE., Soltis, PS. & Doyle, JJ. (eds), *Molecular Systematics of Plants II*. Springer, Boston, MA, pp. 101–131.

Gatesy, J. & Springer, MS. 2017. Phylogenomic red flags: homology errors and zombie lineages in the evolutionary diversification of placental mammals. *Proceedings of the National Academy of Sciences USA* 114: E9431–E9432.

Gauthier, J., Vincent, AT., Charette, SJ. & Derome, N. 2018. A brief history of bioinformatics. *Briefings in Bioinformatics* 2018: 1–16.

Gray, GS. & Fitch, WM. 1983. Evolution of antibiotic resistance genes: the DNA sequence of a kanamycin resistance gene from *Staphylococcus aureus*. *Molecular Biology & Evolution* 1: 57–66.

Hagen, JB. 2000. The origins of bioinformatics. *Nature Reviews* 1: 231–236.

Hagen, JB., 2009. Descended from Darwin? George Gaylord Simpson, Morris Goodman, and primate systematics. In: Cain, J. & Ruse, M. (eds), *Descended from Darwin: Insights into the History of Evolutionary Studies, 1900-1970*. Transactions of the American Philosophical Society, n.s. 99. American Philosophical Society, Philadelphia, pp. 93–109.

Ingram, V. 1961. Gene evolution and the haemoglobins. *Nature* 139: 704–708.

Inkpen, SA. & Doolittle, WF. 2016. Molecular homology and the perennial problem of homology. *Journal of Molecular Evolution* 83: 184–192.

Itano, HA. 1957. The Human hemoglobins: their properties and genetic control. *Advances in Protein Chemistry* 12: 215–268.

Kuraku, S. 2013. Impact of asymmetric gene repertoire between cyclostomes and gnathostomes. *Seminars in Cell & Developmental Biology* 24: 119–127.

Marks, J. 2009. What is the viewpoint of hemoglobin, and does it matter? *History and Philosophy of the Life Sciences* 31: 241–262.

Mindell, DP. & Meyer, A. 2001. Homology evolving. *Trends in Ecology & Evolution* 16: 434–440.

Morgan, GJ. 1988. Emile Zuckerkandl, Linus Pauling, and the molecular evolutionary clock, 1959-1965. *Journal of the History of Biology* 31: 155–178.

Page, RDM. 1989. Comments on component-compatibility in historical biogeography. *Cladistics* 5: 167–182.

Page, RDM. 1994. Maps between trees and cladistic analysis of historical associations among genes, organisms, and areas. *Systematic Biology* 43: 58–77.

Patterson, C. 1982. Morphological characters and homology. In: Joysey, KA. & Friday, AE. (eds.), *Problems of Phylogenetic Reconstruction*. Academic Press, London, pp. 21–74.

Patterson, C. 1987. Introduction. In: Patterson, C. (ed.), *Molecules and Morphology in Evolution: Conflict or Compromise?* Cambridge University Press, Cambridge, UK, pp. 1–22.

Patterson, C. 1988. Homology in classical and molecular biology. *Molecular Biology and Evolution* 5: 603–625.

Box 7.5 *(cont.)*

Pauling, L. & Zuckerkandl, E. 1963. Chemical paleogenetics: molecular 'Restoration Studies' of extinct forms of life. *Acta Chemica Scandinavica* 17 (suppl.): 9–16.

Springer, MS. & Gatesy, J. 2016. The gene tree delusion. *Molecular Phylogenetics and Evolution* 94: 1–33.

Springer, MS. & Gatesy, J. 2018. On the importance of homology in the age of phylogenomics. *Systematics and Biodiversity* 16: 210–228.

Suárez-Díaz, E. 2009. Molecular evolution: concepts and the origin of disciplines. *Studies in the History and Philosophy of the Biological and Biomedical Sciences* 40: 43–53.

Suárez-Díaz, E. 2010. Making room for new faces. Evolution and the growth of bioinformatics. *History and Philosophy of the Life Sciences* 32(1):65–89.

Suárez-Díaz, E. & Anaya-Muñoz, V. 2008. History, objectivity and the construction of molecular phylogenies. *Studies in the History and Philosophy of the Biological and Biomedical Sciences* 38: 451–468.

Williams, DM. 1993. A note on molecular homology: multiple patterns from single datasets. *Cladistics* 9: 233–245.

Zhang, J. 2003. Evolution by gene duplication: an update. *Trends in Ecology & Evolution* 18: 292–298.

Zuckerkandl, M. & Pauling, L. 1965. Molecules as documents of evolutionary history. *Journal of Theoretical Biology* 8: 357–366.

[1] 'It is easy for today's students and researchers to believe that modern bioinformatics emerged recently to assist next generation sequencing data analysis. However, the very beginnings of bioinformatics occurred more than 50 years ago, when desktop computers were still a hypothesis and DNA could not yet be sequenced'.

[2] They meant Colin Patterson.

[3] Mindell & Meyer (2001) summarised various different 'kinds' of homology relating to molecular studies, adding a few more terms: 'Pro-orthology … Semi-orthology … Partial homology … Gametology … Synology …'. None of these need concern us here (Mindell & Meyer 2001, p. 437, table 1).

[4] '*Homonomy* differs from homology in failing the conjunction test because several or many copies of the homologue occur in one individual'. (Patterson 1988, p. 606)

[5] Page recognised the similarity between systematics and biogeography in terms of cladogram analysis, as did Doyle (1992, see also Page 1989, 1994; Williams 1993; Doyle & Davis 1998).

Heterobathmy

The term *Heterobathmy* was introduced by Takhtajan (1959, p. 11, 13, as *heterobathmie*; for the English translation see Takhtajan 1991, p. 227 *et seq.*) and subsequently used by Hennig (*heterobathmie*, Hennig 1965, p. 107).

Heterobathmy literally means 'different steps on a stair', from the Greek *bathmos,* a step or stair. When applied to taxonomy, it means finding the correct level ('step') in the taxonomic hierarchy ('stair') for each character (homologue). The

Table 7.1 Some alternative terms for 'heterobathmy of characters' (Takhtajan 1959, p. 13; Hennig 1965, p. 107)

Chevauchement des specialisations (crossings of specialisations)	Dollo (1895, p. 88)
Heterepistase (heteroepistasy)	Eimer (1898, p. 98)
Law of corresponding stages	Arber & Parkin (1907, p. 35)
Gesetz der Spezialisationskreuzungen	Abel (1912, p. 637)
Spezialisationskreuzungen	Tschulok (1922, p. 205) Hennig (1936, p. 557) Schindewolf (1937, p. 207) Hennig (1950, p. 142)
Evolution entre-croisée	Nierstrasz (1936, p. 674)
Merkmalsphylogenie	Zimmermann (1930, p. 984) Hennig (1950, p. 181)
Mosaic evolution	De Beer (1954, p. 163)
Specialisation-crossings	Hennig (1966, p. 194)
Cross-specialisations	Schindewolf (1993, p. 106)

Data from Takhtajan (1959, p. 13), Nelson (2004, p. 133), Rieppel (2010, p. 482) and Rieppel (2013).

question posed is: where does the character fit? The process of finding the correct place for all characters leads to the rather clumsy phrase 'heterobathmy of synapomorphy' (e.g., Schuh 2000, p. 69), or as Hennig put it, 'heterobathmy of characters', as illustrated in his well-known diagram (Hennig 1965, p. 106, reproduced here as Figure 7.9b) – but the notion, rather than the term, has a long history (see Table 7.1, data for the table is from: Takhtajan 1959, p. 13, Nelson 2004, p. 133, Rieppel 2010, p. 482 and Rieppel 2013).

Thus Hennig wrote,

> ... the mosaic of heterobathmic characters in its distribution over a number of simultaneously living species contains reliable information about the sequence in which the species have evolved from common ancestors at different times – Heterobathmy is therefore a precondition for the establishment of the phylogenetic relationships of species and hence a phylogenetic system. (Hennig 1965, p. 107, quoted in Rieppel 2010, p. 482)

'Heterobathmy of synapomorphy' or 'heterobathmy of characters' was too clumsy a phrase to gain wide usage, but inspection of Hennig's diagram helps explain the general idea (Figure 7.9b). In his diagram, the apomorphic state of character 1 occurs in B–D, the apomorphic state of character 3 occurs in C and D, and the apomorphic state of characters 2, 4, 5 and 6 occur in A, B, C and D, respectively. Each character 'finds' its level in the hierarchy, every character eventually fits.

Additionally, and of some importance, with the exception of autapomorphies, every character is also hierarchical. This is discussed in detail in Chapter 12.

The Cladistic Parameter

What is the optimum, or best, kind of classification?

(Nelson 1979, p. 12)

It may come as no surprise that this question is an old one and can be traced back to Candolle with the idea that, yes, we do need an optimal classification, and, yes, these are synonymous with natural classification. The question 'what bird is that?' concerns identification; the question 'are birds a real taxon found in nature?' concerns natural classification.

Nelson posed a further question: '... is classification to reflect a parameter?' (Nelson 1979, p. 12). There might be many parameters to choose from and much of the debate in the last 200 years can be understood as addressing that issue. One might choose the 'overall similarity' of all homologues, regardless how they are selected. When analysed, or clustered, with a suitable program, the 'overall similarity' of all homologues will always yield a result, one that might be adopted as the optimal classification. Alternatively, one might choose a specific data source, DNA sequences, for example. When analysed, or clustered, with a suitable program, DNA sequences data will always yield a result, one that might be adopted as the optimal classification. Yet, no matter how compelling that choice might be – data or program or both – optimal classification is not about choice, as choice simply depends upon factors outside our empirical reach, or as stated elsewhere (and which stirred up a significant controversy), 'The absence of certain truth represents a philosophical limit of empirical science' (*Cladistics* 32, p.1, 2016, see Chapter 15 for commentary on this controversy).

Nelson opted for what he called the *cladistic parameter*, as all classifications to date possess that aspect in one form or another (see Chapters 2 and 3). The cladistic parameter is 'estimated by the sameness (replication and combinability), if any, exhibited jointly by the two general cladograms' (Nelson 1979, p. 15). In other words, if there is a cladogram (A(CDE)) and another cladogram (B(CDE)), measurement of their similarity to one another is not required to note that together they jointly suggest (AB(CDE)). The 'sameness (replication and combinability)' is a series of homologous relationships, or homologies, and allows the following equation: a homologue is part of homology; homology is part of monophyly; monophyly is part of a natural classification:

{Homology: Monophyly [Taxon]: Natural Classification} = *Cladistic Parameter*

Natural classifications represent observed patterns of relationships, such that:

Patterns of Relationship = Classification

We have also noted that relationship means A(BC), where B and C are more closely related to each other than they are to A. This is the *cladistic parameter*, a property of all classification and all cladograms no matter how constructed: **The *cladistic parameter* is the basis for classification in biological systematics.**

All the terms adopted in the past 200 years – homology, monophyly, taxon – might be accounted for by this general concept of relationship.

In Chapter 2 we discussed the general nature of relationship: two taxa are more closely related to each other than they are to any third. We noted that the relationship can be expressed in a simple diagram. If Taxon B is more closely related to Taxon C than either is to Taxon A, its statement of relationship is:

A(BC)

The fact that each statement of relationship can represent any number of genealogical lineages testifies to their versatility and independence from any explicit evolutionary scenario or mechanism (as noted above and in Nelson & Platnick 1981, figs 3.2–3.9). The nature of statements of relationship and their use in classification is discussed further in Chapter 11.

The cladistic parameter is a statement of relationship; homology is a statement of relationship; a monophyletic group is a statement of relationship; as is a taxon and its natural classification. Thus, the cladistic parameter can be further sum-marised as a form of equivalence[15] between the same types of relationships (i.e., homology and monophyly):

Relationship $\asymp$ Homology $\asymp$ Monophyly $\asymp$ Taxa $\asymp$ Natural Classification

Although each condition is equivalent to the cladistic parameter, the process of discovery begins with evidence. Evidence supports a statement of relationship:

---

[15] A note to philosophers of biology or logic: we are a biologist and a palaeontologist, respectively, and not philosophers. Our usage of the term *equivalence* and its mathematical symbol $\asymp$ is not meant to be understood in any philosophical context. Rather, we use it because the symbol = is often interpreted to mean *identical* or the *sum of*. Homology is a relationship between homologues and monophyly is the relationship between species or genera or families, etc., based on multiple homologies. In this sense a homology is not *identical* to monophyly, but it is part of it and works in the same way. In other words they are different manifestations of the same relationship. We are not concerned if this justifies the term *equivalence* in any philosoph-ical context; we just hope it makes sense to taxonomists and systematists.

Evidence = Statement of Relationship

We noted above evidence for the taxon angiosperms. Such that:

1.   Evidence = Ovules enclosed within a carpel = angiosperms
2.   Evidence = Double fertilisation = angiosperms
3.   Evidence = Stamens with two pairs of pollen sacs = angiosperms
4.   Evidence = Gametophyte structure and development = angiosperms
5.   Evidence = Phloem … sieve tubes and companion cells = angiosperms

Thus, taxon and evidence become one and the same. It should be noted here that while the homologue provides the material evidence, it is homology that supports the relationship. With respect to evidence, if:

Evidence supports = Homology;

Homology supports = Monophyly;

Monophyly supports = Taxon;

Taxon supports = Natural Classification

Then:

Relationship $\asymp$ Homology $\asymp$ Monophyly $\asymp$ Taxa $\asymp$ Natural Classification

and

Cladistic Parameter $\asymp$ Evidence

Not all statements of relationship turn out to be true. Therefore:

Evidence $\asymp$ Statement of Relationship

or

Evidence $\neq$ Statement of Relationship

Evidence may occur as homology, monophyly or taxa. Evidence statements can be expressed as so:

Statement of evidence $\asymp$ Homology $\asymp$ Monophyly $\asymp$ Taxon $\asymp$ Natural Classification

When the evidence conflicts with other evidence, no relationship is formed:

Statement of evidence $\neq$ Homology $\asymp$ Monophyly $\asymp$ Taxon $\asymp$ Natural Classification

An important point is that the cladistic parameter is only of value if evidence is defined as relationship. In this way evidence can be observed in an objective way and empirically scrutinised via cladistic analysis.

Evidence is also absent if any part of the cladistic parameter is poorly defined. If, for instance, we define 'relationship' as overall similarity, as a statement of percentage, distance or by any other quantitative measurement, the cladistic parameter will fail to provide empirical evidence. In such cases the parameter becomes incongruent. For instance:

$$\text{Similarity} \neq \text{Homology} \asymp \text{Monophyly} \asymp \text{Taxa} \asymp \text{Natural Classification}$$

Similarity as non-evidence cannot be used to discover monophyly and therefore is unable to find any other condition of the cladistic parameter. The same applies to all other conditions of the parameter. For instance, if we define 'monophyly' as a group of descendants and their most recent ancestor, we have made a statement of genealogy, rather than relationship (as defined above). The parameter will then still be incongruent:

$$\text{Relationship} \neq \text{Homology} \neq \text{Monophyly} \asymp \text{Taxa} \asymp \text{Natural Classification}$$

Incongruence can only be avoided if we treat all manifestations of the cladistic parameter as evidence, as homologous statements of relationships. Given that cladistic parameter is *equivalent* to natural classification and the study of natural classification is the central tenet of cladistics, we could say, with complete confidence, that:

The cladistic parameter *is* Cladistics – hence the title of this book.

Summary

Cladograms are branching diagrams that summarise current knowledge about organisms. There are different kinds: *Fundamental cladograms* represent information about taxon interrelationships based directly on evidence; *derivative cladograms* are classifications; *general cladograms* are summaries of all sources of data relative to the taxa at hand, relating to *consensus methods* but not necessarily the product of any particular consensus method ('the single most important principle of cladistics is that diverse fundamental cladograms may be combined to form a single general cladogram', Nelson 1979, p. 7).

Trees (*phylograms, phylogenetic trees, phyletic trees*) are not the same as cladograms and are of two sorts: (1) derived directly from a *fundamental cladogram* and interpreted as a phylogeny; and (2) derived from a *fundamental cladogram* but interpreted as a *set of logically possible phylogenetic trees* rather than just one.

Taxon and monophyly refer to the same thing, discovered by evidence. *Paraphyly* is a redundant term in taxonomy, *aphyly* is a more useful term referring to all non-monophyletic groups and is recommended here. The use of aphyly allows others to quickly recognise those taxa deemed appropriate for revision.

Homologues refer to the parts of organisms. Taxonomists refer to these parts as *characters* or *taxonomic characters*. Discovery (elucidation) of homologues is the most important part of comparative biology (taxonomy) and is the arena in which expertise is developed relative to any particular group of organisms; it is the empirical part of comparative biology (taxonomy). Homologues should not, however, be mistaken for homology. *Homology* is the relation implied by *homologues*. This is the primary study of cladistics (systematics), the vehicle with which to arrive at fundamental cladograms. It is succinctly summarised by the formula A (BC), meaning that B and C are more closely related to each other than to A.

Heterobathmy is the process of finding the correct level on the taxonomic hierarchy for every character.

The Cladistic Parameter is the constant in all of modern systematics, modern here meaning from the enlightenment onwards. It too is captured by the formula A (BC), but allows the following conclusions:

$$\text{Homology} \asymp \text{Monophyly} \asymp \text{Natural Classification} \asymp \text{Cladistic Parameter}$$

Thus, the task for taxonomists is clearly set out. It is of some import that the task has been the same since the beginning: study homologues (the parts), reveal relationships.

References

Abel, O. 1912. *Grundzüge der palaeobiologie der wirbeltiere*. E. Schweizerbart, Stuttgart.

Arber, EAN. & Parkin, J. 1907. On the origin of angiosperms. *Journal of the Linnean Society of London, Botany* 38: 29–80.

Benton, MJ. 2000. Stems, nodes, crown clades, and rank-free lists: is Linnaeus dead? *Biological Reviews* 75: 633–648.

Bessey, CE. 1915. The phylogenetic taxonomy of flowering plants. *Annals of the Missouri Botanical Garden* 2(1/2): 109–164.

Brower, AVZ. & Schawaroch, VA. 1996. Three steps of homology assessment. *Cladistics* 12: 265–272.

Candolle, de A-P. 1828. *Memoire sur la famine des Crassulacees*. Treuttel and Wurtz, Paris.

Crawford, RM., Hinz, F. & Gersonde, R. 2013. Reimer Simonsen (10 March 1931 – 9 July 2012). *Diatom Research* 28: 329–331.

Darwin, C. 1859. *On the Origin of Species by Means of Natural Selection, or, the Preservation of Favoured Races in the Struggle for Life*. John Murray, London.

Dayrat, B. 2005. Ancestor-descendant relationships and the reconstruction of the Tree of Life. *Paleobiology* 31: 347–353.

De Beer, G. 1954. *Archaeopteryx* and evolution. *Advances in Science* 11: 160–170.

Delacour, J. & Mayr, E. 1945. The family Anatidae. *Wilson Bulletin* 57: 3–55.

Dollo, L. 1895. Sur la phylogénie des Dipneustes. *Bulletin Sociéte belgique Geologie et Paléontologie et Hydrologie* 9: 79–128.

Ebach, MC. & McNamara, KJ. 2002. A systematic revision of the family Harpetidae (Trilobita). *Records of the Western Australian Museum* 21: 235–267.

Eimer, T. 1898. On species-formation, or the segregation of the chain of living organisms into species. *The Monist* 8: 97–122.

Fortey, RA. 1990. Ontogeny, hypostome attachment and trilobite classification. *Palaeontology* 33: 529–576.

Fortey, RA. 1997. Classification. In: Whittington, HB. (ed.) *Treatise on Invertebrate Paleontology, Part O: Arthropoda 1. Trilobita*. Rev. ed. Geological Society of America/ University of Kansas, Boulder, CO/ Lawrence, KS, pp. 289–302.

Gegenbaur, C. 1870. *Grundzüge der vergleichenden Anatomie. Zweite, umgearbeitete Auflage.*Wilhelm Engelmann, Leipzig.

Ghiselin, MT. 1972. Models in phylogeny. In: Schopf, TJM. (ed.), *Models in Paleobiology*. Freeman, Cooper, San Francisco, pp. 130–145.

Ghiselin, MT. 1976. The nomenclature of correspondence: a new look at 'homology' and 'analogy'. In: Masterton, RB., Hodos, W. & Jerison, H. (eds), *Evolution, Brain, and Behavior: Persistent Problems*. Lawrence Erlbaum Associates, Hillsdale, NJ, pp. 129–42.

Ghiselin, MT. 1984. "Definition," "Character," and other equivocal terms. *Systematic Zoology* 33: 104–110.

Ghiselin, MT. 2016. Homology, convergence and parallelism. *Philosophical Transaction of the Royal Society, B* 371: 20150035. http://dx.doi .org/10.1098/rstb.2015.0035

Harper, CW. Jr. & Platnick, NI. 1978. Phylogenetic and cladistic hypotheses: a debate. *Systematic Zoology* 27: 354–362.

Hawkins, JA., Hughes, CE. & Scotland, RW. 1997. Primary homology assessment, characters and character states. *Cladistics* 13: 275–283.

Hennig, W. 1936. Über einige Gesetzmassigkeiten der geographischen Variation in der Reptiliengattung Draco L.: "Parallele" und "konvergente" Rassenbildung. *Biologisches Zentralblatt* 56: 549–559.

Hennig, W. 1950. *Grundzüge einer Theorie der Phylogenetischen Systematik*. Deutscher Zentralverlag, Berlin.

Hennig, W. 1965. Phylogenetic systematics. *Annual Review of Entomology* 10: 97–116.

Hennig, W. 1966. *Phylogenetic Systematics*. University of Illinois Press, Urbana [reprinted 1979, 1999].

Hughes, NC. 2007. The evolution of trilobite body patterning. *Annual Review of Earth and Planetary Sciences* 35: 401–434.

Kumar, S. Stecher, G. & Tamura, K. 2015. MEGA7: Molecular Evolutionary Genetics Analysis Version 7.0 for Bigger Datasets. *Molecular Biology and Evolution* 33: 1870–1874.

Lam, H. 1957. What is a taxon? *Taxon* 6: 213–215.

Lankester, ER. 1870. On the use of the term homology in modern zoology, and the distinction between homogenetic and homoplastic agreements. *Annals and Magazine of Natural History* 6(4): 34–43.

Levkov, Z., Tofilovska, S. & Mitić-Kopanja, D. 2016. Species of the diatom genus *Craticula* Grunow (Bacillariophyta) from Macedonia. *Contributions, Section of Natural, Mathematical and Biotechnical Sciences (MASA)* 37: 129–165.

Lieberman, BS. & Karim, TS. 2010. Tracing the trilobite tree from the root to the tips: a model marriage of fossils and phylogeny. *Arthropod Structure and Development* 39(2–3): 111–123.

Mavrodiev, EV. 2019. Synapomorphies behind shared derived characters: Examples from the Great Apes' genomic data. *Acta Biotheoretica* doi:10.1007/s10441-019-09368-6.

Mayr, E., Linsley, G. & Usinger, RL. 1953. *Methods and Principles of Systematic Zoology.* McGraw-Hill, New York/London.

Milne, MJ. & Milne, LJ. 1939. Evolutionary trends in caddisworm case construction. *Annals of the Entomological Society of America* 32: 533–542.

Nelson, GJ. 1979. Cladistic analysis and synthesis: principles and definitions, with a historical note on Adanson's *Familles des Plantes. Systematic Zoology* 28: 1–21.

Nelson, GJ. 1985. Outgroups and ontogeny. *Cladistics* 1: 29–45.

Nelson, GJ. 2004. Cladistics: its arrested development. In: Williams, DM. & Forey, PL. (eds.), *Milestones in Systematics.* Taylor and Francis, London, pp. 127—147.

Nelson, GJ. 2011. Resemblance as evidence of ancestry. *Zootaxa* 2946: 137–141.

Nelson, GJ. 2016. What we all learned from Hennig. In: Williams, DM., Schmitt, M. & Wheeler, Q. (eds), *The Future of Phylogenetic Systematics: The Legacy of Willi Hennig.* Cambridge University Press, Cambridge, UK, pp. 200–212.

Nelson, GJ. & Platnick, NI. 1981. *Systematics and Biogeography: Cladistics and Vicariance.* Columbia University Press, New York.

Nierstrasz, HF. 1936. L'Evolution entre-croisée chez les crustacés. *Mémoires de l'Institut Royal des Sciences Naturelles de Belgique,* 2é sér. 3: 667–677.

Owen, R. 1843. *Lectures on the Comparative Anatomy and Physiology of the Invertebrate Animals, Delivered at the Royal College of Surgeons in 1843.* Longman, Brown, Green and Longman, London.

Owen, R. 1847. Report on the archetype and homologies of the vertebrate skeleton. In: *Report of the 16th Meeting of the British Association for the Advancement of Science.* Murray, London, pp. 169–340.

Owen, R. 1848. *On the Archetype and Homologies of the Vertebrate Skeleton.* John van Voorst, London.

Owen, R. 1849. *On the Nature of Limbs.* John van Voorst, London.

Owen, R. 1866. *On the Anatomy of Vertebrates,* Vol. 1: *Fishes and Reptiles.* Longmans, Green and Co., London.

Owen, R. 2007. *On the Nature of Limbs.* University of Chicago Press, Chicago, pp. 119 [Reprint of Owen 1849].

Patterson, C. 1977. The contribution of paleontology to teleostean phylogeny. In: Hecht, MK., Goody, PC. & Hecht, BM. (eds.) *Major Patterns in Vertebrate Evolution.* Plenum, New York, pp. 579–643.

Patterson, C. 1988. Homology in classical and molecular biology. *Molecular Biology and Evolution* 5: 603–625.

Pavlinov, I. 2013. *The Species Problem: Ongoing Issues.* Intech Open. www.intechopen.com/books/the-species-problem-ongoing-issues.

Pietsch, TW. 2012. *Trees of Life. A Visual History of Evolution.* Johns Hopkins University Press, Baltimore, MD.

Pinna de, M. 2014. Species tot sunt diversae quot diversas formas ab initio creavitiv – a dialogue on species. *Arquivos de Zoologia* 45 (esp.): 25–32.

Platnick, NI. 1977. Parallelism in phylogeny reconstruction. *Systematic Zoology* 26: 93–96.

Platnick, NI. & Dupérré, N. 2010. The goblin spider genus *Scaphiella* (Araneae, Oonopidae). *Bulletin of the American Museum of Natural History* 332: 1–156.

Plutynski, A. 2019. Speciation post synthesis: 1960–2000. *Journal of the History of Biology* 52: 569–596.

Prum, RO., Berv, JS., Dornburg, A., Field, DJ., Townsend, JP., Lemmon, EM. & Lemmon, AR. 2015. A comprehensive phylogeny of birds (Aves) using targeted next-generation DNA sequencing. *Nature* 526(7574): 569–573.

Rieppel, O. 2010. Othenio Abel (1875–1946) and "the phylogeny of the parts". *Cladistics* 29: 328–335.

Rieppel, O. 2011. The Gegenbaur transformation: a paradigm change in comparative biology. *Systematics and Biodiversity* 9: 177–190.

Rieppel, O. 2013. The series, the network, and the tree: changing metaphors of order in nature. *Biology & Philosophy* 25: 475–496.

Ronquist, F., Huelsenbeck, J. & Teslenko, M. 2011. *MrBayes version 3.2 Manual: Tutorials and Model Summaries* (November 15, 2011). http://mrbayes .sourceforge.net/mb3.2_manual.pdf

Schindewolf, OH. 1937. Beobachtungen und Gedanken zur Deszendenzlehre. *Acta Biotheoretica Ser. A* 3: 195–212.

Schindewolf, OH. 1993. *Basic Questions in Paleontology: Geologic Time, Organic Evolution, Biological Systematics.* University of Chicago Press, Chicago, IL.

Schuh, RT. 2000. *Biological Systematics: Principles and Applications.* Comstock Publishing Associates, Cornell University Press, Ithaca, NY.

Simonsen R. 1972. Ideas for a more natural system of the centric diatoms. *Beihefte zur Nova Hedwigia* 39: 37–54.

Simonsen, R. 1979. The diatom system: ideas on phylogeny. *Bacillaria* 2: 9–71.

Stamatakis, A. 2016. *The RAxML v8.2.X Manual.* Heidelberg Institute for Theoretical Studies (July 20, 2016). https://sco.h-its.org/exelixis/resource/ download/NewManual.pdf

Takhtajan, A. 1959. *Die Evolution der Angiospermen.* VEB Gustav Fischer, Jena.

Takhtajan, A. 1991. *Evolutionary Trends in Flowering Plants.* Columbia University Press, New York.

Theriot, EC., Cannone, JJ., Gutell, RR. & Alverson, AJ. 2009. The limits of nuclear-encoded SSU rDNA for resolving the diatom phylogeny. *European Journal of Phycology* 44: 277–290.

Thomas, EW., Stepanek, JG. & Kociolek, JP. 2016. Historical and current perspectives on the systematics of the 'enigmatic' diatom genus *Rhoicosphenia* (Bacillariophyta), with single and multi-molecular marker and morphological analyses and discussion on the monophyly of 'monoraphid' diatoms. *PLoS One* 11(4): e0152797. http://doi:10.1371/journal.pone .0152797

Tschulok, S. 1922. *Deszendenzlehre.* Gustav Fischer, Jena.

Wiley, EO. 1981. *Phylogenetics: The Theory and Practice of Phylogentic Systematics.* John Wiley & Sons, New York.

Wiley, EO. 2008. Homology, identity and transformation. In: Arriata, G., Schulze, H-P., Wilson, MVH. (eds), *Mesozoic*

Fishes 4 – Homology and Phylogeny. Verlag Dr. Friedrich Pfeil, Munich, pp. 9–21.

Wiley, EO. & Lieberman, BS. 2011. *Phylogenetics: Theory and Practice of Phylogenetics Systematics.* Wiley-Blackwell, Hoboken, NJ.

Wilkins, JS. 2009. *Species: A History of the Idea.* University of California Press, Berkeley.

Wilkins, JS. 2011. Philosophically speaking, how many species concepts are there? *Zootaxa* 2765: 58–60.

Wilkins, JS. 2018. *Species: A History of the Idea,* 2nd ed. CRC Press, Berkeley.

Williams, DM. 2007. Classification and diatom systematics: the past, the present and the future. In: Brodie, J. & Lewis, J. (eds), *Unravelling the Algae.* CRC Press, Berkeley, pp. 57–91.

Williams, DM. & Ebach, MC. 2008. *Foundations of Systematics and Biogeography.* Springer-Verlag New York Inc., New York.

Williams, DM. & Ebach, MC. 2017. What is intuitive taxonomic practice? *Systematic Biology* 66: 637–643.

Zimmermann, W. 1930. *Die Phylogenie der Pflanzen: Ein Überblick über Tatsachen und Probleme.* Gustav Fischer, Jena.

Further Reading

Nelson, G. 1979. Cladistic analysis and synthesis: principles and definitions, with a historical note on Adanson's *Familles des Plantes. Systematic Zoology* 28: 1–21.

We cannot recommend this paper enough. It is the first attempt to define cladistics as a field *in its own right* and the first to situate it in *the context of systematics and taxonomy* as opposed to the more specialised field of phylogenetics – although over 40 years old, it has lost none of its stimulating power. A similar version can be found in: Nelson, GJ. & Platnick, NI. 1981. *Systematics and Biogeography: Cladistics and Vicariance.* Columbia University Press, New York, xi+567 pp, in the section on 'Classification and General Cladograms' in Chapter 4: 'Systematic Results: Classifications' (p. 305 et seq.).

Also of interest is a series of papers on homology published in the journal *Cladistics*, which are worth examination, as some of the contributions (Nixon & Carpenter *contra* Brower & de Pinna) bring out differences of interpretation under the same analytical paradigm (that of Wagner parsimony), relative to our own viewpoint, which has no commitment to any particular methodology. We present the papers below in chronological sequence:

1. Brower, AVZ. 2011. The meaning of "phenetic". *Cladistics* 28: 113–114.

2. Nixon, KC. & Carpenter, JM. 2011. On homology. *Cladistics* 28: 160–169.

3. Williams, DM. & Ebach, MC. 2012. Confusing homologs as homologies: A reply to "On homology". *Cladistics* 28: 223–224.

4. Nixon, KC. & Carpenter, JM. 2012. More on homology. *Cladistics* 28: 225–226.

5. Williams, DM. & Ebach, MC. 2012. "Phenetics" and its application. *Cladistics* 28: 229–230.

6. Brower, AVZ. & de Pinna, MCC. 2012. Homology and errors. *Cladistics* 28: 529–538.

7.   Nixon, KC. & Carpenter, JM. 2012. More on errors. *Cladistics* 28: 539–544.

8.   Platnick, NI. 2013. Less on homology. *Cladistics* 29: 10–12.

9.   Platnick, NI. 2013. Reification, matrices, and the interrelationships of goblin spiders. *Zootaxa* 3608: 278–280.

10.   Ebach, MC. & Williams, DM. 2013. E quindi uscimmo a riveder le stelle. *Cladistics* 29: 227.

PART IV

How to Study Classification

Modern Artificial Methods and Raw Data

In Chapter 2 we noted some differences between *natural* and *artificial classifications*. To recap: *artificial classifications* are *created* or *imposed* and often constructed so that those who do not know a particular organism are able to identify it. *Natural classification* is about *discovery*; discovering something about the natural world (of which more later). The usual kinds of *artificial classifications* are keys and field guides (see Chapter 2), but here we extend the term to include classifications found by using any specific method, or any specific algorithm, or any specific kind of data, even a combination of the above. This may seem an extreme position to take, one that would eliminate all methods of analysis as having any merit. This is not what we are stating and we will expand on this below, but first we begin by considering 'sets' of *numerical* methods and discussing what we understand to be their underlying philosophy. We do not intend to discuss in detail the technical workings of all those methods. As we have already noted, we are not writing a cookbook.

Our use of the word 'sets' might disappoint or even upset some readers but we crave your indulgence in this matter as it allows us to make clearer our arguments for what we refer to as the *cladistic* approach to taxonomy. We are mindful of the fact that the use of such terms as *phenetics* and *cladistics* etc. are now frowned upon, because they come with conflicting definitions and some persons believe the words themselves serve no useful purpose. It is often said that in some quarters the term phenetics is thought to be a term of abuse; in other quarters cladistics is considered a term of abuse. Such interpretations are *political* rather than scientific. Here we hope to regain a scientific understanding for both phenetics *and* cladistics in an attempt to categorise how, in some areas, systematics has worked: our specific intent here is to outline a *cladistic* approach to taxonomy and as such these terms do serve a useful purpose in grouping sets of methods as well as bodies of thought. Again, we crave your indulgence in these matters so that we can develop our general argument.

Numerical Taxonomic Approaches As Modern Artificial Systems of Classification

Phenetics (Taximetrics, Rogers 1963)

Unweighted phenetics (= overall similarity): Phenetic taxonomy has been variously defined over the years but a relatively recent version is as follows: '. . . a system of classification based on the *overall similarity* of the organisms being classified' (Sokal 1986, p. 423, emphasis ours). The key phrase is 'overall similarity'.

The word 'phenetic' was first defined by Cain and Harrison to mean classification 'by overall similarity, based on all available characters without any weight' (Cain & Harrison 1960, p. 3). Interestingly, Sokal and Sneath, in what was the first book-length treatment of numerical taxonomy, wrote the following: 'For this meaning of similarity [based on the characters of the organisms] we have used the term 'affinity', which was in common use in pre-Darwinian times. We may also distinguish this sort of relationship from relationship by ancestry by calling it *phenetic relationship*, employing the convenient term of H.K. Pusey as used by Cain and Harrison (1960), to indicate that it is judged from the phenotype of the organism and not from its phylogeny' (Sokal & Sneath 1963, pp. 3–4).

One might imagine from that description an intuitively perfect approach to natural classification: abundance of data ('all available characters'), treating all data points as providing equal evidence ('without any weight'), and a means of finding taxa based on 'affinity' that will, as a consequence, find those organisms most similar to one another. The key phrase here is 'most similar to one another'. Later in their book, Sokal and Sneath suggested that 'We would remind the reader before proceeding that the terms 'resemblance', 'similarity', and 'affinity' are used interchangeably throughout this book, and that unless specifically qualified they imply a solely phenetic relationship' (Sokal & Sneath 1963, p. 123). Sokal and Sneath regarded the 'pre-Darwinian' use of the term 'affinity' to be the same as their use of similarity. That assumption may be problematic as the use of affinity in pre-Darwinian times is rather more complex than equating it simply with similarity alone (for a comprehensive discussion on affinity in pre-Darwinian times see, for example, Winsor 2015a–c; also see our summary in Chapter 7).

Nevertheless, phenetics, in its broadest sense, is the formation of taxa by measuring the number of observed shared similarities between taxa – the word 'measuring' is used here in the sense of one or another *clustering algorithm* grouping the sets of similarities. An early example of the kind of results obtained is given in Figure 8.1, presented in the form of a branching diagram (after Sokal & Sneath 1963, p. 199, fig. 7–8).

As time passed, many different clustering techniques were invented, many yielding the same solutions, many differing. The question became: What

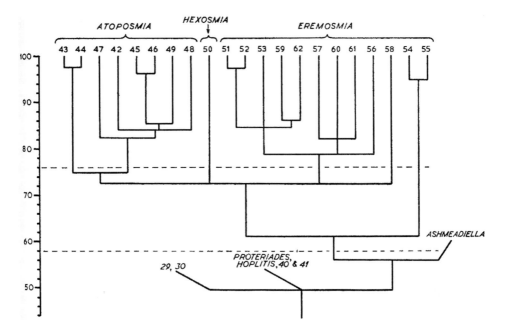

Figure 8.1 Phenogram (after Sokal & Sneath 1963, p. 199, fig. 7–8, with permission).

clustering technique was best? How could we judge that? What should I use? There was no easy answer to these questions and the dilemma of choice may have accounted for the popular view that phenetics died under the weight of its own ingenuity. Viewed differently, it seems that the results of each clustering technique were a reflection, in part, of the method rather than a reflection of the data themselves, hence this is why we refer to these sets of methods as *artificial*. We need to remind readers that when we note a method is *artificial*, it is not necessarily to denigrate it. As Candolle wrote many years ago with respect to classification, it is *purpose* that counts: what is it for? The problem for phenetics was that it was considered to be the method for a general classification, that was its stated purpose.

Somewhat ironically, in our view, phenetics did *not* die but flourished, albeit in a version that began with its creators, supporters and practitioners expressing out-right hostility towards the entire phenetic enterprise: *quantitative phyletics.*

Weighted phenetics (parsimony) ('quantitative phyletics', Kluge & Farris 1969; 'overall similarity of synapomorphy' Nelson 2004, p. 139): We dare say that its proponents would disagree with us that what was first called 'quantitative phy-letics' (= Wagner parsimony) is a version of phenetics, as it is considered by many to be *the* method of cladistics. Quantitative phyletics, which is the same as our *weighted phenetics* or, by its more popular name, Wagner parsimony, began with

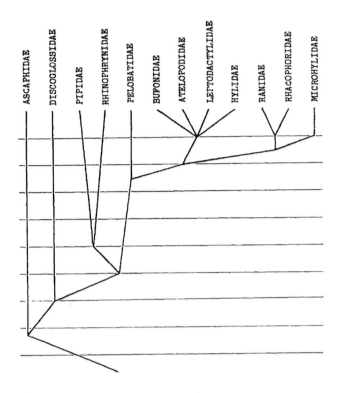

Figure 8.2 Wagner parsimony tree (after Kluge & Farris 1969, p. 14, fig. 5, with permission).

Kluge and Farris (1969)[1]: 'We believe it is worth while to develop still another taxonomic methodology, incorporating the *precision of numerical techniques and the power of evolutionary inference*. We refer to this hybrid methodology as quantitative phyletic taxonomy' (Kluge & Farris 1969, p. 1, our emphasis).

'The precision of numerical techniques' simply meant creating an algorithm so that a computer program could do the sorting (or clustering) of characters to arrive at a 'tree' of taxa (an early example of the results is Kluge & Farris 1969, p. 14, their figure 5, reproduced as our Figure 8.2). But the 'power of evolutionary inference' turned out to be one or another scheme of weighting, disguised, so to speak, as an evolutionary model, a model of the assumed process of change. Rather than dissect each weighting scheme, or each model (but more on that later), it is clear that the data matrix used in any parsimony analysis is seen as essential. The matrix

---

[1] We are aware of the many different beginnings of parsimony methods but here we concentrate on *Wagner parsimony* because it is so often thought to be synonymous with cladistics. We are also mindful of the general use of parsimony, rather than the application of a specific method. Maybe, in the end, all methods utilise some form of parsimony.

Table 8.1 Matrix of four taxa A–D (column 1) and one character (column 2)

A	0
B	0
C	1
D	1

is composed of various scores for various characters, as in any phenetic analysis. The grouping is achieved by weighting some characters more than others. This in itself is not a bad thing – however, it is the clustering algorithm (e.g., Wagner parsimony) that determines the usefulness or otherwise, leading to the view that this method is best characterised as another version of overall similarity, the 'overall similarity of synapomorphy' (Nelson 2004, p. 139). Hence, this is why we refer to these sets of methods as both *phenetic* and *artificial* as well.

Why so? Consider a single binary character distributed among four taxa, A–D. It can be represented in the usual kind of data matrix entry, with one column for data (the zeros and ones) and four rows for taxa, A–D (Table 8.1).

A conventional interpretation of this matrix would simply be that taxon A and B have the pleisomorphic value (as represented by the zero) and C and D have the apomorphic value (as represented by the one). Conventional interpretation would also state that these data relate two taxa more closely to each other than they are to the other two. That is, we have evidence to support the relationship AB(CD) based on the apomorphic value. In other words, this means that C and D are more closely related to each other than either are to A or B. None of this is, or should be, controversial. Every taxonomist is aware of such characters – it is our stock in trade. But it does raise a question: What does conventional really mean? In this sense, conventional simply means after having studied a number of specimens, one is able to identify a defining feature, such that prior to having made any observations relevant to the four taxa A–D, their interrelationships were unknown but now, after a single relevant observation, they can be recognised as related thus: AB(CD). It does not matter what related means at this point – we return to that later. Thus, in this conventional sense 'apomorphic', 'evidence' and 'defining feature' are equivalent terms. But it does not mean it is *true*, in an absolute sense. For further investigation might differ from the view that AB(CD) is the correct relationships for these taxa. Again, none of this should be controversial as it reflects the day-to-day activities of many taxonomists prior to any numerical intervention: we have four known taxa A–D – how might we classify them?

Table 8.2 Matrix of four taxa A–D (column 1) and one character (column 2) with an outgroup added

OUT	0
A	0
B	0
C	1
D	1

When considering *methods*, rather than evidence, to process the matrix above using, for example, a parsimony program, an additional item is required. If the matrix is analysed as above, it would yield no result (a bush A–D) as no distinction would be made concerning the relative evidential value of the zero and the one, and so the character is rendered uninformative, recognising two equal, but conflicting, pairs: AB:0 and CD:1. With respect to *implementation* then, the matrix requires the addition of an *outgroup* such that it becomes as in Table 8.2.

It should be clear that the outgroup is understood as a simple heuristic device to allow 'pleisomorphic' (zero) and 'apomorphic' (one) values to be recognised for what they are – to allow what we have observed (apomorphy) to remain true in the matrix. Again, it should be noted that here 'apomorphic' is equivalent to 'evidence' and 'defining feature'. With the addition of this outgroup, parsimony analysis of the above matrix should find the one result: AB(CD). The data entered into the matrix – AB(CD) – is the same as that found by its analysis. This should not be controversial either.

Controversy begins when some consider these data to be uninformative until *after* analysis rather than before. This is sometimes expressed in terms of 'types' of homology, with data identified prior to analysis referred to as *primary homology*, those after analysis *secondary homology* (de Pinna 1991, Brower & Schawaroch 1996, who elaborate on de Pinna's scheme, see also Hawkins et al. 1997 and Hawkins 2000; see Box 8.1).

We noted above that the single character AB(CD) can be rendered *a priori* uninformative by simply recognising two equal but conflicting pairs, that both the zero value and the one value are *potential* evidence, thus AB:0 and CD:1. This relates to the notion of unrooted trees discussed elsewhere (see Box 3.4).

In this case, all possible permutations might be considered equally relevant at the outset, all equally possible: that is (AB)(CD), (AB)CD, as well as AB(CD), are all potentially supported. Surprisingly, this issue *does* appear controversial, at least to 'modern cladists', the term used by Nixon and Carpenter below:

> Either state might turn out to be a plesiomorphy or synapomorphy on a particular tree, and *as implemented in all modern cladistic analysis*, it is unnecessary to know which state will turn out to group taxa apomorphically and which will not (or, depending on the root, they might both be synapomorphies of collateral groups). (Nixon & Carpenter 2012a, p. 162, our emphasis)

> Only cladistic analysis (e.g. parsimony) can determine where optimized character changes occur and whether the absence may be symplesiomorphic, synapomorphic, or have multiple origins on a given tree. (Nixon & Carpenter 2012b, p. 542)

> Since that revelation, we no longer "group by synapomorphy" nor construct trees by hand; instead, we read synapomorphies from trees that minimize homoplasy over all character states, be they zeros or ones, A, C, G or T. (Nixon & Carpenter 2012c, p. 225)

'Either state [zero or one] might turn out to be a plesiomorphy or synapomorphy on a particular tree'; 'Only … parsimony … can determine where optimized character changes occur'; '… we no longer "group by synapomorphy"' – in other words, rather than allowing the data to speak, the program will do so. Although not stated quite so baldly, other methods, whatever they are, which adopt the matrix as a standard would more or less follow this logic.

Imagine the simple matrix above in Table 8.2 with more data and becoming a matrix of 10 characters, some differing in their implied relationships. What then? Imagine that matrix becoming even larger, say 1000 entries, or larger still with 100 000 entries. What then? Here method is of no help if it is influencing the outcome – no matter how big the dataset, the solution (the result) is likely to contain artefactual relationships – or at least, we would not be able to distinguish the artefactual from the real.

The mechanics of parsimony allows the data to be *distorted* to fit the solution in its search for the shortest cladogram (we discuss this 'distortion' in detail in Chapter 12). For now it should be clear that Wagner parsimony = 'quantitative phyletics' sensu Kluge and Farris = weighted phenetics. Following the logic of artificial classifications, its purpose is to find the shortest tree – an odd purpose in the broader sense of biology.

Box 8.1 Primary Homology and Secondary Homology: More Redundant Concepts

One of the most widely read papers on homology is that by de Pinna (1991[1]), which occupies a special place in the development of cladistics. Among the many topics de Pinna tackled was the idea that there are two stages to the understanding of homology, discussed in terms of the *generation* and *legitimation* of homology propositions:

> I refer to the stages of generation and legitimation of homology propositions as being 'primary' and 'secondary' homology, respectively. This terminology seems more appropriate than the others employed so far, because it emphasizes

Box 8.1 *(cont.)*

that the two stages are interdependent and complementary, and that any homology hypothesis is necessarily tied to both, at least potentially. Also, it indicates that one of the two stages (primary) precedes the other in any analysis (de Pinna, 1991, p. 373)

Primary homology precedes secondary homology because:

A primary homology statement is conjectural, based on similarity, and reflects the expectation that there is a correspondence of parts that can be detected by an observed match of similarities (de Pinna 1991, p. 373)

In contrast, secondary homology is:

... the outcome of a pattern-detecting analysis ... and its search represents a test of the expectation that any observable match of similarities is potentially part of a retrievable regularity indicative of a general pattern. (de Pinna 1991, p. 373)

Here the assumption is that any initial judgements of homology may not be correct because any particular judgement might be overturned by the analysis of other characters. That is, after analysis *some* primary homologies fit the resulting cladogram (= homology) while *others* do not (= homoplasy). We explored this elsewhere using an example of a series of data matrices, the largest being composed of 18 characters (Table B8.1.1, Matrix Total).

With respect to smaller matrices, for matrix a (Table B8.1.1), characters 1–2 support the relationship AB(CD), and thus a parsimony analysis would yield the cladogram AB(CD), with both characters understood as homologous and no character rendered homoplasious (Table B8.1.1, matrix a). For matrix b (Table B8.1.1), characters 1–2 support the relationship AB(CD), while character 3 supports AC(BD). Parsimony analysis yields the cladogram AB(CD), with homoplasy in the third character (Table B8.1.1, matrix b): character 3 fits in either two places, separately for B and D, or its absence from C might be considered a 'reversal'[2]. Note that whenever the solution is not exact for all characters, they still are *made to fit*. For matrix c (Table B8.1.1), characters 1–4 yield two cladograms A(C(BD)) and A(B(CD)) with an uninformative consensus (Table B8.1.1, matrix c).

These results raise two contrasting issues:

(1)   With primary homology, there is no definitive statement that might be made about a characters' homology until such times as analysis shows it to be so – but when do we reach that outcome? One answer might be that this is the nature of science – we never quite know what we have.

Consider the following (see Chapter 3 for more details). If we examine the usual kinds of matrices again (like those in Table B8.1.1), we note that instead of interpreting the entries as putative homologies (relationships) based on similarities, these are *homologues* (the parts) based on our observations – and homologues are not directly equivalent to homology. Thus, the usual kind of data matrix is simply a chart, or table, if you will, of homologues and any subsequent analysis is merely the shifting and sorting of homologues to find a 'tree-of-best-fit'. Homology (relationship) is never really considered. We conclude, then, that de Pinna's *primary homology* is equivalent to the traditional

Box 8.1 (*cont.*)

Table B8.1.1 A series of data matrices for four taxa, A–D. Matrix Total is composed of 18 characters, 12 supporting the relationship AB(CD), 6 supporting the relationship AC(BD); Matrix a is composed of 2 characters, both supporting the relationship AB(CD); Matrix b is composed of 3 characters, 2 supporting the relationship AB(CD), one supporting the relationship AC(BD); Matrix c is composed of 4 characters, 2 supporting the relationship AB(CD), 2 supporting the relationship AC(BD)

Matrix Total

	1	2	3	4	5	6	7	8	9	10	11	12	13	14	15	16	17	18
A	0	0	0	0	0	0	0	0	0	0	0	0	0	0	0	0	0	0
B	0	0	1	1	1	1	1	1	0	0	0	0	0	0	0	0	0	0
C	1	1	0	0	0	0	0	0	1	1	1	1	1	1	1	1	1	1
D	1	1	1	1	1	1	1	1	1	1	1	1	1	1	1	1	1	1

Matrix a

	1	2
A	0	0
B	0	0
C	1	1
D	1	1

Matrix b

	1	2	3
A	0	0	0
B	0	0	1
C	1	1	0
D	1	1	1

Matrix c

	1	2	3	4
A	0	0	0	0
B	0	0	1	1
C	1	1	0	0
D	1	1	1	1

Box 8.1 *(cont.)*

understanding (e.g., Owen) of *homologue* and that *secondary homology* is an unnecessary concept created by the idea that *analysis of characters* is necessary for their ultimate understanding (our argument is similar to that made by Grand, save some of the vocabulary (Grand, 2013); see also Williams 2004).

Interestingly, Nixon and Carpenter also rejected the use of primary and secondary homology: 'The use of the terminology "primary homology" and "secondary homology" is unfortunate, and we instead use "hypothesis of homology" and "homology" in their places' (Nixon & Carpenter 2011, p. 168). But this is little improvement as it implies de Pinna's 'primary homology' is simply a 'hypothesis of homology'; secondary homology is the substantiated 'primary homology'. It still confuses homologues with homology.

The alternative is:

(2)   That we have spent a great deal of time studying our organisms, so we can have some confidence in our judgements of what is and is not a character, analysis is somewhat irrelevant. It would be perverse to spend so much time assessing characters by detailed observation on specimens to then allow a program (of whatever kind) to sort out the useful from the useless in terms of their fit to a tree.

We believe (2) is the most appropriate for systematics and taxonomy, for organism biologists. In fact, we believe that is what is usually done. Option (1) is really the use of phenetic data interpreted via a model (parsimony, or whatever alternative is implemented).

This view necessarily leads into discussion of synapomorphy and symplesiomorphy and whether synapomorphy is equivalent to homology or homologue. We have dealt with this elsewhere (Chapter 7), but here we note Richter's (2017) recent contribution on the subject. He drew attention to an earlier comment of ours on the difference between homologue and homology (Williams & Ebach 2012), suggesting we thought there was 'semantic difficulty in applying theses [sic] terms' (Richter 2017, p. 1, footnote 1). Instead, we were stating that *homologue* and *homology* were conceptually different, as did Richard Owen (Owen 1849). For us, a homologue is *a part of an organism*, something that might be examined and studied, often called the same thing (Owen's 'namesakes'), while homology *is the relation determined by those comparisons*. One might represent this as: homologue: A + B; homology: CD(AB).

For Richter 'the actual question is whether synapomorphy and homologue are synonymous' (Richter, 2017, p. 1, footnote 1). At present, we find nothing wrong with that equation, only our wish to reinforce the idea that if synapomorphies and homologues are synonymous, then they both refer to *material parts of organisms*, things that can be studied (Nelson 2011). Does this imply two kinds of homologues: synapomorphic homologues and sympleisomorphic homologues? Homologues are not just parts – they are the parts that *imply* relationships; each synapomorphy implies a relationship; symplesiomorphy does not: the phrase 'sympleisomorphic homologues' is, then, an oxymoron. That one might still understand there to be synapomorphic homologues and sympleisomorphic homologues is a remnant of yet another aspect of phenetics, coupled with the view that such things must be determined (or inferred) from some (or every) computer program starting with a matrix of similarities. Ours is a slight departure from the view that *homology* is equivalent to synapomorphy.

Box 8.1 *(cont.)*

With further reflection, treating homologues and homology in the way we suggest yields the sought after marriage between comparative biology (the study of homologues, how organisms are *constructed*) and taxonomy (the study of homology, how organisms are *related*), something Owen, and countless others, grappled with on numerous occasions. It remains to be seen if the differences we understand between homologues and homology translate into something useful for molecular data. This is an open question – but we expect it will.

What is required for taxonomic studies is the sorting of 'good' characters from 'bad' characters. By 'good' and 'bad' we mean those characters we understand properly.

In summary: Primary homology is not a statement of relationship but an assessment of what homologues (parts) we are dealing with. Homology statements, derived from those homologues, are the relationship of the organisms. Homoplasy is only necessary when attempting to account for *all* similarities and as such has been co-opted into the analytical terminology – it is merely a term to account for all mismatches between dataset and tree. It might be worthwhile pointing out that de Pinna's paper appears to be a response to Patterson (1982), the latter focused on a compatibility approach to understanding homology, the former – de Pinna – a defence of parsimony (i.e., Wagner parsimony).

Lest we are misunderstood: Primary homology, secondary homology, homoplasy and the data matrix are all *redundant concepts* that belong to the phenetic era of taxonomy. All that is required are homologues (an understanding of the parts of organisms) and homology (an understanding of relationships) – from that, knowledge emerges concerning both the organisms and how they are constructed, and their relationships to one another, and to the rest of life.

References

de Pinna, MCC. 1991. Concepts and tests of homology in the cladistic paradigm. *Cladistics* 7: 367–394.

Grand, A. 2013. Représentation sémantique des phénotypes: métamodèle et ontologies pour les caractères taxonomiques et phylogénétiques. Unpublished Ph.D. thesis, MNHN Paris.

Nelson, G. 2011. Resemblance as evidence of ancestry. *Zootaxa* 2946: 137–141.

Nixon, KC. & Carpenter, JM. 2011. On homology. *Cladistics* 28: 160–169.

Owen, R. 1849. *On the Nature of Limbs.* John van Voorst, London.

Patterson, C. 1982. Morphological characters and homology. In: Joysey, KA. & Friday, AE. (eds), *Problems of Phylogenetic Reconstruction*. Academic Press, London, pp. 21–74.

Richter, S. 2017. Homology and synapomorphy – symplesiomorphy – neither synonymous nor equivalent but different perspectives on the same phenomenon. *Cladistics* 33: 540–544.

Williams, DM. 2004. Homologues and homology, phenetics and cladistics: 150 years of progress. In: Williams, DM. & Forey, PL. (eds) *Milestones in Systematics*. CRC Press, Florida, pp. 191–224.

Williams, DM. & Ebach, MC. 2012. Confusing homologs as homologies: a reply to "On homology". *Cladistics* 28: 223–224.

[1] Google scholar reports over 1000 citations since publication, a remarkable number given that it is largely a theoretical paper.

[2] We have placed 'reversal' in quotes as it is a form of data distortion; for more details see Chapter 12.

Table 8.3 Matrix of four taxa A–D (column 1) and one character (column 2) with outgroup added and the implied relationship of the one character specified (column 4)

OUT	0		
A	0		
B	0	=	AB(CD)
C	1		
D	1		

Weighted Phenetics (Compatibility)

The generalities of compatibility analysis are discussed more fully in Chapter 9. Briefly, compatibility can be understood as a kind of parsimony method that deals with the data in a different way to Wagner parsimony. In short, compatibility analysis views data such that they are constrained to consist of only 'unique' derivations of the apomorphic state. That is, the binary character A–D will be constrained as much as possible to reflect only the relationship AB(CD), with the view that state 1 is uniquely derived for C and D (Table 8.3).

With relationships specified as 'unique origins', the analytical process is simply to find the most characters compatible with one another. For AB(CD), for example, A(BCD) is absolutely compatible, yielding A(B(CD)) – other characters, like B (ACD), for example, are partially compatible, such that AB(CD) and B(ACD) have the CD group in common. (This partial compatibility turns out to be useful. We return to it in Chapter 10.) D(ABC), however, conflicts with AB(CD) in an absolute sense. While we see the compatibility approach as an improvement over the use of the conventional binary matrix, it is likely to underestimate potential relationships as conflict is likely to be greater when characters are viewed this way.

This method might also be understood as employing an evolutionary model, one that uses unique derivations and no 'reversals' (an example of the results of a compatibility analysis is presented in our Figure 8.3, after Meacham 1980, figs 16–17).

Weighted Phenetics (Phylogeny – Model Methods)

... our results imply that model selection may be unnecessary when one is interested in inferring ancestral sequences or *in revealing the cladistic relationships [= classification]* among genes and organisms.

(Abadi et al. 2019, p. 8, our emphasis)

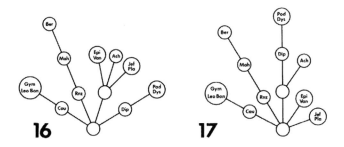

Figure 8.3 Compatibility trees (after Meacham 1980, figs 16–17, with permission).

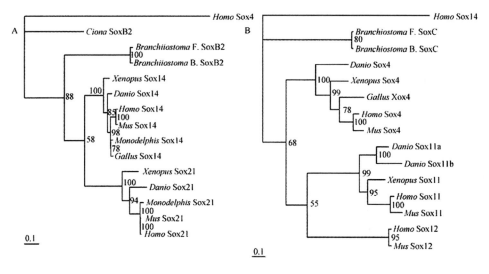

Figure 8.4 Modelled tree (after Lin et al. 2009, fig. 2, with permission). 'A, Maximum Likelihood trees of the entire sequence SoxB2 . . . B, Maximum Likelihood trees of the entire sequence SoxC . . .'.

We can understand all current model-based systems as really *kinds of weighting schemes*, favouring certain kinds of 'change' over other kinds of change with respect to particular characters. For example, examination of the instruction manual for MrBayes reveals descriptions of numerous models of change for different kinds of DNA data (Ronquist et al. 2011, 'MrBayes is a program for Bayesian inference and model choice across a wide range of phylogenetic and evolutionary models', http://mrbayes.sourceforge.net/). It should be clear that these models are implemented by ever complex weighting schemes to take account of all the various forms of change deemed possible (an example of results from these kinds of analyses is captured in our Figure 8.4, after Lin et al. 2009, figure 2).

Crucial to this kind of implementation, as for those above, is to view matrix characters as representing linear series of changing states such that 0 →1 would be the *interpretation* of data in Tables 8.1–8.3 (for further commentary see Chapter 7 and Box 7.2).

We closed our piece on phenetics above with these words: As time passed numerous different clustering techniques were invented, many yielding the same solutions, many differing. Questions began to emerge: What clustering technique was best? What should we use? Should we use many different methods or is one better than another? And so it is with *phylogeny – model methods*: What model is best? What weighting scheme should I apply? Should we use many different methods or is one better than another? Or is one better at one time, and another at another time? A general solution is not obvious.

And so journals fill up with yet more and more methods, with more and more subtle changes to existing methods (see any issue of *Systematic Biology*, for example) – yielding more methods than can never be successfully evaluated in any meaningful way: to us, this represents the *second wave of phenetics* – and has yielded a plethora of new, but artificial, methods of character analysis and the resulting taxonomy[2].

Assessment of Artificial Methods of Classification

We began by suggesting that modern versions of *artificial classifications* are the results presented when using any particular method, or any particular algorithm, or any particular kind of data, or any combination of the above. Our reasons are simple: for any and every method utilised (or, more accurately, invented), for any and every data source collected, one has to make the assumption that the method or data source (or their combination) are appropriate and accurate enough to discover 'true' relationships. For the moment, it does not matter what 'true' means

---

[2] Most methods that include models were originally designed to deal with the then new DNA sequence data. Since that time, the modelling approach has been applied to morphological data (e.g., Wright & Hillis 2014). Discussion of these matters is now current (e.g., Schrago et al. 2018) – we have not dwelt on these issues but offer the following two quotes from Goloboff et al.: 'The homogeneous Markov model assumes (when applied to morphology) that the characters are like units that can simply switch into one or another state at any point in the tree' (Goloboff et al. 2018, p. 625). That is, once again, characters are viewed as sequences of 'units'. The second quote is of more significance: 'But the very fact that taxonomists beginning to investigate a group first need to learn the relevant characters speaks against the idea that all groups can be classified by looking at the same sets of characters randomly changing over all the tree; that is why someone who has worked extensively on spider morphology needs to learn a whole new suite of anatomical characters if starting now to work on, say, beetles' (Goloboff et al. 2018, p. 625). We need add no more.

only that the method or data source (or some combination) discovers something that reflects the natural world rather than something that is simply an artefact, or includes artefacts, of the method applied.

To be sure, over the last few decades, a veritable avalanche of metrics and measures have been invented with the intention of establishing some degree of 'certainty' or 'confidence' in the results obtained but, in truth, no measure or metric could possibly exist that can establish that the nodes (relationships) of any resulting cladogram are actually real relationships – all these metrics and measures can do is confirm that the particular method in question works well enough according to some specified criteria rather than confirm that it actually discovers something about the natural world. Therefore, we see all these methods as producing *artificial classifications*. Of course, any method or any data source (or any combination) may indeed find 'true' relationships, but how could that be established? How can one sort the real from the artefactual?

Two options have been suggested to support a method's veracity, both very different: (1) simulation studies and (2) philosophical claims.

A simulation study applies a particular method to some artificial dataset generated according to a model of evolution. One might judge how well that method performs relative to that model. One might even judge how well that method performs relative to many models (e.g., Kuhner & Felsenstein 1994). Yet, overall, if the method works well (by finding the expected result), it simply confirms that the particular method in question works well enough (or not well enough) according to that particular model – that it is a *successful artificial method*. It still cannot inform us whether the results are accurate reflections of the real world rather than artefact. One of the earliest simulation studies put it like this: 'The main limitation of simulation studies is that, taken by themselves, *they cannot indicate how methods will perform in the real world*' (Hulsenbeck 1995, p. 17, our emphasis). This is our view– quite simply there is no possible way to judge how methods work in this respect[3].

If it is accepted that ultimate justification cannot be found via simulation, no matter how sophisticated that is, one might instead attempt to justify a method (or data source) by appealing to philosophy, a tactic that has been utilised by invoking a variety of different philosophers and their viewpoints, Karl Popper being a

---

[3] As Nixon and Carpenter wrote in a footnote addressing the issue of the *True Tree*: '. . . whether simulated . . . or merely assumed . . . these elusive/illusory commodities loom large in certain circles as arbiters of goodness for phylogenetic methodology' (Nixon & Carpenter 2000, p. 311). We agree.

particular favourite of systematists (Rieppel 2008)[4]. Yet this approach to justification is also flawed, as one might pick and choose a philosopher or philosophy and defend it as the most significant or most representative or whatever, but that justification cannot reach beyond the empirical.[5]

Summary

Our views on methods might seem unusually pessimistic – nihilistic even – as it may appear that we are suggesting each and every method, each and every data source, is flawed and there is little we can do to discover the flaws with respect to documenting the natural world. But this would be an inaccurate understanding of our position: we do not mean to imply that these various efforts do not produce something useful, accurate even – accurate in the sense of reflecting something about the real world, as we have enough faith in the cumulative achievements of the taxonomy of our forefathers, pre-Darwinian and after, that in aggregate, seem fairly much in agreement.

We note two things here of importance. All methods described above have in common the requirement of a *data matrix*, this is what we refer to as its *phenetic aspect*; all methods described above employ some form of *weighting*, however that weighting is determined. On the optimistic side, if readers inspect the results in all four figures illustrating this chapter (Figures 8.1–8.4), they will note that they have one thing in common: every diagram is a *fundamental cladogram*, every diagram has a *cladistic parameter*.

We commented above, in the weighted phenetics section, on the approach known as *compatibility analysis*, or *clique analysis*. It might be unfair to include this approach under the general name of phenetics[6] as it has, or appears to have, a different approach to characters such that instead of the usual linear approach, it treats them hierarchically[7]:

---

[4] The literature on Popper and systematics is extensive and complex. Perhaps the best we can offer is to refer interested readers to Ward Wheeler's summary, which begins: 'It is hard to overstate the influence of Popper on practising scientists today, especially in systematics' (Wheeler, 2012, p. 71). As Rieppel noted: 'As the case of the "cladistic revolution" demonstrates, scientists who turn to philosophy in defense of a research program read philosophers with an agenda in mind. That agenda is likely to distort the philosophical picture, as happened to Popper's philosophy of science at the hands of cladists' (Rieppel 2008, p. 316). If we might be blunt: we are tempted to recommend avoiding the subject.

[5] Again, interested readers might like to explore the many papers of Kirk Fitzhugh or Olivier Rieppel, to name two prominent contributors to this philosophical literature in systematics. It would be difficult to suggest any one paper typical for either of these people so, *caveat lector*, any of their philosophical contributions would suffice. If we might again be blunt: we recommend avoiding the whole subject.

[6] At least under its interpretation by Patterson (1982).

[7] '... characters are *already* hypotheses of evolutionary relationships' (Meacham & Estabrook 1985, p. 431, abstract, their emphasis).

$$A \rightarrow B \rightarrow C$$

vs.

A(BC)

We have noted elsewhere that compatibility analysis has been reinvented or rediscovered on numerous occasions, not least among the suite of methods that deal with consensus techniques. In the following chapter we discuss some generalities of consensus methods. Before we discuss our understanding of the *cladistic* approach to taxonomy (natural classification), we will briefly examine some of the issues in a pivotal debate on systematics, which, to some extent, still lingers in various forms (for example, see the recent exchange between Simões et al. 2017a, 2017b and Laing et al. 2017 – albeit this time their debate is shrouded in the context of 'Big Data'): the 'total evidence' versus 'consensus' debate (Chapter 11).

References

Abadi, S., Azouri, D., Pupko, T. & Mayrose, I. 2019. Model selection may not be a mandatory step for phylogeny reconstruction. *Nature Communications* 10(934). https://doi.org/10.1038/s41467-019-08822-w

Brower, AVZ. & Schawaroch, V. 1996. Three steps of homology assessment. *Cladistics* 12: 265–272.

Cain, AJ. & Harrison, GA. 1960. Phyletic weighting. *Proceedings of the Zoological Society of London* 135: 1–31.

de Pinna, MCC. 1991. Concepts and tests of homology in the cladistic paradigm. *Cladistics* 7: 367–394.

Goloboff, PA., Torres, A. & Arias, JS. 2018. Parsimony and model-based phylogenetic methods for morphological data: a comment on O'Reilly et al. *Palaeontology* 61: 625–663. https://doi.org/10. 1111/pala.12353

Hawkins, J. 2000. A survey of primary homology assessment: different botanists perceive and define characters in different ways. In: Scotland, RW. & Pennington, RT. (eds), *Homology & Systematics: Coding*

Characters for Phylogenetic Analysis. Taylor and Francis, London, pp. 22–53.

Hawkins, JA., Hughes, CE. & Scotland, RW. 1997. Primary homology, characters and character states. *Cladistics* 13: 275–283.

Hulsenbeck, JP. 1995. Performance of phylogenetic methods in simulation. *Systematic Biology* 44: 17–48.

Kluge, AG. & Farris, JS. 1969. Quantitative phyletics and the evolution of Anurans. *Systematic Zoology* 18: 1–32.

Kuhner, MK. & Felsenstein, J. 1994. A simulation comparison of phylogeny algorithms under equal and unequal evolutionary rates. *Molecular Biology and Evolution* 11: 459–468.

Laing, AM., Doyle, S., Gold, MEL., Nesbitt, SJ., O'Leary, MA., Turner, AH., Wilberg, EW. & Poole, KE. 2017. Giant taxon-character matrices: the future of morphological systematics. *Cladistics* 34: 333–335.

Lin, YS., Chen, DY., Fan, QS. & Zhang, HW. 2009. Characterization of SoxB2 and SoxC genes in Amphioxus (*Branchiostoma belcheri*): implications for their evolutionary conservation. *Science in China, Series C: Life Sciences* 52: 813–822.

Meacham, CA. 1980. Phylogeny of the Berberidaceae with an evaluation of classifications. *Systematic Botany* 5: 149–172.

Meacham, CA. & Estabrook, GF. 1985. Compatibility methods in systematics. *Annual Review of Ecology and Systematics* 16: 431–446.

Nelson, GJ. 2004. Cladistics: its arrested development. In: Williams, DM. & Forey, PL. (eds) *Milestones in Systematics*. CRC Press, Florida, pp. 127–147.

Nixon, KC. & Carpenter, JM. 2000. On the other "Phylogenetic Systematics". *Cladistics* 16: 298–318.

Nixon, KC. & Carpenter, JM. 2012a. On homology. *Cladistics* 28: 160–169.

Nixon, KC. & Carpenter, JM. 2012b. More on errors. *Cladistics* 28: 539–544.

Nixon, KC. & Carpenter, JM. 2012c. More on homology. *Cladistics* 28: 225–226.

Patterson, C. 1982. Morphological characters and homology. In: Joysey, KA. & Friday, AE. (eds), *Problems of Phylogenetic Reconstruction*. Academic Press, London, pp. 21–74.

Rieppel, O. 2008. Re-writing Popper's philosophy of science for systematics. *History and Philosophy of the Life Sciences* 30: 293–316.

Rogers, DJ. 1963. Taximetrics – new name, old concept. *Brittonia* 15: 285–290.

Ronquist, F., Huelsenbeck, J. & Teslenko, M. 2011. *MrBayes version 3.2 Manual: Tutorials and Model Summaries* (November 15, 2011). http://mrbayes .sourceforge.net/mb3.2_manual.pdf

Schrago, CG., Aguiar, BO. & Mello, B. 2018. Comparative evaluation of maximum parsimony and Bayesian phylogenetic reconstruction using empirical morphological data. *Journal of Evolutionary Biology* 31: 1477–1484.

Simões, TR., Caldwell, MW., Palci, A. & Nydam, RL. 2017a. Giant taxon-character matrices: quality of character constructions remains critical regardless of size. *Cladistics* 33: 198–219.

Simões, TR., Caldwell, MW., Palci, A. & Nydam, RL. 2017b. Giant taxon-character matrices II: a response to Laing et al. (2017). *Cladistics* 34: 702–707.

Sokal, RR. 1986. Phenetic taxonomy: theory and methods. *Annual Review of Ecology and Systematics* 17: 423–442.

Sokal, RR. & Sneath, PHA. 1963. *Principles of Numerical Taxonomy*. W.H. Freeman, San Francisco.

Wheeler, WC. 2012. *Systematics: A Course of Lectures*. Wiley & Co, Chichester.

Williams, DM. & Ebach, MC. 2005. Drowning by numbers: rereading Nelson's "Nullius in Verba". *Botanical Review* 71: 415–477.

Williams, DM., Ebach, MC. & Wheeler, QD. 2010. Beyond belief. In: Williams, DM. & Knapp, S. (eds), *Beyond Cladistics*. University of California Press, Berkeley, pp. 169–197.

Winsor, MP. 2015a. Considering affinity: an ethereal conversation (part one of three). *Endeavour* 39: 69–79.

Winsor, MP. 2015b. Considering affinity: an ethereal conversation (part two of three). *Endeavour* 39: 116–126.

Winsor, MP. 2015c. Considering affinity: an ethereal conversation (part three of three). *Endeavour* 39: 179–187.

Wright, AM. & Hillis, DM. 2014. Bayesian analysis using a simple likelihood model outperforms parsimony for estimation of phylogeny from discrete morphological data. *PLoS One* 9: e109210.

Further Reading

Readers who are interested in 'point-and-click' guidance should seek other sources in addition to this book. In all honesty, we cannot recommend any of the methods discussed by these authors as they are all approaches to artificial classification rather than natural classification, the latter being the subject we are interested in. We are, therefore, a little reluctant to suggest any single book (and there are very many), but perhaps those below are at least representative of what to expect. In any case, we suggest reading them with a critical eye – if so, it should not take too long to come to the conclusion the subject, as conceived via these contributions, is 'drowning in number' (Williams & Ebach 2005).

Felsenstein, J. 2004. *Inferring Phylogenies.* Sinauer Associates, Sunderland, MA.

Although comprehensive, now a remarkable 15 years old, it lacks some of the developments of the past decade. As far as we are aware, Felsenstein is not planning any revised version.

Hall, B. 2017. *Phylogenetic Trees Made Easy: A How-To Manual.* 5th ed. Oxford University Press, Oxford.

Each edition was reviewed extensively with mixed reception. We suggest reading a few reviews then proceed with caution.

As a further note of caution, neither of these two books discuss classification (or taxonomy) in the sense we explore in this book.

Phenetics

The best we can offer on this subject is Sokal and Sneath's two books. These are now quite old but both worth dipping into for a comprehensive view of how phenetics began, why it began, what it hoped to achieve, how it developed and flourished under the 'numerical taxonomy' umbrella, and – in part – its apparent slide into 'phylogenetics'. Above we suggested that Felsenstein's *Inferring Phylogenies* should be considered the third in this series of books, a view we still hold (Williams et al. 2010). None of these books is without value, if read critically – even if that value is to understand how the first wave of numerical

taxonomy lost its way, and, more crucially, how the second wave persists in misleading others.

Steussy's book has a more up-to-date summary of phenetics.

Sneath, PHA. & Sokal, RR. 1973. *Numerical Taxonomy.* Freeman, San Francisco.

Sokal, RR. & Sneath, PHA. 1963. *Principles of Numerical Taxonomy.* W.H. Freeman, San Francisco.

Steussy, T. 2010. *Plant Taxonomy: The Systematic Evaluation of Comparative Data.* Columbia University Press, New York.

Weighted Phenetics (Parsimony)

There are several books that focus on parsimony, as understood and implemented by the Wagner Parsimony

algorithm (e.g., Farris 1983), such as *Biological Systematics: Principles and Applications*, which is the best available

(Schuh and Brower 2009). We have listed all editions below – the 3rd edition is due in 2021 (Brower and Schuh 2021). Another recent contribution is Caetano-Anollés et al. (2018).

There are a few books that follow Willi Hennig's original version of *Phylogenetic Systematics* more closely, and differ considerably from what has been called 'modern cladistics' (by Nixon and Carpenter, 2012a, for example). It is perhaps incorrect (and maybe inaccurate) to include them here, but we do not want these books to disappear from sight. These authors are critical of 'modern cladistics' as discussed above under 'weighted phenetics' (parsimony). It is worth noting, somewhat inexplicably, that neither Mikoleit nor Wiesemüller et al. have been translated into English. They both deserve an English translation.

Wagner Parsimony

Brower, A. & Shuh, T. 2021. *Biological Systematics: Principles and Applications.* Cornell University Press, Ithaca, NY.

Caetano-Anollés, G., Nasir, A., Kim, KM. & Caetano-Anollés, D. 2018. Rooting phylogenies and the tree of life while minimizing ad hoc and auxiliary assumptions. *Evolutionary Bioinformatics* 14: 1–21.

Farris, JS. 1983. The logical basis of phylogenetic analysis. In: Platnick, NI. & Funk, VA. (eds), *Advances in Cladistics II.* Columbia University Press, New York, pp. 7–36.

Schuh, T. 1999. *Biological Systematics: Principles and Applications.* Comstock Publishing Associates, New York.

Schuh, T. & Brower, A. 2009. *Biological Systematics: Principles and Applications.* Comstock Publishing Associates, New York.

Hennig

Borkent, A. 2018. The state of phylogenetic analysis: narrow visions and simple answers – examples from the Diptera (flies). *Zootaxa* 4374(1): 107–143.

Disney, RHL. 2003. Is not Hennig's method of producing cladograms as defensible as those derived from parsimony algorithms? *Bonner zoologische Beiträge* 50: 305–311.

Hennig, W. 1966. *Phylogenetic Systematics.* University of Illinois Press, Champaign [reprinted in 1979 and 1999 with a new Preface].

Kraus, O. 1998. Elucidating the historical process of phylogeny: Phylogenetic Systematics versus cladistic techniques. In: Selden, PA. (ed.), *Proceedings of the 17th European Colloquium of Arachnology, Edinburgh 1997.* British Arachnological Society, Burnham Beeches, pp. 1–7.

Mikoleit, G. 2004. *Phylogenetische Systematik der Wirbeltiere.* Dr. Friedrich Pfeil, Munich.

Wägele, W. 2001. *Grundlagen der Phylogenetischen Systematik.* Dr. Friedrich Pfeil, Munich.

Wägele, W. 2005. *Foundations of Phylogenetic Systematics.* Dr. Friedrich Pfeil, Munich.

Wiesemüller, B., Rothe, H. & Hencke, W. 2003. *Phylogenetische Systematik: Eine Einführung.* Springer, Berlin, Heidelberg.

Weighted Phenetics (Compatibility)

Sadly, the development of compatibility methods was truncated after it was ferociously (and somewhat unfairly) attacked in the mid-1980s, mostly in the pages of *Taxon* (primarily between 1984 and 1986). Many empirical studies were published in *Systematic Botany*, and it is worth perusing back issues of that journal for examples. From the vast literature on the subject, we recommend the papers by Le Quesne (1979), Meacham (1980) and the summary in Felsenstein (1982, pp. 389–393), as they are all, if a little dated, clearly written. The Meacham & Estabrook review (1985) is also a bit dated but still worth a glance, as is Scotland & Steel (2015), for its recent discussion, and Williams & Ebach (2017), who generalise the issue. Patterson explored compatibility in terms of homology testing (1982, 1988).

It is worth noting that when reading papers on compatibility one must distinguish between the method of analysis and some of the ideas its proponents expressed about how to and what to classify. For example, their discussions concerning the necessity of 'convex groups' in classification is really a defence of paraphyly and promoted as support for what was referred to as 'traditional evolutionary classification' (Meacham & Duncan 1987, for commentary see Wiley 1981 and 2009). In this they were mistaken (see Chapters 3–5) but it should not detract from the general usefulness of the method.

Felsenstein, J. 1982. Numerical methods for inferring evolutionary trees. *Quarterly Review of Biology* 57: 379–404.

Le Quesne, WJ. 1979. Compatibility analysis and the uniquely derived character concept. *Systematic Zoology* 28: 92–94.

Meacham, CA. 1980. Phylogeny of the Berberidaceae with an evaluation of classifications. *Systematic Botany* 5: 149–172.

Meacham, CA. & Duncan, T. 1987. The necessity of convex groups in biological classification. *Systematic Botany* 12: 78–90.

Meacham, CA. & Estabrook, GF. 1985. Compatibility methods in systematics. *Annual Review of Ecology and Systematics* 16: 431–446.

Patterson, C. 1982. Morphological characters and homology. In: Joysey, KA. & Friday, AE. (eds), *Problems of Phylogenetic Reconstruction*. London: Academic Press, pp. 21–74.

Patterson, C. 1988. Homology in classical and molecular biology. *Molecular Biology and Evolution* 5: 603–625.

Scotland, RW. & Steel, M. 2015. Circumstances in which parsimony but not compatibility will be provably misleading. *Systematic Biology* 64: 492–504.

Williams, DM. & Ebach, MC. 2017. What is intuitive taxonomic practice? *Systematic Biology* 66: 637–643.

Wiley, EO. 1981. Convex groups and consistent classifications. *Systematic Botany* 6: 346–358.

Wiley, EO. 2009. Patrocladistics, nothing new. *Taxon* 58: 2–6.

Weighted Phenetics (Phylogeny – 'Model' Methods)

We have already mentioned the books by Felsenstein and Hall. We could add Baum and Smiths's *Tree Thinking: An*

Introduction to Phylogenetic Biology (Baum & Smith 2012) and Ward Wheeler's *Systematics: A Course of Lectures* (Wheeler

2012). All four of these books are quite different from one another but each has extended discussions on the modelling approach and some delve into theory here and there – but *caveat lector*: these books primarily focus on phylogeny to the exclusion of classification. There are some philosophical contributions such as Sober (1988 and 2015, but see Brower 2017).

There are many highly technical (that is mathematical) books available, and it is beyond our ability (and stamina!) to make recommendations from this veritable mountain of literature. These are just a selection (we have resisted noting any of the numerous books on *phylogenomics*, see Box 10.3 for our brief comments on that topic). Interested readers who feel the urge to explore phylogenetic modelling might dip into any of the following:

Baum, DA. & Smith, SD. 2012. *Tree Thinking: An Introduction to Phylogenetic Biology*. Roberts, Greenwood Village, CO.

Brower, AVZ. 2017. "Parsimony be damned!". *Cladistics* 33: 667–670.

Cadette, MW. & Davies, TJ. 2016. *Phylogenies in Ecology: A Guide to Concepts and Methods*. Princeton University Press, Princeton, NJ.

Huson, DH., Regula, R. & Scornavacca, C. 2010. *Phylogenetic Networks*. Cambridge University Press, Cambridge, UK.

Lemey, P. 2009. *The Phylogenetic Handbook*. 2nd ed. Cambridge University Press, Cambridge, UK.

Semple, C. & Steel, M. 2003. *Phylogenetics*. Oxford University Press, Oxford.

Sober, E. 1988. *Reconstructing the Past: Parsimony, Evolution and Inference*. MIT Press, Cambridge, MA.

Sober, E. 2015. *Ockham's Razors: A User's Manual*. Cambridge University Press, Cambridge, UK.

Steel, M. 2016. *Phylogeny: Discrete and Random Processes in Evolution*. Society for Industrial and Applied Mathematics, Philadelphia.

Warnow, T. 2017. *Computational Phylogenetics: An Introduction to Designing Methods for Phylogeny Estimation*. Cambridge University Press, Cambridge, UK.

Wheeler, W. 2012. *Systematics: A Course of Lectures*. Wiley-Blackwell, Chichester. Books of this kind are numerous, reading them is exhausted only by your purse, your enthusiasm or your patience – whichever is the greater. We are compelled to note that, from our perspective, there is not much to recommend here for the taxonomist. Our advice: spend your money on a field trip!

9

How to Study Classification: Consensus Techniques and General Classifications

In 1972, Edward N. Adams III published what might be the first paper on consensus techniques for use in biological classification. He addressed the following question: '... can we combine the information from rival classifications into a new, hopefully more accurate classification? Such a consensus of the rivals is useful both in tree comparison and tree discovery' (Adams 1972, p. 390). Since Adams' paper, nearly half a century ago, numerous consensus tree techniques have been proposed, numerous critiques of each have been published and an almost infinite number of suggestions have been made as to how to use one of them, some of them, any of them, all of them, or none of them (Bininda-Emonds 2004a, 2004b). Alongside this avalanche of technical detail are discussions concerning *supertrees* (which are a form of consensus analysis) and *supermatrices*, the latter being an extension of the 'combining data' debate (Sanderson et al. 1998, see the following Chapter 10). Again, as with the methods of data analysis described in Chapter 8, we do not intend to discuss each and every consensus technique in detail but deal with what we understand to be the basic issues (on the details of consensus methods, we make some suggestions in the Further Reading section below).

The world of consensus techniques, like the world of phylogeny reconstruction, has developed several almost impenetrable jargon-filled vocabularies. We have tried to use just one vocabulary in this book: the component terminology *sensu* Nelson and Platnick (1981). For us, this is the most appropriate as we have no desire to bog readers down in mathematical jargon when it can be better explained in plain language (other comments on mathematical jargon can be found in the opening paragraphs of Chapter 7 and Box 7.2[1]).

---

[1] Obviously we could be accused of preferring our own jargon. It might be useful, then, while reviewing the details of this chapter to refer to Chapter 3 for brief summaries of the terminology we have adopted in this book and to Chapter 7 for more detailed discussion of that terminology.

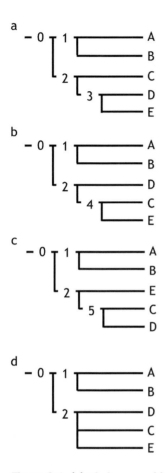

Figure 9.1 (a) Cladogram of five taxa, A–E, with four components: ABCDE (0), AB (1), CDE (2) and DE (3). (b) As (a) but with one difference: component 4 (D+E). (c) Cladogram with four components, but instead of either component 3 or 4, it has component 5 (C + D). (d) An optimal solution of (AB)(CDE).

Cladograms and Components: Every cladogram consists of a set of *components* (branch points) and *terms* (terminal taxa are occasionally referred to as *tips*). A branch point (i.e., a node) is defined by its inclusive taxa. For example, the cladogram in Figure 9.1a is a simple example of five taxa, A–E, with four components: ABCDE (0), AB (1), CDE (2) and DE (3). The cladogram in Figure 9.1b is similar to that in Figure 9.1a but with one difference: component 3 (D+E) is not included but instead new component 4 (C+E): ABCDE (0), AB (1), CDE (2) and CE (4). The cladogram in Figure 9.1c also has four components, but instead of either component 3 or 4, it has component 5 (C + D).

For all three cladograms (Figures 9.1a–c) component 0 can be considered *trivial* or *uninformative* as it simply describes the contents of the entire cladogram (A–E),

Table 9.1 Three cladograms from Figures 9.1a–c and their respective components (1–5)

	(AB)(C(DE)) (Figure 9.1a)	(AB)(D(CE)) (Figure 9.1b)	(AB)E(CD)) (Figure 9.1c)
Component			
0	ABCDE	ABCDE	ABCDE
1	AB	AB	AB
2	CDE	CDE	CDE
3	DE		
4		CE	
5			CD

which remains true even if there was no further resolution. Components 1–5 describe all the groups found in all three cladograms and are informative in that respect (Table 9.1).

In Figure 9.1a–c, excluding component 0 (the trivial component), each cladogram has component 1(A+B) and component 2 (C+D+E) (Table 9.1). The three cladograms differ with respect to the interrelationships of CDE (Figure 9.1a–c). As the individual components 3, 4 and 5 conflict, no decision on the resolution of CDE can be made. In this case, to find a 'hopefully more accurate classification' *sensu* Adams (above), application of most consensus techniques would yield the same optimal solution: (AB)(CDE) (Figure 9.1d). But a fundamental issue can now be identified: how is conflict resolved?

Component analysis, then, is a way of decomposing any cladogram (fundamental or derivative, see Chapter 7) into their constituent parts. In one sense the components above are incomplete: for Figure 9.1, components 1–3 describe three groups, AB (1), CDE (2) and DE (3) (Figure 9.1a); components 1, 2 and 4 describe three groups, AB (1), CDE (2) and CE (4) (Figure 9.1b); and components 5–7 describe three groups, AB (1), CDE (2) and CD (5) (Figure 9.1c). The component representation is *incomplete* because it describes *the contents of the terminal group only*. If they were to describe the data more fully to include the *relationship* then that would involve more than just the terminal group, as one would have to answer the question: what is the group related to? Implicitly, then, component 1 of Figure 9.1a describes the relationship CDE(AB), component 2 describes the relationship (CDE)AB and component 3 describes the relationship ABC(DE). The meaning should be clear: for component 1, A and B are more closely related to each other than they are to either C, D and/or E. This is important – we return to it later in Chapter 10.

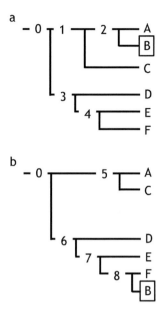

Figure 9.2 Two fundamental cladograms, both (a and b) with six taxa (A–F), each having an entirely different set of components, differing only in the placement of taxon B.

Methods of cladogram analysis (consensus) usually differ in their approach to the question of conflict. It should be noted here that conflict between a suite of fundamental cladograms and conflict between a set of characters share similar problems. We discuss this issue for characters in Chapter 10.

It is not difficult, however, to find examples where different consensus techniques yield different solutions. For example, in Figure 9.2a and b, there are two fundamental cladograms each with six identical taxa (A–F). Each has an entirely different set of components, differing only in the placement of taxon B (Table 9.2).

A *strict consensus solution*[2], for example, would be an unresolved polychotomy (polytomy) (Figure 9.3a) as there are no shared nodes common to the two cladograms. Using alternative consensus techniques, for example Adams' original suggestion, a more or less fully resolved solution can be obtained merely by treating errant taxon B as a *rogue*[3], and thus leaving the rogue as an unresolved basal taxon (unplaced) in the optimal solution (Figure 9.3b). In any case, it can be easily appreciated from a simple visual inspection of the two cladograms in

---

[2] A method that includes only nodes (components) common to all cladograms.

[3] The notion of rogue taxa was first suggested by Wilkinson (1996, p. 439) for these kinds of circumstances.

Table 9.2 Component information from the two cladograms in Figure 9.2a and b and that in Figure 9.3c

Cladogram 1: ((AB)C)(D(EF)) (Figure 9.2a = Figure 9.5a)	Cladogram 2: ((AC)(D(E(FB)))) (Figure 9.2b = Figure 9.5b)	Cladogram 3: ((AC)(DE(FB))) (Figure 9.2c)
Component 0 = ABCDEF	Component 0 = ABCDEF	Component 0 = ABCDEF
Component 1 = ABC		
Component 2 = AB		
Component 3 = DEF		
Component 4 = EF		
	Component 5 = AC	Component 5 = AC
	Component 6 = BDEF	Component 6 = BDEF
	Component 7 = BEF	
	Component 8 = BF	Component 8 = BF

Figure 9.2a and b that if taxon B was removed, then the Adams consensus solution is the obvious one (Figure 9.4a–c). This identifies a second fundamental issue: what to do with taxon B and the rogues? Surprisingly, this has elicited such a vast volume of response it is hard to make sense of it all beyond recognising that technical solutions to finding the 'correct' or 'best' placement of taxon B are numerous, their advantages and disadvantages documented and discussed in detail (Bininda-Emonds 2004a) – not unlike the current literature documenting methods for analysing raw data (Chapters 8 and 10).

Once again, as with methods of data analysis, one might find justification in any one of the available consensus techniques, but this simply makes the problem a technical one, solved only by the ingenuity of those focused on that task. The example in Figure 9.2 demonstrates the major shortcomings of a technique-driven approach: two fundamental cladograms are in fact insufficient – a third, at least, is required. Suppose the third cladogram acquired is as in Figure 9.5c. The solution now becomes obvious: the cladogram in Figure 9.5b (=Figure 9.2b). This is an *empirical* solution not a *technical* solution and is to be preferred. It also highlights the view that taxa are only 'rogue' inasmuch as their position is not determined by the cladograms (or data) to hand.

Supertrees and supermatrices: The language of *supertrees* is usually couched in that of mathematics and phylogenetics, but its definition is more or less

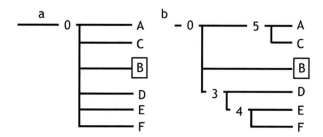

Figure 9.3 (a) A *strict consensus solution* with an unresolved polychotomy (polytomy); (b) Adams's solution with a more or less fully resolved solution 'isolating' the *rogue* taxon B.

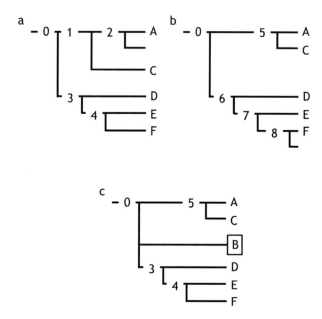

Figure 9.4 Explanation of the Adams consensus by omitting taxon B from (a) and (b), with (c) being the result.

the same as that of consensus trees in relation to classification, as above. For example:

> Supertree construction [can be defined] as the generation of one or more output trees (the supertrees) from a set of source trees that possess fully or partially overlapping sets of taxa. Because the source trees need only overlap – minimally each source tree must share at least two taxa with the rest of the source trees – the supertree can be more inclusive than any individual source contributing to it. The supertree ordinarily contains all taxa found in the source trees. Our definition distinguishes between supertree and consensus techniques, the latter of which we hold to combine fully overlapping source trees ... We recognize that this distinction is arbitrary in that many consensus techniques can be adapted for a supertree (Bininda-Emonds et al. 2002, p. 266)

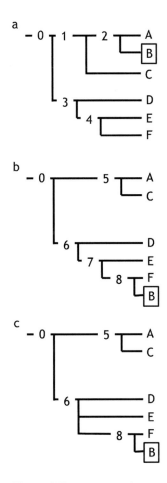

Figure 9.5 Figure 9.2 demonstrates the shortcomings of a *technique-driven* approach: two fundamental cladograms are insufficient – a third, at least, is required. The third is part (c). The solution is part (b) (=Figure 9.2b), an *empirical* rather than *technical* solution.

The primary difference, then, between consensus trees and supertrees is the latter does not need all of its fundamental cladograms to have identical taxa, as long as there is some overlap. An example taken from Gordon (1986) is reproduced as our Figure 9.6a: there are two cladograms, one with terminals 1–9, the second with terminals 1–5 of the first cladogram and a further five, 10–14. Terminals 1–5 are common to both cladograms, terminals 6–9 occur only in the first cladogram, terminals 10–14 occur only in the second cladogram. As there is an overlap (terminals 1–5), it is still possible to combine them into one summary cladogram. As above, any set of fundamental

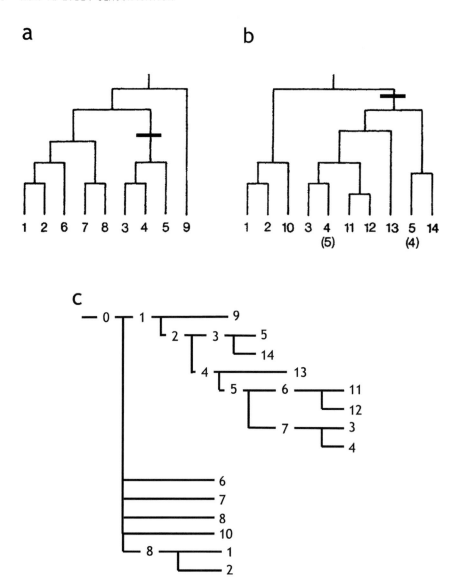

Figure 9.6 Example from Gordon (1986, figure 1, with permission): two cladograms, (a) with terminals 1–9; (b) with terminals 1–5 of the first cladogram and an alternative five novel terminal, 10–14; (c) consensus cladogram from analysis of all 16 terminals using Wagner parsimony (consensus of five most parsimonious cladograms). The black bar on parts a and b indicates the relationship 3(4, 5). (See text for further details.)

cladograms[4] can be decomposed into their components and added together following some criteria.

While not all supertree construction is achieved in the same way, here we focus only on the coding of such cladograms via conventional matrices as this is useful for the points we wish to make later.

For supertree (and consensus) construction each component in the fundamental cladograms can be coded as a binary matrix character with the resulting pooled data analysed using Wagner parsimony or some similar program. This coding procedure was originally called *Matrix Representation with Parsimony* (MRP, Ragan 1992; Baum 1992). Since its original description there have been a number of variants. These need not concern us here.

MRP for the cladograms in Figure 9.1 would be as follows:

Figure 9.1a: Component 0 includes all five taxa, A–E, hence all are coded with a 1; component 1 includes only A and B, hence only those two are coded with a 1; component 2 includes three taxa, C, D and E, which are coded with a 1; and component 3 includes only two taxa, C and E, only those are coded with a 1. An all-zero outgroup is added so that the score of 1 is read (initially) as the 'grouping state'. The final matrix will look like this:

	(AB)(C(DE)) Figure 9.1a			
	0	**1**	**2**	**3**
OUT	0	0	0	0
A	1	1	0	0
B	1	1	0	0
C	1	0	1	0
D	1	0	1	1
E	1	0	1	1

The matrix, when analysed using Wagner parsimony, unsurprisingly yields one cladogram, (AB)(C(DE)), equivalent to the original.

---

[4] Most contributions to supertree terminology refer to any set of original cladograms used for finding the supertree as *source trees* (e.g., Bininda-Emonds et al. 2002). These need not be fundamental cladograms (i.e., based on original data) but can, for example, be derived from a classification. We have chosen to use only the term *fundamental cladogram* here for *all* of the original cladograms. Strictly speaking, the term *derivative cladogram* (see Chapter 7) could be applied to cladograms derived from a classification but to keep matters simple we use just one term here. Terminology to one side, it is the *set of relationships* that is important.

Figure 9.1a-c: If this coding procedure is applied to all three cladograms in Figure 9.1, then it results in a matrix containing all the component information of all three cladograms:

	(AB)(C(DE)) Figure 9.1a				(AB)(D(CE)) Figure 9.1b				(AB)E(CD)) Figure 9.1c			
	0	1	2	3	0	1	2	4	0	1	2	5
OUT	0	0	0	0	0	0	0	0	0	0	0	0
A	1	1	0	0	1	1	0	0	1	1	0	0
B	1	1	0	0	1	1	0	0	1	1	0	0
C	1	0	1	0	1	0	1	1	1	0	1	1
D	1	0	1	1	1	0	1	0	1	0	1	1
E	1	0	1	1	1	0	1	1	1	0	1	0

The resulting matrix has a total of 12 binary variables. Nine are replicated in all three cladograms and uninformative (components 0, 1 and 2), only three are informative (components 3, 4 and 5). The matrix might be represented differently with the weight assigned to the coded component (bottom row, *w*):

	0	**1**	**2**	**3**	**4**	**5**
OUT	0	0	0	0	0	0
A	1	1	0	0	0	0
B	1	1	0	0	0	0
C	1	0	1	1	1	0
D	1	0	1	0	1	1
E	1	0	1	1	0	1
w	3	3	3	1	1	1

When either matrix, weighted or unweighted, is processed using Wagner parsimony, for example, it yields three cladograms of which the consensus is: (AB)(CDE) (Figure 9.1d).

MRP for the cladograms in Figure 9.2a and b is as follows:

Figure 9.2a–b (=Figure 9.4a–b): Yields a matrix of 10 binary variables of which only 1 is replicated, the uninformative component 0.

	Figure 9.2a = Figure 9.5a					Figure 9.2b = Figure 9.5b				
	0	1	2	3	4	0	5	6	7	8
OUT	0	0	0	0	0	0	0	0	0	0
A	1	1	1	0	0	1	1	0	0	0
B	1	1	1	0	0	1	0	1	1	1
C	1	1	0	0	0	1	1	0	0	0
D	1	0	0	1	0	1	0	1	0	0
E	1	0	0	1	1	1	0	1	1	0
F	1	0	0	1	1	1	0	1	1	1

Parsimony analysis of this matrix yields six cladograms; the strict consensus of the six is unresolved. MRP for the cladograms in Figure 9.5a and b is as above.

Figure 9.6a and b: The coding procedure is as above but with nodes not included in the original cladograms coded as missing (?). The cladogram in Figure 9.6a is:

Cladogram 1 (Gordon 1986, fig. 1a = Figure 9.6a)								
	Nodes							
	1	2	3	4	5	6	7	8
Terminals								
1	1	1	0	1	0	0	1	1
2	1	1	0	1	0	0	1	1
3	0	0	0	0	1	1	1	1
4	0	0	0	0	1	1	1	1
5	0	0	0	0	0	1	1	1
6	0	1	0	1	0	0	1	1
7	0	0	1	1	0	0	1	1
8	0	0	1	1	0	0	1	1
9	0	0	0	0	0	0	0	1

The cladogram in Figure 9.6b is:

Cladogram 2 (Gordon 1986, fig. 1b = Figure 9.6b)								
Nodes								
	1	**2**	**3**	**4**	**5**	**6**	**7**	**8**
Terminals								
1	1	1	0	0	0	0	0	0
2	1	1	0	0	0	0	0	0
3	0	0	1	0	1	1	0	1
4	0	0	1	0	1	1	0	1
5	0	0	0	0	0	0	1	1
10	0	1	0	0	0	0	0	0
11	0	0	0	1	1	1	0	1
12	0	0	0	1	1	1	0	1
13	0	0	0	0	0	1	0	1
14	0	0	0	0	0	0	1	1

The combined matrix for Figures 9.6a and b is:

Cladogram 1 (Gordon 1986, fig. 1a = Figure 9.6a)								+	Cladogram 2 (Gordon 1986, fig. 1b = Figure 9.6b)							
	1	**2**	**3**	**4**	**5**	**6**	**7**	**8**		**1**	**2**	**3**	**4**	**5**	**6**	**7**	**8**
T.									T.							
1	1	1	0	1	0	0	1	1	1	1	1	0	0	0	0	0	0
2	1	1	0	1	0	0	1	1	2	1	1	0	0	0	0	0	0
3	0	0	0	0	1	1	1	1	3	0	0	1	0	1	1	0	1
4	0	0	0	0	1	1	1	1	4	0	0	1	0	1	1	0	1
5	0	0	0	0	0	1	1	1	5	0	0	0	0	0	0	1	1
6	0	1	0	1	0	0	1	1		?	?	?	?	?	?	?	?
7	0	0	1	1	0	0	1	1		?	?	?	?	?	?	?	?
8	0	0	1	1	0	0	1	1		?	?	?	?	?	?	?	?

(*cont.*)

	Cladogram 1 (Gordon 1986, fig. 1a = Figure 9.6a)								+	Cladogram 2 (Gordon 1986, fig. 1b = Figure 9.6b)							
	1	2	3	4	5	6	7	8		1	2	3	4	5	6	7	8
9	0	0	0	0	0	0	0	1		?	?	?	?	?	?	?	?
10	?	?	?	?	?	?	?	?		0	1	0	0	0	0	0	0
11	?	?	?	?	?	?	?	?		0	0	0	1	1	1	0	1
12	?	?	?	?	?	?	?	?		0	0	0	1	1	1	0	1
13	?	?	?	?	?	?	?	?		0	0	0	0	0	1	0	1
14	?	?	?	?	?	?	?	?		0	0	0	0	0	0	1	1

The resulting matrix of 16 binary variables, when analysed using Wagner parsimony, yields five cladograms of which the strict consensus is 6,7,8,10 (1,2) ((9((5, 14)(13(3,4)(11,12)))))) (Figure 9.6c). Inspection of the components in the solution, when compared to the two fundamental cladograms, yields the following:

	Figure 9.6c	Figure 9.6a	Figure 9.6b
Component 0	**1–14**	**1–14**	**1–14**
Component 1	3–5, 9, 11–14		
Component 2	3–5, 11–14		
Component 3	5, 14		
Component 4	3–4, 11–13		
Component 5	3,4, 11,12		3,4, 11,12
Component 6	**11, 12**		**11, 12**
Component 7	**3,4**	**3,4**	**3,4**
Component 8	**1,2**	**1,2**	**1,2**

Of the eight components found in the solution, only three are found in the two fundamental cladograms (components 6–8, excluding component 0) (Figure 9.7). The significance of this is largely due to the missing taxa in the fundamental cladograms, but other parts of the cladogram are found in both:

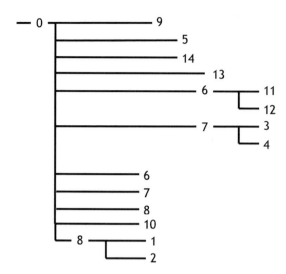

Figure 9.7 Solution for example in Figure 9.6. Of the eight components found only three are in the three fundamental cladograms (6, 7, and 8).

for example, the *relationship* 3(4, 5) (Figure 9.6a). One might argue, then, that to find the optimal solution requires refining the methodology so that we might include more information in the resulting cladogram. Mickevich and Platnick, for example, noted

> . . . for a consensus estimate, the best fit criterion requires us to choose a method which combines all, rather than some, of the taxonomic statements found in common; hence, the Adams consensus can be preferred over a Nelson consensus. (Mickevich & Platnick 1989, p. 46)

Given that one might search each fundamental cladogram and list the *relationships* ('taxonomic statements') instead of the components, the solution might again be sought in the data rather than the method(s) (Chapter 11).

Supermatrices need not detain us long as they are relatively straightforward. As can be seen from Table 9.3a (after de Queiroz & Gatesy 2006, figure 1), separate matrices derived from different studies are simply added together to form one larger matrix which is then analysed using one or another program. Differences arise when results of the analysis of a supermatrix differ from the analysis of cladograms (supertrees). In the de Queiroz and Gatesy (2006, figure 1) example, Dataset #1 yields the cladogram E(B(A(CD); Dataset #2 yields the cladogram E(A(B (CD); Dataset #3 yields the cladogram E(AB)(CD) (Table 9.3a). The supermatrix of Dataset #1 + Dataset #2 + Dataset #3 yields the cladogram E(D(C(AB) (Table 9.3b). The strict consensus of cladograms E(B(A(CD), E(A(B(CD) and E(AB)(CD) yields E (AB(CD)).

Table 9.3a Cladogram coding for three cladograms after de Queiroz & Gatesy (2006, figure 1)

	Dataset #1: E(B(A(CD))						Dataset #2: E(A(B(CD))							Dataset #3: E(AB)(CD)				
E	0	0	0	0	0	0	0	0	0	0	0	0	0	0	0	0	0	0
A	1	1	1	1	1	0	1	1	1	0	0	0	0	1	1	1	0	0
B	1	1	1	0	0	0	1	1	1	1	1	0	0	1	1	1	0	0
C	1	1	1	1	1	1	1	1	1	1	1	1	1	1	0	0	1	1
D	0	0	0	1	1	1	0	0	0	1	1	1	1	0	0	0	1	1

Table 9.3b Cladogram coding for three datasets after de Queiroz & Gatesy (2006, figure 1)

	Dataset #1+Dataset #2+Dataset #3: E(D(C(AB))																	
	0	1	2	3	4	5	6	7	8	9	10	11	12	13	14	15	16	17
E	0	0	0	0	0	0	0	0	0	0	0	0	0	0	0	0	0	0
A	1	1	1	1	1	0	1	1	1	0	0	0	0	1	1	1	0	0
B	1	1	1	0	0	0	1	1	1	1	1	0	0	1	1	1	0	0
C	1	1	1	1	1	1	1	1	1	1	1	1	1	1	0	0	1	1
D	0	0	0	1	1	1	0	0	0	1	1	1	1	0	0	0	1	1

Given these different results, one might investigate ways of refining the method, as if one method or one kind of information content can yield a 'better' estimate of the data than another method or another kind of information. Rather, one might better investigate where the support comes from for each solution. That is: *investigate the data*.

Summary

Consensus methods have been designed to find a summary cladogram from a set of cladograms. This seems a useful approach to the problem of differing cladograms from differing datasets. Exploration of consensus methods has descended into a search for a method, or a search for some methodology, that will justify one approach or another. This is flawed reasoning as it reflects methods not data. *Supermatrices* have all the problems associated with ordinary matrices and should be abandoned.

It is the data that requires exploration, not the methodology.

References

Adams, EN. 1972. Consensus techniques and the comparison of taxonomic trees. *Systematic Zoology* 21: 390–397.

Baum, BR. 1992. Combining trees as a way of combining data sets for phylogenetic inference, and the desirability of combining gene trees. *Taxon* 41: 3–10.

Bininda-Emonds, ORP. (ed.) 2004a *Phylogenetic Supertrees: Combining Information to Reveal the Tree of Life.* Kluwer/Academic Press, Dordrecht.

Bininda-Emonds, ORP. 2004b. The evolution of supertrees. *Trends in Ecology & Evolution* 19: 315–322.

Bininda-Emonds, ORP., Gittleman, J. & Steel, M. 2002. The (Super)Tree of Life: procedures, problems, and prospects. *Annual Review of Ecology and Systematics* 33: 265–289.

de Queiroz, A. & Gatesy, J. 2006. The supermatrix approach to systematics. *Trends in Ecology & Evolution* 22: 34–41.

Gordon, AD. 1986. Consensus supertrees: the synthesis of rooted trees containing overlapping sets of labeled leaves. *Journal of Classification* 3: 335–348.

Mickevich, MF. & Platnick, NI. 1989. On the information content of classifications. *Cladistics* 5: 33–47.

Nelson, G. & Platnick, NI. 1981. *Systematics and Biogeography: Cladistics and Vicariance.* Columbia University Press, New York.

Ragan, MA. 1992. Matrix representation in reconstructing phylogenetic relationships among the eukaryotes. *Biosystems* 28: 47–55.

Sanderson, MJ., Purvis, A. & Henze, C. 1998. Phylogenetic supertrees: assembling the trees of life. *Trends in Ecology and Evolution* 13: 105–109.

Wilkinson, M. 1996. Majority-rule reduced consensus methods and their use in bootstrapping. *Molecular Biology and Evolution* 13: 437–444.

Further Reading

Bininda-Emonds, ORP, Gittleman, JL. & Steel, MA. 2002. The (Super)Tree of Life: procedures, problems, and prospects. *Annual Review of Ecology and Systematics* 33: 265–289.

Bininda-Emonds, ORP. (ed.) 2004. *Phylogenetic Supertrees: Combining Information to Reveal the Tree of Life.* Kluwer/Academic Press, Dordrecht.

These are two useful summaries of the methods available at that time. Bear in mind that most contributions to Bininda-Emonds (2004) are still immersed in finding better *methods* rather than exploring the actual data.

10

How to Study Classification: 'Total Evidence' vs. 'Consensus', Character Congruence vs. Taxonomic Congruence, Simultaneous Analysis vs. Partitioned Data

A significant debate in systematics that began in the late-1970s, developed in the mid-1980s and still with us today is the discussion on the use of what was initially called the 'Total evidence versus Consensus' debate. The essence of the debate can be captured with two contrasting approaches to systematics, whether to combine evidence or keep it partitioned:

(1)   All available data, regardless of their source or 'kind', should be combined into one large binary data matrix, analysed using a method or methods noted in Chapter 8, and the cladogram(s) resulting from that analysis should be the basis for any proposed classification ('Total Evidence' or 'Combined Evidence').

   Or

(2)   Data should be partitioned into discrete sets or 'kinds' – such as morphology and molecules – and analysed separately using a method or methods from those noted in Chapter 8, and a consensus of those results, using a method or methods from those noted in Chapter 9, is the basis for any proposed classification (Consensus, Taxonomic Congruence or 'Partitioned Evidence').

   The two approaches are illustrated as flow charts in Figure 10.1 (after Kluge & Wolf 1993, p. 184, figure 1, modified from Kluge 1989, p. 8, figure 1). It would take considerable time and effort to cover every aspect of what was, and has been, involved in this debate. Suffice to say that the issues never really disappeared and

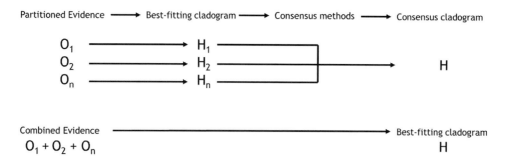

Figure 10.1 The two approaches to data analysis are illustrated as flow charts (redrawn from Kluge & Wolf 1993, p. 184, figure 1, modified from Kluge 1989, p. 8, figure 1); see text for explanation.

are present today, albeit, in some cases, with a different vocabulary (concatenation ≈ combined data, e.g., Zharkikh & Li, 1993; supermatrices ≈ combined data, e.g., Gatesy et al. 2002; supertrees ≈ consensus methods, e.g., Bininda-Emonds 2004, etc., see Chapter 9) and in some cases with different aims than arriving at an optimal classification (e.g., 'A *total-evidence* approach to dating with fossils, applied to the early radiation of the Hymenoptera' (Ronquist et al. 2012) and 'Bayesian *total-evidence* dating reveals the recent crown radiation of penguins' (Gavryushkina et al. 2017), emphasis in each title ours) – research ambition to one side, the substantive issues remain the same: what method or methods are best to find the optimal cladogram with respect to varied sources of data[1].

We discussed in Chapter 8, albeit briefly, 'sets' of methods that can be applied to the analysis of raw data to generate fundamental cladograms – these methods cover the 'total evidence' aspect. We discussed in Chapter 9, also briefly, the basics of consensus methods that can be applied to the analysis of fundamental clado-grams to generate a general cladogram.

Below we explore some aspects of the 'total-evidence' debate and will offer a solution (in Chapters 11 and 12) different to that of the generation of more and more methods (and accumulation of more and more data of varying quality).

'Total Evidence' vs 'Consensus'

Although consensus techniques were first applied to comparisons of trees resulting from differing methods (such as parsimony analyses vs. phenetic

---

[1] One might gain an appreciation of the 'rabbit-hole' these studies disappear into by reviewing the 'concatenation versus coalescence versus concatalescence' dispute, one we do not attempt to examine or summarise here; see Gatesy & Springer (2013, 2014) and Gatesy et al. (2017) for details.

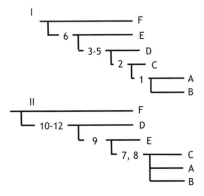

Figure 10.2 Two different fundamental cladograms: (I) (AB)C)D)E)F and (II) (ABC)E)D)F (redrawn from Miyamoto 1985, figure 1). The first cladogram (I) is supported by characters 1–6, the second (II) by characters 7–12.

analyses, e.g., Mickevich 1978), the beginning of the 'Total evidence/Consensus' debate', in a cladistic context at least, began with Miyamoto (1985), whose study was a critique of consensus methods in the context of deriving a general cladogram or general classification (*sensu* Nelson 1979; see Chapter 7). Miyamoto's (1985) example went like so:

Miyamoto (1985, figures 1–3, reproduced as our Figures 10.2 and 10.3): He examined two different fundamental cladograms: (AB)C)D)E)F and (ABC)E)D)F (Figure 10.2, cladograms indicated by I and II, reproduced from Miyamoto 1985, figure 1). The first cladogram (I) is supported by characters 1–6, the second (II) cladogram by characters 7–12. Miyamoto noted that:

> The two fundamental cladograms (I and II) support two stable (A-C and A-E, excluding the universal set A-F), two unstable (A-D vs. A-E excluding D), and one ambiguous (A-B) components [sic, component]. (Miyamoto 1985, p. 186)

Miyamoto then applied two consensus methods to find a general cladogram. The Nelson (1979) consensus resulted in cladogram (AB)C)DE)F and the Adams (1972) consensus resulted in cladogram (ABC)DE)F (Figure 10.3a, the Adams consensus, Figure 10.3b the Nelson consensus, redrawn from Miyamoto 1985, figure 2). He then plotted the original 12 characters back on to both consensus cladograms to show that neither were most parsimonious solutions relative to the original data as they required extra steps to account for all 12 characters on each of the consensus cladograms (4 extra steps for the Adams consensus, 5 for the Nelson consensus). In contrast, a Wagner parsimony analysis of all 12 original characters yielded the cladogram (AB)C)D)E)F, which needed only one extra step, and was

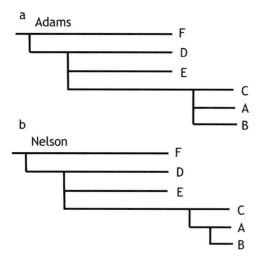

Figure 10.3 Two consensus results for the two fundamental cladograms in Figure 10.2. (a) The Adams consensus, (ABC)DE)F; (b) the Nelson consensus, (AB)C)DE)F.

the cladogram Miyamoto preferred. The data are summarised as in Table 10.1a; the number of components per solution is summarised in Table 10.1b.

Miyamoto concluded:

> Consensus techniques may be useful for comparing the relative stabilities of cladistic and phenetic approaches ... However, these methods should not be extended to the construction of general cladograms and classifications. (Miyamoto 1985, p. 188)

That is, direct analysis of all the data was to be preferred rather than a consensus of partitioned data analyses[2]. Note that Miyamoto was using his parsimony results as the standard with which to compare all other cladograms.

A more detailed critique of consensus methods by Arnold Kluge appeared a few years later (Kluge, 1989). Kluge's paper followed Miyamoto's in its implied critique of the derivation of general cladograms by consensus (i.e., Nelson, 1979). Part of Kluge's abstract is sufficient to capture his intent:

> Character congruence, the principle of using all the relevant data, and character independence are important concepts in phylogenetic inference, because they relate directly to the evidence on which hypotheses are based. Taxonomic congruence, which is agreement among patterns of taxonomic relationships, is less important, because its connection to the underlying character evidence is indirect and often imperfect. (Kluge, 1989)

---

[2] It should be noted here that later Miyamoto changed his mind, preferring consensus solutions (Miyamoto & Fitch, 1995).

Table 10.1a Summary of components (Cps) required for the two fundamental cladograms and the two consensus cladograms in Figures 10.2 and 10.3 (after Miyamoto 1985)

	(I): Characters 1–6; Characters 1–12	**(II): Characters 7–12**	**Consensus Adams**	**Consensus Nelson**
Cladogram	(AB)C)D)E)F	(ABC)E)D)F	(ABC)DE)F	(AB)C)DE)F
Cps	AB [ch. 1]			AB
	ABC x 3 [ch. 2, 7, 8]	ABC x 2 [ch. 7, 8]	ABC	ABC
	ABCD x 3[ch. 3, 4, 5]			
		ABCE [ch. 9]		
	ABCDE [ch. 6]	ABCDE x 3 [ch. 10, 11, 12]	ABCDE	ABCDE
	ABCDEF	ABCDEF	ABCDEF	ABCDEF

Table 10.1b Summary of components relative to characters for the two fundamental cladograms and the two consensus cladograms in Figures 10.2 and 10.3 (after Miyamoto 1985)

	(I): Characters 1–6; Characters 1–12	**(II): Characters 7–12**	**Consensus Nelson**	**Consensus Adams**
	(AB)C)D)E)F	**(ABC)E)D)F**	**(AB)C)DE)F**	**(ABC)ED)F**
AB	+	−	+	−
ABC	+++	++	+	+
ABCD	+++	−	−	−
ABCE	−	+	−	−
ABCDE	+++++	+++	+	+
ABCDEF	+	+	+	+
Total (components = nodes)	5	4	4	3

The issue was one of congruence – but congruence of what – cladograms or characters? Here then, it was clearly stated in an alternative way: *character congruence* (total evidence) versus *taxonomic congruence* (consensus).

According to Kluge, character congruence referred to agreement among the original characters (the raw data), which '. . . is usually evaluated in phylogenetic studies with *cladistic parsimony* . . .' (Kluge, 1989, p. 9, our emphasis). By cladistic parsimony, Kluge meant Wagner parsimony (see Chapter 8). He contrasted this with taxonomic congruence which 'involves deriving a consensus object from a comparison of two or more fundamental topologies (Nelson, 1979), and measuring the agreement of the taxonomic groups in the fundamental branching patterns' (Kluge 1989, p. 8). Kluge went on to note that 'A *large number* of consensus methods and indexes have been used to judge taxonomic congruence' (Kluge 1989, p. 8, emphasis ours) suggesting that it would be almost impossible to make a rational choice between these many methods – as it happens, a not unreasonable point.

Kluge set out his views on how to determine taxon relationships:

1. Character congruence, which could be found *only* by using cladistic parsimony, a specific method that, according to him, had considerable justification, and using all available evidence, hence the original term 'total evidence';

    Contrasted with:

2. Taxonomic congruence, which corresponded to a *large number* of different consensus methods, none of which had any comprehensive justification, and was always once removed from the raw data.

Thus, cladistics *in this form* became inextricably linked with (1) using all available data, but more crucially (2) with a specific algorithm: Wagner parsimony.

A number of papers appeared directly after Kluge's initial attempt at defining these approaches – some were in support of a consensus approach (e.g., de Queiroz, 1993), while others offered a more 'balanced' view (e.g., de Queiroz et al., 1995), but most (uneasily, in some cases) veered towards character congruence with weighting of some kind or another applied, usually in the guise of models of evolution (e.g., Huelsenbeck et al., 1994, 1996, of which more later).

From a specifically cladistic perspective, the next comprehensive survey was Nixon and Carpenter (1996). They modified some of Kluge's original terminology (for reasons that we need not discuss here) with 'total evidence' becoming 'simultaneous analysis'[3] and 'taxonomic congruence' becoming 'partitioned data' (Nixon & Carpenter 1996, p. 223[4]). It should be noted in passing that Nixon and Carpenter had earlier discussed 'simultaneous analysis' in the context of outgroup comparison (Nixon & Carpenter 1993), where their meaning was much the same: that all *relevant* data should be analysed simultaneously using Wagner parsimony. With a stab at humour (Are '. . . all truly sacred truths rich in comedy'?[5]), Nixon and Carpenter summarised their position thus:

> In conclusion, we can state the following: if there is reason to suspect that two datasets are independent estimates of a phylogeny based on samples from different underlying distributions, the two datasets should be combined and analysed simultaneously; conversely, if there is reason to suspect that the two datasets are not independent estimates, they should not be analysed separately. (Nixon & Carpenter 1996, p. 237)

This was followed by the view that 'The position that datasets should be analysed separately is clearly based on a *rejection of the principle of parsimony in cladistics*' (Nixon & Carpenter 1996, p. 237) – by that they meant a rejection of the view that Wagner parsimony is *the* method of (cladistic) analysis. Once again, one might appreciate a further connection, or *hardening*[6] if you will, of an attitude to one particular perspective: *cladistics = Wagner parsimony*. But as we noted in Chapter 8, Wagner parsimony is but one method among many[7] for which there is no special justification for its use beyond some measure of its own internal logic based on a simple model of character transformation: there is no clear way to understand if all its results reflect the natural world or are caused by

[3] Nixon and Carpenter understood differences between the meaning of 'total evidence' and their own 'simultaneous analysis': 'Given that some available data may be useful in ways other than a parsimony analysis, and some available data may be excluded through compartmentalization or incomplete sampling, the term "total evidence" is perhaps best restricted to the use of all data, while "simultaneous analysis" refers to the analysis of multiple data sets with as much data included as is appropriate and tractable' (Nixon & Carpenter 1996, p. 223). Discussion of what is and is not 'appropriate and tractable' ensued (e.g., Lecointre & Deleporte 2005).

[4] With respect to the phrase 'total evidence', Nixon and Carpenter wrote: 'The term is probably not appropriate to contrast the method of separate analyses of partitioned data followed by consensus of results with the method of simultaneous analysis of multiple combined datasets. In both methods, one might argue that all of the evidence is considered (e.g., Miyamoto & Fitch, 1995); the two methods differ not in how much evidence is considered, *but in the procedure by which evidence is evaluated*. We therefore prefer to use the terminology "simultaneous analysis" in place of "total evidence" following Nixon and Carpenter (1993)' (Nixon & Carpenter 1996, p. 223, emphasis ours).

[5] George Bernard Shaw – but as a statement; we converted it into a question.

[6] We have borrowed the word *hardening* from Stephen Jay Gould (1983) who applied it to the 'hardening of the modern synthesis', a move rejecting an original pluralistic view of evolutionary change (see Witteveen 2011).

[7] It should be noted that Wagner parsimony was not the only implementation of a parsimony approach.

methodological artefact (we take the latter up in the section 'Character Conflict: Representation and Data-Distortion' in Chapter 12).

As we noted above, further commentary on methods focused almost entirely on exploration of models of change – weighting, in fact (e.g., Bull et al., 1993; Huelsenbeck et al., 1994; Huelsenbeck et al., 1996) – along with tables presenting 'Why' or 'Why Not To' combine datasets (de Queiroz et al. 1995, p. 665; Miyamoto & Fitch 1995, p. 68; Huelsenbeck et al. 1996, p. 153).

Before we proceed, recall once again how phenetics was, and perhaps still is, perceived: classification 'by overall similarity, based on all available characters without any weighting'. Suppose the term 'overall similarity' is replaced with a weighted version, giving us 'overall similarity of synapomorphy' (Nelson, 2004, p. 139), we then arrive at weighted phenetics (Chapter 8). The proponents of the modelling approach clearly accepted that they were introducing weighting to their characters. Nixon and Carpenter were seemingly oblivious to the fact that as time passed their version of cladistics (i.e., Wagner parsimony, or as they later began to write, 'modern cladistics') was looking more and more like phenetics (see Chapter 8), a transformation almost fully realised in their later discussions on homology (Nixon & Carpenter 2012a, 2012b, more details in Chapter 7).

One critique that emerged in support of combining datasets deserves discussion. Nixon and Carpenter, following Barrett et al. (1991), noted that 'Simultaneous analysis can allow "secondary signals" to emerge because it measures strength of evidence supporting disparate results' (Nixon & Carpenter 1996, p. 237). What are *secondary signals*? What does that mean? A simple example was proposed by Barrett et al. (1991).

Barrett et al. (1991, p. 488): Barrett et al. offered two datasets for consideration with the view that: '. . . if we can show that a tree based on the combined data is incompatible with a strict consensus tree, then, in effect, we will have shown that all consensus methods have the same problem' (Barrett et al. 1991, p. 488). Their two datasets are as follows:

	Dataset 1				Dataset 2			
	1	**2**	**3**	**4**	**5**	**6**	**7**	**8**
	1	2	1	3	1	2	1	3
A	0	1	0	0	0	1	0	0
B	1	0	1	0	1	0	1	1
C	1	1	0	1	1	1	0	1
D	1	0	1	1	1	0	1	0

Table 10.2 Results from analysis of the two datasets above (after Barrett et al. 1991, p. 488)

	Dataset 1	Dataset 2	Dataset 1+2	Dataset 1: cladogram + Dataset 2: cladogram
	Characters 1–4	Characters 5–8	Characters 1–8	Strict consensus
Result:	A(B(CD))	A(D(BC))	(AC)(BD)	A(BCD)

For Dataset 1 there are four characters (numbered 1–4 in the first row) in varying numbers (character 1 is represented by one character; character 2 is represented by two characters; character 3 is represented by one character; character 4 is represented by three characters, the figures in the second row of the matrix above). Parsimony analysis of Dataset 1 yields just one cladogram: A(B(CD)).

For Dataset 2 there are four characters (numbered 5–8, in the first row) in varying numbers (character 5 is represented by just one character; character 6 is represented by two characters; character 7 is represented by just one character; character 8 is represented by three characters, the figures in the second row of the matrix above). Parsimony analysis of Dataset 2 yields just one cladogram: A(D (BC)). The strict consensus of the pair A(B(CD)) + A(D(BC)) is A(BCD). If Dataset 1 and 2 are combined and analysed as one dataset, parsimony analysis yields another cladogram: (AC)(BD). The results are summarised in Table 10.2 and in terms of characters and components, the results are in Table 10.3.

Chippindale and Wiens (1994, p. 284, figure 2): Chippindale and Wiens (1994) offered a similar example to that of Barrett et al. (1991) where separate datasets yield contrasting solutions while combining the data yields a third but different solution. They considered the third cladogram to be the correct tree – and later that it does not matter if it is correct or not, just that it differs (Wiens & Chippindale, 1994, p. 565). The example is as follows:

	Dataset 1					Dataset 2				
	1	2	3	4	5	6	7	8	9	10
	50	75	25	50	25	25	25	25	50	75
A	0	1	1	0	0	0	0	1	0	1
B	0	1	1	1	0	0	0	1	1	1
C	0	0	0	1	1	1	1	0	1	1
D	1	1	0	1	1	0	1	0	1	0
E	1	1	0	1	1	1	1	0	1	1

Table 10.3 Results from analysis of the two datasets above with respect to cladograms and characters (after Barrett et al. 1991, p. 488)

	Dataset 1	Dataset 2	Dataset 1 +2	Solution #1	Solution #2	Solution #3	Consensus
A(BCD)	+	+	++	+	+	−	+
BD(AC)	++	++	++++	−	−	++++	−
(BD)AC	+	+	++	−	−	++	+
AB(CD)	+++	−	+++	+++	−	−	+++
AD(BC)	−	+++	+++	−	+++	−	+++
Totals	7	7	14	4	4	6	8

For dataset 1 there are five characters (numbered 1–5, first row) in varying numbers (character 1 is represented by 50 characters; character 2 is represented by 75 characters; character 3 is represented by 25 characters; character 4 is represented by 50 characters; and character 5 is represented by 25 characters, second row). Parsimony analysis yields one cladogram: C(AB)(DE).

For dataset 2 there are five characters (numbered 6–10, second row) in varying numbers (character 6 is represented by 25 characters; character 7 is represented by 25 characters; character 8 is represented by 25 characters; character 9 is represented by 50 characters; and character 10 is represented by 75 characters, second row). Parsimony analysis yields one cladogram: D(AB)(CE). The strict consensus of C(AB)(DE) + D(AB)(CE) is (AB)CDE. If dataset 1 and 2 are analysed as one combined dataset, parsimony analysis yields just one cladogram: A(B(C(DE))). The results can be summarised as in Table 10.4 and in terms of characters and components, as in Table 10.5.

Resolution and false nodes: How can these results be evaluated? In a commentary on Chippindale and Wiens' example, Huelsenbeck et al. made the not unreasonable point that '... the 'correct' tree is arbitrarily defined as the most-parsimonious tree for the combined data' (Huelsenbeck et al. 1994, p. 290). Of more interest would be to establish what evidence supports each resolved node. The result from the combined data yields the cladogram A(B(C(DE))), which has three components BCDE, CDE and DE in spite of apparent conflict between the DE and CE component.

Characters that support A(B(C(DE))) are drawn from the 50 implying the relationship ABC(DE) (dataset 1, character 1), the 100 implying the relationship A (BCDE) (dataset 1, character 4, dataset 2, character 9) and the 50 implying the

Table 10.4 Results from analysis of the two datasets above (after Chippindale & Wiens 1994)

	Dataset 1	Dataset 2	Dataset 1+2	
	Characters 1–5	Characters 6–10	Characters 1–10	Consensus strict
Result	C(AB)(DE)	D(AB)(CE)	A(B(C(DE)))	(AB)CDE

Table 10.5 Results from analysis of the two datasets above with respect to cladograms and characters (after Chippindale & Wiens 1994)

	Dataset 1	Dataset 2	Dataset 1+2	Solution #1	Solution #2	Solution #3	Consensus
ABC(DE)	50	–	50	50		50	
C(ABDE)	75	–	75	75			
(AB)CDE	25	25	50	50	50		50
A(BCDE)	50	50	100			100	
AB(CDE)	25	25	50			50	
D(ABCE)	–	75	75		75		
ABD(CE)	–	25	25		25		
Totals	225	200	425	175	150	200	50

relationship AB(CDE) (dataset 1, character 5, dataset 2, character 7). Yet, in all there is a total of 100 characters supporting AB(CDE), achieved from a combination of the 50 supporting (AB)CDE (dataset 1, character 3, dataset 2, character 8) and 50 that *inadvertently* support AB(CDE) (dataset 1, character 5, dataset 2, character 7) – support in the sense that they do not contradict the CDE component; those characters have two steps on the cladogram rather than one. There are various ways in which these results could be interpreted given different optimisations of the characters in question. In the combined results, the (AB)CDE characters may contribute to the CDE component when interpreted as *reversals*, as the 0 state would be understood as secondarily derived and state 1 is actually primitive. Alternatively, state 1 could be considered to have arisen in A and B separately. Interpretations aside, the (AB)CDE characters fit to more than one node on the cladogram found from the analysis of the pooled datasets and hence component AB is not found.

This example (and the two before) illustrates what we mean by *artefactual resolution*. The pooled data in each case resolve nodes that are *created* (they are not part of the original data) because of the need to satisfy the demands of the program – in the case of Wagner parsimony it *needs* to find the shortest tree (Chapter 8). The shortest tree is justified by the model underlying the execution of the program, with artefactual resolution being explained away as characters that are 'secondarily derived', or 'reversals', homoplasy, or whatever. We refer to these as *false nodes* rather than 'secondary signal', as that is what they are: *false*, a product of the program.

Chippindale and Wiens observed the following:

> A nonexhaustive survey of the literature ... [25 studies] ... indicates that combined analyses generate trees that are incongruent with each of those from the separately analyzed data sets in more than half the cases. Thus, combination of data can allow discovery of relationships (and therefore sets of congruent and discordant characters) that would have been missed had the data sets only been analyzed separately. (Chippindale & Wiens 1994, p. 281)

In our view, the cases they examined reveal an abundance of *false nodes* being found – artefacts everywhere. Huelsenbeck et al. went on to note that '[I]t is difficult to imagine an explanation for such striking conflict between two datasets that does not involve different underlying histories, strong selection, or extreme non-independence' (Huelsenbeck et al. 1994, p. 290). In other words, one might need to know more about 'how evolution actually worked', and propose models of character change to find the 'best' tree (a proposal initially made by Bull et al. 1993).

Wiens and Chippindale responded to Huelsenbeck et al., and their (Wiens & Chippindale) final conclusion is of interest:

> We believe that systematists should not be discouraged from combining their data, and we suggest that a more useful direction for future studies of data integration might be to emphasize consideration of the *processes of character evolution* (through *character weighting*) rather than agreement among partitions of the data. (Wiens & Chippindale 1994, p. 566, emphasis ours)

Wiens and Chippindale are effectively suggesting *more* weighting of characters. But this simply ended up enveloping the topic of data analysis (and as a consequence any resulting classifications) in varying kinds of complex (as well as simple) models for estimating changes in characters, a procedure that now dominates the entire field of phylogenetic studies (systematics) – and it is instructive that Wiens and Chippindale understood this as the application of character weighting, echoing Kluge and Farris, way back in 1969 (Kluge & Farris 1969, p. 1).

Once again, we note that it requires no deep analysis of the history of the first wave of phenetics to learn that at first it was '... based on all available characters without any weight' (Cain & Harrison 1960, p. 3), and then, when results seemed almost arbitrary, weights were applied (Sneath & Sokal 1973) – and so it seems for this second wave of phenetics, whether it is disguised in parsimony (as in Wagner parsimony, which has a simple model) or with the incorporation of more complex models to represent 'processes of character evolution' (Wiens & Chippindale 1994). For other versions of 'Total Evidence' see Boxes 10.1–10.3.

Box 10.1   Other Versions of 'Total Evidence' (= Phenetics): Integrative Taxonomy

Note that we are not defending 'traditional taxonomy' here, but instead we argue that the real cutting-edge future for systematics and biodiversity research is *integrative taxonomy*, which uses a large number of characters including DNA and many other types of data, to delimit, discover, and identify meaningful, natural species and taxa at all levels.

(Will et al. 2005, p. 845)

'Integrative taxonomy' is defined as the science that aims to delimit the units of life's diversity from multiple and complementary perspectives (phylogeography, comparative morphology, population genetics, ecology, development, behaviour, etc.).

(Dayrat 2005, p. 407)

The basis behind an 'integrative taxonomy' was first outlined in two separate contributions, written independently of one another. Both differ in certain respects, but both agree on the basic idea as captured in the two quotations above: large amounts of diverse information (data) are required for modern taxonomic studies (Will et al. 2005; Dayrat 2005). The two accounts differ in that Will et al. makes a plea for using a 'large number of characters' for taxon discovery, while Dayrat focuses mostly on a mix of processes ('... population genetics, ecology, development, behaviour ...') and theories ('... phylogeography ...'), rather than solely on organismal characters ('comparative morphology...' or DNA sequences).

Inspiration for Dayrat's integrative taxonomy came, in part, from a paper by Marvelee Wake (Wake 2003), who addressed the issue of 'Integrative Biology' in the renamed journal *Integrative and Comparative Biology* (formerly known as *American Zoologist*), the journal of the *American Society of Zoologists,* itself renamed the *Society for Integrative and Comparative Biology*. Wake commented on the difficulty in precisely defining 'Integrative Biology' noting that:

> There are almost as many conceptions of 'integrative biology' as there are people interested in the idea; this results in those people considering themselves to be 'integrative biologists' without any clarification of or agreement upon the central themes of the concept. (Wake 2003, p. 240)

Dayrat suggested that, for him, 'Integration means multi-disciplinary ...' (Dayrat 2005, p. 409), one of the many perspectives offered by Wake. What might that mean?

Box 10.1 *(cont.)*

> Under the new paradigm of integrative taxonomy, morphologists could describe morphodiversity, analyse the variation of morphological features among individuals and propose hypothetical morphospecies. Then, those morphospecies – or some of them – could be submitted to the filter of other approaches and additional data (molecular, behavioural, developmental, ecological, etc.). (Dayrat 2005, p. 409)

History – again – is instructive. More than 80 years ago William Bertram Turrill (1890–1961, see Hubbard 1971) wrote the following:

> Those who, having been trained to an appreciation of modern discoveries in ecology, cytology, genetics, &c., are trying to widen the basis of taxonomy, have undertaken a long, slow and perhaps thankless task. They have, however, a vision of a revivified taxonomy in which an important place is found for all observational and experimental data relating, even if indirectly, to the constitution, subdivision, origin and behaviour of species and other taxonomic groups. (Turrill 1935, pp. 104–105)

Turrill's paper is unique inasmuch as he later went on to describe, for the first time, the concept of *alpha taxonomy*[1]. Turrill's paper, and much of the discussion it generated, was a major influence on the development of what became known as the *New Systematics* (Huxley 1942). Turrill's thrust, nevertheless, was the desire to help taxonomists become 'multi-disciplinary' – in fact, to make them explore experimental methods in population biology and genetics alongside their descriptive accounts.

Turrill was not alone in his 'multi-disciplinary' quest for taxonomists. George Gaylord Simpson (1902–1984, see Olson 1991), an evolutionary palaeontologist and leading light in the creation of the *Modern Synthesis*, said of taxonomy that:

> ... in its various guises and branches [taxonomy] gathers together, utilizes, summarizes, and implements everything that is known about animals, whether morphological, physiological, psychological, or ecological. (Simpson 1945, p. 1)

Willi Hennig (1913–1976), more or less outside the confines of either the *New Systematics* or the *Modern Synthesis*, created the term *holomorphology* specifically to describe the totality of comparative data (see Higgins 2005; Yeo et al. 2018):

> We will call the totality of all these characters [... physiological, morphological, and physiological ...] simply the total form (or the holomorphy) of the semaphoront, which thus is to be regarded as a multidimensional construct. (Hennig 1966, p. 7)

Herbert H. Ross (1908–1978) defined *biosystematics* as a subset of systematics

> ... based on any scientific information that can be brought to bear on the problems of the evolution of species, whether they concern speciation or phylogeny. (Ross 1974)

Whatever Dayrat's 'Integrative Taxonomy' might be (and it has become extremely popular), and all subsequent developments, it is not so much a new paradigm inasmuch

Box 10.1   *(cont.)*

as it requires, or calls for, *vast amounts of data*, and, under certain circumstances, a plethora of evolutionary models are required to understand and interpret the data.

In the same issue of *Integrative and Comparative Biology* that included Marvelee Wake's essay referred to above, John Pearse began his summary of the collection of papers thus: 'Integrative biology has been practiced for a very long time; it is in our roots' (Pearse 2003, p. 276).

The notion that taxonomy is based on the integration of multiple data sources is, of course, not new, nor is it confined to any particular body of thought, doctrine or philosophy, as can be seen by contrasting Simpson's approach with that of Hennig's, for example.

Regardless of the obvious plea for more data, even for very large datasets, it seems that, in its most predictive sense, taxonomy/systematics has more or less functioned with small, even alarmingly small, datasets.

References

Dayrat, B. 2005. Towards integrative taxonomy. *Biological Journal of the Linnean Society* 85: 407–415.

Hennig, W. 1966. *Phylogenetic Systematics*. University of Illinois Press, Urbana [reprinted 1979, 1999].

Higgins, W. 2005. Holomorphology: the total evidence approach to phylogenetic reconstruction. In: Raynal-Roques, A., Rogeuenant, A. & Prat, D. (eds), *Actes du 18e Congrès mondial et exposition d'orchidées. 18th Proceedings of the World Orchid Conference*. Naturalia, Turriers, pp. 268–276.

Hubbard, CE. 1971. William Bertram Turrill 1890-1961. *Biographical Memoirs of Fellows of the Royal Society* 17: 688–712.

Huxley, J. 1942. *Evolution. The Modern Synthesis*. George Allen & Unwin, London.

Olson, EC. 1991. George Gaylord Simpson: June 16, 1902-October 6, 1984. *Biographical Memoirs, National Academy of Sciences (U.S.)* 60: 331–353.

Pearse, JS. 2003. The promise of integrative biology: resurrection of the naturalist. *Integrative and Comparative Biology* 43: 276–277.

Ross, HH. 1974. *Biological Systematics*. Addison-Wesley, Reading, MA.

Simpson, GG. 1945. The principles of classification and a classification of mammals. *Bulletin of the American Museum of Natural History* 85.

Turrill, WB. 1935. The investigation of plant species. *Proceedings of the Linnean Society of London* 147: 104–105.

Wake, MH. 2003. What is "Integrative Biology"? *Integrative and Comparative Biology* 43: 239–241.

Will, KW., Mishler, BD. & Wheeler, QD. 2005. The perils of DNA barcoding and the need for integrative taxonomy. *Systematic Biology* 54: 844–851.

Yeo, D., Puniamoorthy, J., Ngiam, RW. & Meier, R. 2018. Towards holomorphology in entomology: rapid and cost-effective adult–larva matching using NGS barcodes. *Systematic Entomology* 43: 678–691.

[1] 'The aim of the alpha taxonomist must be to complete the preliminary and mainly morphological survey of plant-life' (Turrill 1935).

Box 10.2 Other Versions of 'Total Evidence' (= Phenetics): Polyphasic Taxonomy

Polyphasic taxonomy, a phrase first used by bacteriological taxonomists but in more recent times adopted by a few phycologists, is yet another version of 'Integrative Taxonomy' (Vandamme & Peeters 2014, e.g., Das et al. 2014 and Komárek 2016). The ideas first appeared in a 1970 paper written by bacteriologist Rita Colwell[1]:

> Thus, a taxonomy is required which assembles and assimilates the many levels of information, from the molecular to the ecological, and incorporates the several distinct, and separable, portions of information extractable from a nonhomogeneous system to yield a multidimensional taxonomy. Such a taxonomy has been termed 'polyphasic' … (Colwell 1970, p. 410)

As noted in a later review, 'In principle, all genotypic, phenotypic, and phylogenetic information may be incorporated in polyphasic taxonomy' (Vandamme et al. 1996, p. 409):

> Polyphasic classification is purely empirical, follows no strict rules or guidelines, may integrate any significant information on the organisms, and results in a consensus type of classification, satisfying most but not necessarily all users of taxonomic results. Polyphasic taxonomy is not hindered by any conceptual prejudice, except that the more information that can be integrated on a group of organisms, the better the outcome might reflect its biological reality. (Vandamme et al. 1996, pp. 430–431)

For us, this is yet again another version of phenetics (and total evidence), if a rather poorly defined one (see Sneath 1995).

References

Colwell, RR. 1970. Polyphasic taxonomy of the genus *Vibrio*: numerical taxonomy of *Vibrio cholerae*, *Vibrio parahaemolyticus*, and related *Vibrio* species. *Journal of Bacteriology* 104: 410–433.

Das, S., Dash, HR., Mangwani, N., Chakraborty, J. & Kumari, S. 2014. Understanding molecular identification and polyphasic taxonomic approaches for genetic relatedness and phylogenetic relationships of microorganisms. *Journal of Microbiological Methods* 103: 80–100.

Komárek, J. 2016. A polyphasic approach for the taxonomy of cyanobacteria: principles and applications. *European Journal of Phycology* 51: 346–353.

Sneath, PHA. 1995. Thirty years of numerical taxonomy. *Systematic Biology* 44: 281–298.

Vandamme, P. & Peeters, C. 2014. Time to revisit polyphasic taxonomy. *Antonie van Leeuwenhoek* 106: 57–65.

Vandamme, P., Pot, B., Gillis, M., de Vos, P., Kersters, K. & Swings, J. 1996. Polyphasic taxonomy, a consensus approach to bacterial systematics. *Microbiological Reviews* 60: 407–438.

[1] Among other things, https://en.wikipedia.org/wiki/Rita_R._Colwell

Box 10.3   Phylogenomics

Omics: The suffix *-ome* as used in molecular biology refers to a *totality* of some sort . . .
https://en.wikipedia.org/wiki/Omics

The other twist to 'omics' may be associated with the 'Om' (pronounced 'Aum'), an ancient Sanskrit intonation, which, like music, transcends the barriers of age, race, culture, and even species. . . . It is believed to be the basic sound of the world and to contain all other sounds, and is a mantra in itself. If repeated with the correct intonation, it can resonate throughout the body, so that the sound penetrates to the center of one's being, the atman or soul. There is harmony, peace, and bliss in this simple but deeply philosophical sound.

(Yadav 2007)

Although phylogenomics began as a way to understand and explore gene function (Eisen 1998), it now primarily embraces the study of taxon relationships (evolutionary relationships) using vast amounts of data derived from whole genome studies (DeSalle & Rosenfeld 2012). It is occasionally contrasted (maybe somewhat unjustly) with *molecular phylogenetics*, the latter supposedly dealing with one-gene (single-gene, or few-gene) trees (cladograms) rather than whole genomes or multi-gene trees (cladograms) – supposedly phylogenomics was to usher in 'the end of incongruence' (Gee 2003, but see Jeffroy et al. 2006 and Philippe et al. 2017). Of course, incongruence did not disappear but phylogenomics is founded on the notion that *more data yields better results regardless of quality* – leaving aside what 'better' might mean, the notion that vast amounts of data are required is a phenetic idea, or closely associated with the phenetic enterprise. This is not necessarily a bad thing.

The complexities behind the many techniques involved in phylogenomics is beyond the purview of this book, but in general the approach owes a great deal to the 'total evidence' argument, in spirit at least – although it is never pitched that way (a brief examination of Bleidorn 2017, for example, especially his chapter 9, 'Sources of error and incongruence in phylogenomic analyses', which explores nearly all the issues raised in the earlier total evidence discussions). The major thrust of our book is to ask: what nodes on any tree (cladogram) (however derived) can be considered 'true' for the world we are hoping to document – how much artefact is in the tree, and how much is caused by the method itself. We need not make those arguments all over again here (see Chapter 7). We do wish to stress, however, that in the development of phylogenomics little has been said of homology[1], the core principle of comparative biology (Bleidorn does discuss the concept: Bleidorn 2017, p. 130 et seq., as do DeSalle & Rosenfeld 2012, chapter 5). Springer and Gatesy (and Gatesy & Springer 2017[2]) offer a timely and eloquent discussion of the issues involved:

> Homology is perhaps the most central concept of phylogenetic biology. Molecular systematists have traditionally paid due attention to the homology statements that are implied by their alignments of orthologous sequences, but some authors have suggested that manual gene-by-gene curation is not sustainable in the phylogenomics era (Springer & Gatesy 2017, abstract)

Box 10.3   *(cont.)*

Their conclusions are worth recording here in full:

> If automated pipelines are inadequate, the unintended introduction of hom-
> ology problems at the molecular level is equivalent to morphologists scoring the
> 'same' character in the tooth of one species and in the tail of another species.
> In the absence of high-quality phylogenomic data, neither concatenation nor
> coalescence methods should be expected to resolve difficult phylogenetic
> problems that turn on razor-thin margins. Rather, attention to detail is crucial
> for resolving short internodes in the Tree of Life. *The availability of vast quan-*
> *tities of data should never be an excuse for lack of careful examination of*
> *phylogenomic data.* (Springer & Gatesy 2017, p. 17, emphasis ours)

But hope springs eternal (Du & Hahn 2019).

References

Bleidorn, C. 2017. *Phylogenomics: An Introduction*. Springer, Berlin.

DeSalle, R. & Rosenfeld, J. 2012. *Phylogenomics: A Primer*. CRC Press, Boca Raton, FL.

Du, P. & Hahn, MW. 2019. Species tree inference under the multispecies coalescent on data with paralogs is accurate. *BioRxiv* 498378. https://doi.org/10.1101/498378

Eisen, JA. 1998. Phylogenomics: improving functional predictions for uncharacterized genes by evolutionary analysis. *Genome Research* 8: 163–167.

Gatesy, J. & Springer, MS. 2017. Phylogenomic red flags in placental radiation. *Proceedings of the National Academy of Sciences USA* 114(45): E9431–E9432.

Gee, H. 2003. Evolution: ending incongruence. *Nature* 425: 782.

Jeffroy, O., Brinkmann, H., Delsuc, F. & Philippe, H. 2006. Phylogenomics: the beginning of incongruence? *Trends in Genetics* 22: 225–231.

Liu, L., Zhang, J., Rheindt, FE., Lei, F., Qu, Y., Wang, Y., Zhang, Y., Sullivan, C., Nie, W., Wang, J., Yang, F., Chen, J., Edwards, SV., Meng, J. & Wu, S. 2017. Reply to Gatesy and Springer: claims of homology errors and zombie lineages do not compromise the dating of placental diversification. *Proceedings of the National Academy of Sciences USA* 114(45): E9433–E9434.

Philippe, H., de Vienne, DM., Ranwez, V., Roure, B., Baurain, D. & Delsuc, F. 2017. Pitfalls in supermatrix phylogenomics. *European Journal of Taxonomy* 283: 1–25.

Sanderson, MJ., Nicolae, M. & McMahon, MM. 2017. Homology-aware phylogenomics at gigabase scales. *Systematic Biology* 66: 590–603.

Springer, MS. & Gatesy, J. 2017. On the importance of homology in the age of phylogenomics. *Systematics and Biodiversity* 16: 1–19.

Yadav, SP. 2007. The Wholeness in suffix -omics, -omes, and the word Om. *Journal of Biomolecular Techniques* 18(5): 277.

[1] Sanderson et al. (2017), for example, appear to treat homology as if it means similarity.
[2] There was a reply: Liu *et al.* (2017).

Summary

A major discussion took place in systematics during the late-1970s, through the 1980s and remains a topic of discussion today. Two approaches to systematics were contrasted as: (1) all available data, regardless of their source or 'kind', are combined into one large data matrix ('combined evidence'), analysed using a method or methods noted in Chapter 8, and any cladogram(s) resulting from that analysis should be the basis for any proposed classification. Or: (2) data should be partitioned into discrete sets or 'kinds' ('partitioned evidence)' and analysed separately with a consensus of those results being the basis for any proposed classification. We have traced the development of these ideas and although it appears that the 'total evidence' approach (or versions of it) prevails, there is no agreement on which method to apply to all these data. We, however, see it differently. We see the general argument – combining or not – as irrelevant if one views character data differently. We discuss this in Chapters 11 and 12.

References

Adams, EN. 1972. Consensus techniques and the comparison of taxonomic trees. *Systematic Zoology* 21: 390–397.

Barrett, M., Donoghue, M. & Sober, E. 1991. Against consensus. *Systematic Zoology* 40: 486–493.

Bininda-Emonds, ORP. 2004. The evolution of supertrees. *Trends in Ecology & Evolution* 19: 315–322.

Bull, JJ., Huelsenbeck, JP., Cunningham, CW., Swofford, DL. & Waddell, PJ. 1993. Partitioning and combining data in phylogenetic analysis. *Systematic Biology* 42: 384–397.

Cain, AJ. & Harrison, GA. 1960. Phyletic weighting. *Proceedings of the Zoological Society of London* 135: 1–31.

Chippindale, PT. & Wiens, JJ. 1994. Weighting, partitioning, and combining characters in phylogenetic analysis. *Systematic Biology* 43: 278–287.

de Queiroz, A. 1993. For consensus (sometimes). *Systematic Biology* 42: 368–372.

de Queiroz, A., Donoghue, MJ. & Kim, J. 1995. Separate versus combined analysis of phylogenetic evidence. *Annual Review of Ecology and Systematics* 26: 657–681.

Gatesy, J. & Springer, MS. 2013. Concatenation versus coalescence versus "concatalescence". *Proceedings of the National Academy of Sciences USA* 110: E1179.

Gatesy, J. & Springer, MS. 2014. Phylogenetic analysis at deep timescales: unreliable gene trees, bypassed hidden support, and the coalescence/concatalescence conundrum. *Molecular Phylogenetics & Evolution* 80: 231–266.

Gatesy, J., Matthee, C., DeSalle, R. & Hayashi, C. 2002. Resolution of a Supertree/Supermatrix paradox. *Systematic Biology* 51: 652–664.

Gatesy, J., Meredith, RW., Janecka, JE., Simmons, MP., Murphy, WJ. & Springer, MS. 2017. Resolution of a concatenation/coalescence kerfuffle: partitioned coalescence support and a robust family-level tree for Mammalia. *Cladistics* 33: 295–332.

Gavryushkina, A., Heath, TA., Ksepka, DT., Stadler, T., Welch, D. & Drummond, AJ. 2017. Bayesian total-evidence dating reveals the recent crown radiation of penguins. *Systematic Biology* 66: 57–73.

Gould, SJ. 1983. The hardening of the modern synthesis. In: Grene, M (ed.), *Dimensions of Darwinism*. Cambridge University Press, Cambridge, UK, pp. 71–93.

Huelsenbeck, JP., Swofford, DL., Cunningham, CW., Bull, JJ. & Waddell, PJ. 1994. Is character weighting a panacea for the problem of data heterogeneity in phylogenetic analysis? *Systematic Biology* 43: 288–291.

Huelsenbeck, JP., Bull, JJ. & Cunningham, CW. 1996. Combining data in phylogenetic analysis. *Trends in Ecology and Evolution* 11: 152–158.

Kluge, AG. 1989. A concern for evidence and a phylogenetic hypothesis of relationships among *Epicrates* (Boidae, Serpentes). *Systematic Zoology* 38: 7–25.

Kluge, AG. & Farris, JS. 1969. Quantitative phyletics and the evolution of Anurans. *Systematic Zoology* 18: 1–32.

Kluge, AG. & Wolf, AJ. 1993. Cladistics: what's in a word? *Cladistics* 9: 183–199.

Lecointre, G. & Deleporte, P. 2005. Total evidence requires exclusion of phylogenetically misleading data. *Zoologica Scripta* 34: 101–117.

Mickevich, MF. 1978. Taxonomic congruence. *Systematic Zoology* 27: 143–158.

Miyamoto, MM. 1985. Consensus cladograms and general classifications. *Cladistics* 1: 186–189.

Miyamoto, M. & Fitch, W. 1995. Testing species phylogenies and phylogenetic methods with congruence. *Systematic Biology* 44: 64–76.

Nelson, G. 1979. Cladistic analysis and synthesis: principles and definitions, with a historical note on Adanson's *Familles des Plantes*. *Systematic Zoology* 28: 1–21.

Nelson, GJ. 2004. Cladistics: its arrested development. In: Williams, DM. & Forey, PL. (eds), *Milestones in Systematics*. CRC Press, Boca Raton, FL, pp. 127–147.

Nixon, KC. & Carpenter, JM. 1993. On outgroups. *Cladistics* 9: 413–426.

Nixon, KC. & Carpenter, JM. 1996. On simultaneous analysis. *Cladistics* 12: 221–241.

Nixon, KC. & Carpenter, JM. 2012a. On homology. *Cladistics* 28: 160–169.

Nixon, KC. & Carpenter, JM. 2012b. More on errors. *Cladistics* 28: 539–544.

Ronquist, F., Klopfstein, S., Vilhelmsen, L., Schulmeister, S., Murray, DL. & Rasnitsyn, AP. 2012. A total-evidence approach to dating with fossils, applied to the early radiation of the Hymenoptera. *Systematic Biology* 61: 973–999.

Sneath, PHA. & Sokal, RR. 1973. *Numerical Taxonomy*. Freeman, San Francisco.

Wiens, JJ. & Chippindale, PT. 1994. Combining and weighting characters and the prior agreement approach revisited. *Systematic Biology* 43: 564–566.

Witteveen, J. 2011. [Book review] The softening of the Modern Synthesis, Julian Huxley: Evolution: The Modern Synthesis; The Definitive Edition. Massimo Pigliucci and Gerd B. Müller (eds): Evolution–The Extended Synthesis. *Acta Biotheoretica* 59: 333–345.

Zharkikh, A. & Li, W-H. 1993. Inconsistency of the maximum-parsimony method: the case of five taxa with a molecular clock. *Systematic Biology* 42: 113–125.

11

How to Study Classification
Natural Methods I: Consensus Revisited

To Tournefort who anticipated it; to Bernard de Jussieu who proved it; to Adanson who developed it; and to Antoine-Laurent de Jussieu who subjected it to fixed laws . . .

(A.-P. de Candolle 1804, p. 6, our translation)

. . . almost all [,] systematists seem to be, and seem ever to have been, 'cladists.' At least cladistic elements are discernible in systematic work of whatever age.

(Nelson & Platnick 1981, p. 327)

Towards Understanding Natural Classification

In Chapter 7 we noted that 'the single most important principle of cladistics is that diverse fundamental cladograms may be combined to form a single general cladogram' (Nelson 1979, p. 7) and that while the construction of *general cladograms* may related to consensus methods they are not necessarily the product of any particular kind of consensus method. We have discussed a few elements of consensus methods in Chapter 9. Below we pursue one version in a little more detail.

Cladistic Analysis and Synthesis: General Cladograms

In this book we have repeatedly noted that *cladistic analysis* 'is the analysis of branching. Specifically, the analysis is of cladograms (hierarchically branching diagrams), or classifications that may be represented by cladograms' (Nelson 1979, p. 3). This is important as it means that *cladistic analysis* is synonymous with the *discovery of relationships* rather than the *analysis of groups*.

As noted in Chapter 9, every fundamental cladogram, however found, will consist of a set of components (branch points) and terms (terminal taxa). Given a set of fundamental cladograms, and as a consequence a set of components that may differ amongst themselves, and given the aim of *cladistic synthesis* is to find 'a single general cladogram', or an optimal classification (Nelson 1979, p. 7), what, then, is that general cladogram? Nelson's reasoning was straightforward.

Suppose we argue that, for whatever reason, one method of analysis in particular is preferred over all others in its ability to find the general cladogram from some raw data. Suppose, for example, we argue that a cladogram derived from a measure of 'overall similarity' of all available data best reflects the requirements for a general cladogram. To support such a view, philosophical arguments would be required. Suppose, instead, we prefer to use Wagner parsimony to find a cladogram that best reflects the requirements for a general cladogram. To support such a view, different philosophical arguments would be required. Suppose we take a more contemporary approach, and set about reconstructing a phylogeny using some statistical approach that incorporates some model of evolution. To support that view, even more philosophical arguments would be required (see Chapter 8). The same might apply to different kinds of data. For example, we might argue that to reconstruct a phylogeny morphology is *the* preferred source of data, adopting some philosophical arguments to support this (e.g., '. . .morphology *is* the phenotype so it *must* be fundamental to discovering phylogeny'; 'morphology has development, ontogeny, which is the key to homology'; 'and anyway, DNA is far too *noisy*', etc.). Or we might argue that to reconstruct a phylogeny DNA is the preferred source of data, adopting some philosophical arguments to support this (e.g., 'DNA is the basic unit of Life, so it *must* be fundamental to phylogeny'; 'DNA comes in large quantities, and we *need* large amounts of data'; 'and anyway, morphology is far too *noisy*', etc.). It should be clear that arguments to support any of these preferences – methods or data – are, in the court of last resort, philosophical, as any realistic empirical investigation would require knowledge of the true relationships among the organisms, which, of course, we do not have.

We could attempt something empirical rather than philosophical. If we change the measure of 'overall similarity' and the structure of our cladogram remains the same, can we then assume this indicates that real relationships have been discovered? Or maybe if we are so certain that Wagner parsimony is the best technique and add more data, and the cladogram remains the same, does this indicate that real relationships have been discovered? Suppose we change the model and our phylogeny remains the same, can we assume this indicates that in this case real phylogenetic relationships have been discovered? Suppose for every permutation we try the results differ in every case, what then? More crucially, how can we learn if the result is a true representation of real relationships, or that the result is merely an artefact of the chosen method or methods, or a reflection of that particular data source?

The last 30 years of development in phylogenetic methods have seen journals and books fill up with arguments supporting one or another method, or supporting one or another source of data, or supporting some permutation of methods, or all the data and one method, or all the methods and one source of data – the arguments have been, are and always will be, endless as there appears to be no empirical way to eliminate the problem of method dependency: that the method adopted, for whatever reason, is causing the results, either in the case of agreement between or disagreement between various analyses by different methods of identical (or even different) datasets.

Nelson (1979) suggested the problem was a relatively old one but, nevertheless, one that has never been completely solved: understanding the distinction between artificial and natural approaches to classification. In Chapter 8 we documented some modern methods of artificial classification. Here, we extended our understanding of artificial classifications to include any particular method in use today as each is one of *imposition* rather than *discovery* – for example, Wagner parsimony minimises steps; model-based methods depend upon the accuracy of the model, etc.; and both depend on some underlying philosophy, that is, some metaphysics. Approaches to natural classification focus as much as possible on what is true of the data, without interference from the method, to enable *discovery*. Nelson suggested the following:

> A general solution requires not combining data according to some particular clustering technique, nor even combining various clustering techniques (which would result merely in a hybrid technique), but rather the combining of the results of clustering techniques in general. (Nelson 1979, p. 7)

Nelson's approach, then, was devised to combine results from various fundamental cladograms, to find signal amidst the noise by seeking replicated components (see below). The process was to find the best fit of as many components from an array of fundamental cladograms, with the resulting cladogram being *the* general cladogram and hence *the* general classification (see Liebherr 1986, for an example). Of course, as in all scientific investigation, everything is to a certain extent provisional.

First we need to add a note on Nelson's approach, which is a form of compatibility analysis (Page 1988, 1990, see Chapter 8). With respect to the mechanics of compatibility, it has been invented, reinvented and re-introduced on many different occasions (Williams & Ebach 2008, p. 195), beginning, in its modern guise at least, with E.O. Wilson's 'consistency test for phylogenies' (Wilson 1965; Patterson 1982; Nordal 1987; Scotland & Steel 2015). Compatibility can be simply explained as: characters or cladograms will either: (1) agree in an absolute sense; (2) agree via overlapping and non-conflicting sub-sets; (3) or will completely disagree (conflict). Previously, we presented a table of a few versions of compatibility analysis (Williams & Ebach 2008, p. 204). Here we present a table modified from Williams & Ebach (2017) (Table 11.1).

Table 11.1 Various versions of compatibility analyses

Nelson (1979)	Patterson (1982)	Scotland (1992)	Nelson & Ladiges (2001)
Cladistic synthesis	Homology testing	Character analysis	Biogeographical congruence
Exclusion	(a) a different group	Consistent	Consistent
Inclusion	(b) a larger group (c) a smaller group	Consistent	Consistent
Replication	(d) the same group	Congruent	Corroborate
Non-combinable	(e) conflicting groups	Conflict	Conflict

After Williams & Ebach (2008, p. 204) and Williams & Ebach (2017).

Example: Figure 11.1a is a diagram of the relationships of some Bicosoecid genera[1] (after Kim et al. 2010, p. 190, fig. 11). The diagram was found using maximum likelihood (ML) with SSU rDNA sequences of 65 taxa with 1490 unambiguously aligned sites (Kim et al. 2010, p. 190). In the sense used here, the diagram is a *fundamental cladogram*. Figure 11.1b is the same cladogram redrawn to indicate the individual components, of which there are nine (excluding some more basal taxa and component 0) (Figure 11.1b). Figure 11.2a is an alternative cladogram for a similar set of Bicosoecid genera (after Kim et al. 2010, p. 192, fig. 13). This cladogram was found using an ME (maximum evolution) distance tree from 1560 nucleotides (Kim et al. 2010, p. 192). It too can be regarded as a *fundamental cladogram*. Figure 11.2b is the same as Figure 11.2a except that the uncultured Bicosoecid and *Symbiomonas scintillans* have been removed for ease of comparison with the cladogram in Figure 11.2a. Figure 11.2b is redrawn to indicate the components, of which there are nine (excluding component 0). The components differ between these two fundamental cladograms: the cladogram in Figure 11.1b has components 1–9, with components 3, 4 and 9 being unique to it; the cladogram in Figure 11.2b has components 1, 2, 5, 6, 7, 8, 10–12, with the last three being unique to it. Thus, components 1, 2, 5, 6, 7 and 8 are common to both.

Figure 11.3a is a recent classification of subclass Rictidia, Superorder Cyathobodiniae, which includes the Bicosoecids (Cavalier-Smith & Scoble 2013). The classification has been modified to allow comparison with the fundamental cladograms in Figures 11.1 and 11.2. The cladogram derived from the classification in Figure 11.3b has four components: 2, 6, 8, and 12 (*Bicosoecida* is only a terminal in the two fundamental cladograms so was not labelled as such and is of no

---

[1] Bicosoecids are heterotrophic flagellates – https://eol.org/pages/52511512

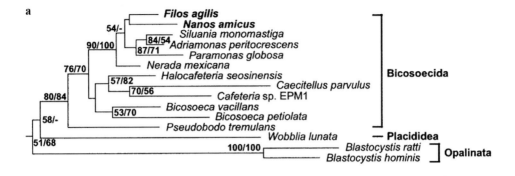

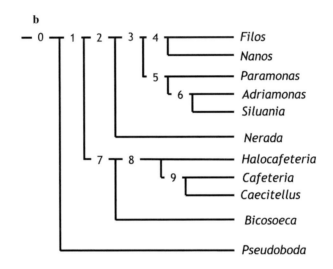

Figure 11.1 (a) Diagram of relationships (cladogram) of some Bicosoecid genera (after Kim et al. 2010, p. 190, fig. 11, with permission); (b) redrawn as a *fundamental cladogram* with nine individual components

significance here). Following from Chapter 7, this is a *derivative cladogram* (derived from the classification). It can easily be seen that the classification in Figure 11.3 is composed of components from both of the fundamental cladograms in Figures 11.1 and 11.2: components 2, 6 and 8 are shared (Figures 11.1 and 11.2). As can be seen from this example, the two fundamental cladograms, based on different datasets and different analyses, differ from each other but have enough common information to be able to construct a classification in broad agreement with much of the original data. This is summarised in Table 11.2.

Interestingly enough, component 2 is equivalent to Order Pseudodendromonadida, recognised long before DNA data was available; component 6 (Siluaniidae, in some classifications) has potential morphological synapomorphies; and, with respect to the relationships in component 12, *Halocafeteria* is a monotypic genus.

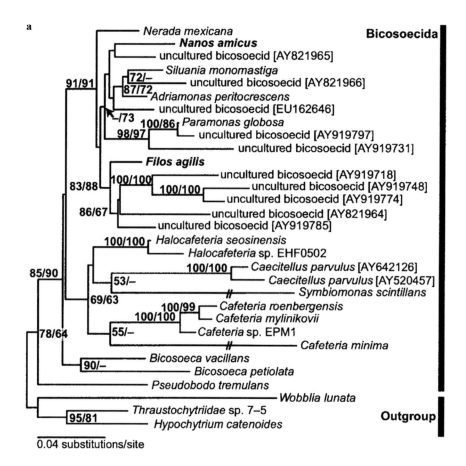

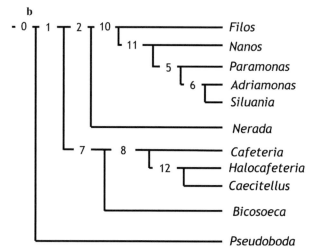

Figure 11.2 (a) An alternative cladogram for a similar set of Bicosoecid genera (after Kim et al. 2010, p. 192, fig. 13, with permission); (b) redrawn from part a, except that the uncultured Bicosoecid and *Symbiomonas scintillans* have been removed for ease of comparison with the cladogram in part a.

a Superorder Cyathobodiniae
   Order Anoecida
      Family Caecitellidae
         Genus *Halocafeteria*
         Genus *Caecitellus*
      Family Cafeteriaceae
         Genus *Cafeteria*
   Order Bicosoecida
      Family Bicosoecaceae
         Genus *Bicosoeca*
   Order Pseudodendromonadida
      Family Filidae
         Genus *Filos*
      Family Nanidae
         Genus *Nanos*
      Family Neradidae
         Genus *Neruda*
      Family Paramonadidae
         Genus *Paramonas*
      Family Siluaniaceae
         Genus *Adriamonas*
         Genus *Siluania*

b

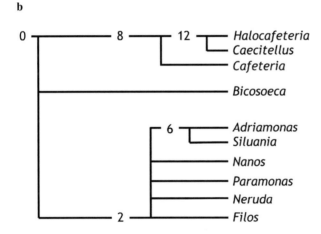

Figure 11.3 (a) Recent classification of subclass Rictidia, Superorder Cyathobodiniae (modified from Cavalier-Smith & Scoble 2013), which includes the Bicosoecids, modified to allow comparison with the fundamental cladograms in Figures 11.1 and 11.2; (b) cladogram derived from the classification in part a with four components.

Given these replicated components and the general cladogram it implies, does this suggest that SSU rDNA sequences are *the* data to use to find general clado-grams? No, how can it? The two fundamental cladograms disagree. Does it suggest that ML is *the* most efficient method? No, how can it? The two fundamental

Table 11.2 Summary of data for some Bicosoecids

Components	Fundamental cladogram 1 (Figure 11.1) (Kim et al. 2010, p. 190, fig. 11)	Fundamental cladogram 2 (Figure 11.2) (Kim et al. 2010, p. 192, fig. 13)	Classification of subclass Rictidia, Superorder Cyathobodiniae
1	+	+	−
2	**+**	**+**	**+**
3	+	−	−
4	+	−	−
5	+	+	−
6	**+**	**+**	**+**
7	+	+	−
8	**+**	**+**	**+**
9	+	−	−
10	−	+	−
11	−	+	−
12	−	+	+

Bold type indicates three common components.

cladograms disagree. An experiment could be made using permutations of these methods and data, but nothing could definitively point to *the* method or *the* data source. We can only assume that, for these data, never mind the method used, components 2, 6 and 8 are signal as they are replicated. This assumption may, of course, be false but only further study will tell.

Different methods do indeed produce different results for the same data, but all results (or at least most) have some aspects in common. Nelson's original approach depended on the assumption that 'the differences are artefact, and the samenesses method-independent [so] the problem is only distinguishing signal from noise' (Nelson 1979, p. 14). While there is much to be said for this approach, an additional question has since arisen, admittedly one that has not been addressed by many commentators: is there a way of finding a method-independent approach to the analysis of data? Re-phrased: is there a way to pursue natural classification outside mere replications?

We began this chapter by noting that *cladistic analysis* 'is the analysis of branching'. Suppose, then, that raw data *itself* could be represented by some form of branching instead of the usual linear series? That the information in characters was best represented by A(BC) rather than A→B→C. The former states only that B and C are more closely related to each other than they are to A; the latter assumes that C evolved from B, which evolved from A but, interestingly, also that there is a relationship A(BC). The former version states what the evidence implies (suggests); the latter invokes, in addition, a model of character change.

This is not about seeking a theory-free, or almost theory-free, mathematical manipulation of data, or some form of instrumentalism (cf., Rieppel 2007a, 2007b). That effort was part of the phenetic enterprise in numerical taxonomy. Ours is an exploration of 'empiricism' in the context of seeking a natural classification.

Nelson's original 1979 outline was partly in reaction to a lengthy paper published by Charles D. Michener (1918–2015, Beckemeyer 2015) entitled 'Discordant evolution and the classification of allodapine bees' (Michener 1977). Twenty years earlier, Michener, an expert on bee systematics, played an influential role in the first explorations in numerical (phenetic) taxonomy (Michener & Sokal 1957), the story of his collaboration with Robert Sokal (1926–2012, Futuyma 2012) having been told a number of times (e.g., Hull 1988[2]). Many years later Michener offered a summary of these experiments:

> The method [phenetics] would soon be largely abandoned by systematists, but it accomplished useful objectives: It often stimulated discussion and, for the first time, it forced systematists to record and tabulate the status of all known characters for all the taxa under consideration. (Michener 2007, p. 11)

Those words have significance for some of the ideas we develop below. Michener continued:

> Thus when phylogenetic methods became well known (about 1966) and computerized methods were developed later, collection of data largely appropriate for use was already being done. (Michener 2007, p. 11)

We consider the 'collection of data largely appropriate for use' with 'computerized methods' to be flawed in most, if not all, taxonomic methods that have followed from the numerical 'phenetic-cladistic' experiments of the last half century, matters we now follow up and will be the subject of the following chapter.

---

[2] For a view from the horse's mouth see Sokal (1985) and Michener (2007).

Cladistic Analysis and Synthesis: Cladogram Structure

There is one further property of cladograms that needs consideration at this stage. Many, if not most, cladograms will have multiple branching – that is, some parts of the cladogram remain unresolved. What of potential further resolution? There are two cases worth considering: *terminal trichotomies* and *basal trichotomies*.

Terminal trichotomies: For the simplest possible example, consider a clado-gram with four taxa A–D and a *terminal trichotomy* such that the four taxa are related as D(ABC). There are three possible resolutions, all found by resolv-ing the ABC component: D(C(AB)), D(B(AC)) and D(A(BC)) (Table 11.3, upper row).

Basal trichotomies: For the simplest possible example, consider a cladogram with four taxa A–D and a *basal trichotomy* such that the four taxa are related as DC (AB). There are three possible resolutions, which retain the AB component and have a node added to resolve the relationships of C and D: D(C(AB)), C(D(AB)) and (CD)(AB) (Table 11.3, lower row). For the complete suite of resolutions, one is common to both D(ABC) and DC(AB) (Table 11.3).

These resolutions were referred to as *Interpretation 1* by Nelson and Platnick, based on the general assumption that for the basal trichotomy example, A and B are more closely related to each other than either of them is to C *and* D (Nelson & Platnick 1980, p. 87). In addition, the cladogram DC(AB) is com-posed of two three-item statements, C(AB) and D(AB). This turned out to be a useful observation as an alternative interpretation, *Interpretation 2*, was dis-cussed, in which A and B are more closely related to each other than either of them are to taxon C *and/or* taxon D. In this interpretation each three-item statement can be considered *separately* allowing five resolutions per three-item statement (Table 11.4).

What is the significance of this? Let us return to the example of a simple single binary character AB(CD). For four taxa there are a total of 26 possible solutions (Table 11.6). Of the 26 solutions, 15 are fully resolved (2-nodes), 4 have a terminal trichotomy (1-node), 6 have a basal trichotomy (1-node) and one is the unresolved bush (0-nodes) (Table 11.6). If we consider character data (evidence) then the

Table 11.3 Summary of resolutions for multiple branching cladograms using Interpretation 1

Multiple branching	Resolutions: Interpretation 1				
	$1^{1\ 2}$	2^1	3^1	2^2	3^2
Terminal D(ABC):	D(C(AB))	D(B(AC))	D(A(BC))		
Basal DC(AB):	D(C(AB))			C(D(AB))	(CD)(AB)

Table 11.4 Summary of resolutions for multiple branching cladograms using Interpretation 2

3-item statement	Possible resolutions: Interpretation 2						
	1	2	3	4	5	4	5
C(AB):	D(C(AB))	(AB)(CD)	C(D(AB))	C(B(AD))	C(A(BD))		
D(AB):	D(C(AB)	(AB)(CD)	C(D(AB))			D(B(AC))	D(A(BC))

Note that the three solutions 1–3 are common to both statements (they include the AB component), so there are a total of seven solutions when both statements are considered together. If both statements were found to be true, then only the first three solutions are the possible resolutions, which are those found using *Interpretation 1* (Table 11.5). If only one three-item statement turns out to be true, then one set of five resolutions is possible, *Interpretation 2*. If one statement is deemed true, but unknown, then seven solutions are possible.

Table 11.5 Comparison between resolutions for Interpretations 1 and 2

		1	2	3	4	5
Interpretation 1	DC(AB):	D(C(AB))	(AB)(CD)	C(D(AB))		
Interpretation 2	C(AB):	D(C(AB))	(AB)(CD)	C(D(AB))	C(B(AD))	C(A(BD))
Interpretation 2	D(AB):	D(C(AB)	(AB)(CD)	C(D(AB))	D(B(AC))	D(A(BC))

character AB(CD) can be fit to all 26, which divide into two sets: four that the character fits exactly (with one step, as it were) and the remaining 22 fit inexactly (with two steps, marked with * in Table 11.6).

The character AB(CD) can *also* be viewed from the perspective of their three-item statements, of which there are two, C(AB) and D(AB). The 26 cladograms can then be divided into *three* sets instead of two: the 4 fully resolved cladograms, where both statements fit exactly, but of the remaining 22, 6 differ as one of the two statements will fit these (marked with + on Table 11.6) and of the remaining 20 no statements fit. Regardless of method, AB(CD) is the best of the 26, meaning that C and D are more closely related to each other than either of them is to A and B, the Interpretation 1 option. But if available data were not so clear cut and one statement happened to be true for that particular solution, then precision is achieved.

Cladograms with basal trichotomies under this interpretation allow for three further resolutions with respect to accumulating more relevant data (Table 11.5). Under Interpretation 2, however, in which C and D are more closely related to each other than either of them are to taxon A *and/or* taxon B, a further seven are possible, again with respect to accumulating relevant data (Table 11.5). Both

Table 11.6 All 26 solutions for four taxa with the fit of the character (evidence) AB(CD)

(AB)(CD)	*(AC)(BD)	*(AD)(BC)	A(B(CD))	*+A(C(BD))	*+A(D(BC))	B(A(CD))	*+B(C(AD))	*+B(D(AC))	*C(A(BD))	*C(B(AD))	*C(D(AB))	*D(A(BC))	*D(B(AC))	*D(C(AB))
*+A (BCD)	*+B (ACD)	*C (ABD)	*D (ABC)											
AB(CD)	*CD (AB)	*BD (AC)	*BC (AD)	*AD (BC)	*AC (BD)									
*(ABCD)														

interpretations allow for non-dichotomous cladograms, so Interpretation 1 allows a total of three cladograms, Interpretation 2 allow a total of 10 (Table 11.5).

As noted above, these resolutions can be understood in terms of actual evidence.[3] The terminal trichotomy is equivalent to a character shared among all three taxa, as well as a particular component. Other characters could be considered such as any combination of pairs.

Interestingly, Interpretation 1 is analogous to Assumption 1 (A1) in biogeographical analysis when dealing with widespread taxa where the 'plesiomorphic' state is equivalent to the 'widespread' taxon, such as AB(CD), with AB as the widespread taxa. Similarly, Interpretation 2 is analogous to Assumption 2 (A2) for biogeographical analysis (Nelson & Platnick 1981). We discuss this subject further in Chapter 12.

Summary

Cladistics focuses on the construction of general cladograms, summaries of 'common' knowledge. 'Common' knowledge refers to the relationships discovered. The analysis of cladograms (relationships) is an entirely different process than the analysis of change (graph theory).

References

Beckemeyer, RJ. 2015. In remembrance: Charles Duncan Michener (Sept 22, 1918–Nov 1, 2015). *Journal of the Kansas Entomological Society* 88: 457–464.

Candolle, A-P. de. 1804. *Essai sur les propriétés médicales des plantes, comparées avec leurs formes extérieures et leur classification naturelle.* Didot Jeune, Paris.

Cavalier-Smith, T. & Scoble, JM. 2013. Phylogeny of Heterokonta: *Incisomonas marina*, a uniciliate gliding opalozoan related to *Solenicola* (Nanomonadea), and evidence that Actinophryida evolved from raphidophytes. *European Journal of Protistology* 49: 328–353.

Farris, JS. 1997. Cycles. *Cladistics* 13: 131–144.

Farris, JS. & Kluge, AG. 1998. A/The brief history of three-taxon analysis. *Cladistics* 14: 349–362.

Farris, JS., Kluge, AG. & De Laet, JE. 2001. Taxic revisions. *Cladistics* 17: 79–103.

---

[3] Farris, and associates, misunderstood, or did not appreciate, the full implications behind Nelson & Platnick's Interpretations 1 and 2. They understood them 'as rules for resolving multifurcations on given trees. They concern trees only as abstract diagrams, not the connection of trees to characters' (Farris and Kluge, 1998, p. 350, also Farris 1997, p. 131, Farris et al. 2001, p. 95). On the contrary, Interpretations 1 and 2 are not so much *rules* as options relative to *actual data* (evidence) when viewed from a relationship point of view, as clearly outlined in the original paper describing the two interpretations (Nelson & Platnick 1980).

Futuyma, DR. 2012. Robert R. Sokal (1926–2012). *Science* 336: 816.

Hull, DL. 1988. *Science As a Process. An Evolutionary Account of the Social and Conceptual Development of Science*. The University of Chicago Press, Chicago and London.

Kim, E., Yubuki, N., Leander, BS. & Graham, LE. 2010. Ultrastructure and 18S rDNA Phylogeny of *Apoikia lindahlii* comb. nov. (Chrysophyceae) and its epibiontic Protists, *Filos agilis* gen. et sp. nov. (Bicosoecida) and *Nanos amicus* gen. et sp. nov. (Bicosoecida). *Protist* 161: 177–196.

Liebherr, JK. 1986. Cladistic analysis of North American Platynini and revision of the *Agonum extensicolle* species group (Coleoptera: Carabidae). *University of California Publications in Entomology* 106: 1–198.

Michener, CD. 1977. Discordant evolution and the classification of allodapine bees. *Systematic Zoology* 26: 32–56.

Michener, CD. 2007. The professional development of an entomologist. *Annual Review of Entomology* 52: 1–15.

Michener, CD. & Sokal, RR. 1957. A quantitative approach to a problem of classification. *Evolution* 11: 130–162.

Nelson, G. 1979. Cladistic analysis and synthesis: principles and definitions, with a historical note on Adanson's *Familles des Plantes. Systematic Zoology* 28: 1–21.

Nelson, GJ. & Ladiges, PY. 2001. Gondwana, vicariance biogeography, and the New York School revisited. *Australian Journal of Botany* 49: 389–409.

Nelson, G. & Platnick, NI. 1980. Multiple branching in cladograms: two interpretations. *Systematic Zoology* 29: 86–91.

Nelson, G. & Platnick, NI. 1981. *Systematics and Biogeography: Cladistics and Vicariance*. Columbia University Press, New York.

Nordal, I. 1987. Cladistics and character weighting: a contribution to the compatibility versus parsimony discussion. *Taxon* 36: 59–60.

Page, RDM. 1988. Quantitative cladistic biogeography: constructing and comparing area cladograms. *Systematic Zoology* 37: 254–270.

Page, RDM. 1990. Component analysis: a valiant failure? *Cladistics* 6: 119–136.

Patterson, C. 1982. Morphological characters and homology. In: Joysey, KA. & Friday, AE. (eds), *Problems of Phylogenetic Reconstruction*. Academic Press, London, pp. 21–74.

Rieppel, O. 2007a. The nature of parsimony and instrumentalism in systematics. *Journal of Zoological Systematics and Evolutionary Research* 45: 177–183.

Rieppel, O. 2007b. Parsimony, likelihood, and instrumentalism in systematics. *Biology and Philosophy* 22: 141–144.

Scotland, RW. 1992. Cladistic theory. In: Forey, PL., Humphries, CJ., Kitching, IJ., Scotland, RW. & Williams, DM. *Cladistics. A Practical Course in Systematics*. Oxford University Press, Oxford, pp. 3–13.

Scotland, RW. & Steel, M. 2015. Circumstances in which parsimony but not compatibility will be provably misleading. *Systematic Biology* 64: 492–504.

Sokal, RR. 1985. The principles of numerical taxonomy: Twenty five years later. In: Goodfellow, M., Jones, D., & Priest, EG. (eds), *Computer-Assisted Bacterial Systematics*. Academic Press, London, pp. 1–20.

Williams, DM. & Ebach, MC. 2008. *Foundations of Systematics and Biogeography*. Springer, New York.

Williams, DM. & Ebach, MC. 2017. What is intuitive taxonomic practice? *Systematic Biology* 66: 637–643.

Wilson, EO. 1965. A consistency test for phylogenies based on contemporaneous species. *Systematic Zoology* 14: 214–220.

12

How to Study Classification

Natural Methods II: Beyond Method, the Philosophy of Three-Item Analysis

From the preceding chapters – and the wealth of literature on the subject – one thing seems clear: different solutions to systematic problems are possible from different methods, whether those methods are directed toward the analysis of raw data or the analysis of cladograms (consensus). Once again the issue is whether the solutions found (by whatever means, data or method, or combination thereof) represent aspects of the natural world or include artefacts of the methods used.

With respect to the original 'Character congruence vs. Taxonomic congruence' debate, it seems that the only concrete outcome was to accept a wide variety of methods for consensus/supertree construction alongside a similarly wide variety of methods for the analysis of raw data. Negotiating those minefields is a daunting task. For example, after reading the review on consensus techniques written by David Bryant (2003, especially his figure 2, 'A classification of consensus methods', also see Bryant et al., 2017) one might still have no clear idea about which one, if any, to use and when[1]. The same might be said for raw data analysis, where one could inspect, for example, Ward Wheeler's recent book *Systematics* and while terribly informative, still not really be sure what one is supposed to do or how one is supposed to choose from such an array of algorithms (Wheeler 2012). Wheeler's book is intended as a

---

[1] Bryant writes in the abstract to his review: 'The first consensus method was proposed by Adams in 1972. Since then a large variety of different methods have been developed, and there has been considerable debate over how they should be used'. More recently, and – to us at least – rather oddly, maximum likelihood and Bayesian methods have now been applied to supertree construction (Steel & Rodrigo 2008; Akanni et al. 2015). Oddly, for quite why a method that attempts to 'model' supposed evolutionary character transformations (character change) should be applied to finding common parts of differing cladograms (structural representations of relationships) is unclear to us. It seems to be a case of using a procedure simply because it exists.

series of lecture notes describing the many different kinds of methods available. He writes in the Preface: 'These notes are not meant to be the last word in systematics, but the first'. Perhaps they might possibly be the last word on artificial methods in systematics relative to *phylogeny reconstruction* (but we doubt it!). For practising taxonomists, a way out of this analytical quagmire is desperately needed as '... it is not consensus versus "combined data sets" but rather the *nature of data* and how they are combined' (Nelson & Ladiges 1992a, p. 481, emphasis ours).

Three-item analysis treats data from a relational point of view. It is a direct representation of the *cladistic parameter* (see Chapter 7). Given that cladistic analysis is *the analysis of branching*, each three-taxon statement directly represents the cladistic parameter and in our view represents *the unit of systematics* (Williams & Ebach 2008). By that we mean it is the basic building block for any scheme of relationships. In Chapter 9 we very briefly addressed the incompleteness of components as groups, such that any *group* defined in a cladogram also expresses a *relationship*. Thus, five taxa A–E that include the group AB is better represented as (AB)CDE, meaning that A+B are more closely related to each other than they are to either C, D or E.

When the rudiments of three-item analysis were first published in the context of character analysis (Nelson & Platnick 1991, rather than for biogeography, Nelson & Ladiges 1991, 1992a) its initial reception was not entirely favourable, eliciting a flurry of responses of varying quality and relevance, most emanating from self-professed cladists[2] and primarily from the perspective of a defence of Wagner Parsimony (e.g., Harvey 1992; Kluge 1993; for more details see Box 12.1). With more than 25 years of reflection, those critiques can now perhaps be understood more accurately, although in our view such a lengthy passage of time should never have really been necessary. It is instructive, too, that during this period, several other versions of three-item analysis emerged, entirely unrelated or uninfluenced by what we refer to here as *the first critique*[3]. Rather than document the early mechanics of three-item analysis, we have provided a few details in Box 12.1. Our intention below is to document some generalities rather than getting bogged down with issues concerning implementation (again, see Box 12.3). Implementation, perhaps, accounts for most of the negative contributions to the *first critique*.

---

[2] It is worth noting that early opposition to 'pattern cladistics' not only came from cladists themselves (e.g., Brooks & Wiley, 1985) but also from philosophers of biology (Beatty, 1982; Hull, 1988), evolutionary taxonomists (Charig, 1982), geneticists (Dawkins, 1986), and evolutionary biologists (Ridley, 1986), among others. Oddly, the term 'pattern cladistics' was first defended in print by the philosopher Ronald H. Brady (Brady, 1982, see Ebach & Williams, 2020).

[3] Alongside those we note in our Box 12.2, additional exploration of three-item data have been made in the mathematical literature, for example: Wu (2004), Jansson et al. (2006), Lin et al. (2009), Thu-Hien & Habib (2009), Tazehkand et al. (2013), and others (see references in these papers). These relations are sometimes discussed as 'Rooted Triplets' rather than three-item statements. There is some doubt that triplets are the same as three-item statements but that is a discussion we leave for another time.

Box 12.1   The First Critiques of Three-Item Analysis (1991–2002)

After Nelson and Platnick (1991) published their first account of three-item analysis as applied to character data[1], a number of commentaries followed, most published in the journal *Cladistics*. Several issues emerged during what we call *the first critique*, issues that now, with hindsight, seem relatively trivial: at best basic misunderstandings, at worse, misrepresentations – perhaps now of interest only to social scientists.

It would exhaust the patience of our readers (not to say our own) to detail all the various criticisms that were made during those first 10 years. Below we sketch a few that may have influenced non-partisans preventing them from taking three-item analysis seriously (but see Box 12.2). The issues raised in these critiques dealt mostly with *Methodology* and to a lesser extent *Philosophy*; the former easier to deal with, the latter is addressed in the body of this chapter.

Methodology

'Transforming' the binary matrix: Every binary character has an equivalent suite of three-item characters. For example, the three-item equivalent of the simple binary character AB (CD) is two statements of relationships: A(CD) and B(CD). How can these be represented in a conventional matrix? That is, how can they be entered into a conventional matrix for subsequent analysis? This is merely a *practical* detail, one of *implementation*, but nevertheless, caused some discussion (initially by Harvey 1992, later by Kluge 1993, 1994). To accurately represent these two three-item statements A(CD) and B(CD) in a conventional matrix it would have to be like so:

	1	2
A	0	
B		0
C	1	1
D	1	1

If a parsimony program is used for subsequent analysis (which it was in these early explorations), the so-called primitive state (zero) needs to be identified, so the matrix requires the addition of an all-zero outgroup:

	1	2
OUT	0	0
A	0	
B		0
C	1	1
D	1	1

Box 12.1 *(cont.)*

Note that two of the cells are empty: there is no value for taxon A and statement 2 and no value for taxon B and statement 1. As each three-item statement is, by definition, composed of only three taxa, no entries are required for the taxa excluded from any particular statement – there is *literally no relevant evidence*. Yet now, as then (and somewhat to our surprise), parsimony programs *still* cannot deal with matrices that have entries that mean 'no data', cells that should be left blank. To overcome this, question marks (?) were used as place holders for those blank (empty) cells to indicate that for these there are in actual fact *no data*:

OUT	0	0
A	0	?
B	?	0
C	1	1
D	1	1

This procedure was merely one of implementation – *to make it work*. Of course, this became controversial as the addition of question marks suggested to some that data was being *invented* or, worse, *altered* (Harvey 1992; Kluge 1993, p. 250: 'The matrix resulting from the three-taxon conversion has considerable missing data, *where none existed before*', italics in the original), rather than understanding that the question marks were simply acting as *place holders* to indicate 'no data'. It should have been abundantly clear that the process of converting a binary matrix into a three-item matrix was merely *operational*. Critics, as they are wont to do, fell upon the word *transform*. This was unfortunate. Farris, for example, and somewhat oddly, offered this rather convoluted critique:

> Matrices of three-taxon statements (3ts) represent a transformation of original data in that they are calculated from normal data matrices. Normal matrices cannot be considered transformational (or 3ts matrices original) in that sense, because the 3ts transformation cannot be reversed. That the 3ts transformation cannot be reversed also means that the transformation discards information. (Farris 2000a, , p. 403, abstract)

Oddly, because the three-item matrix is derived from a dataset of which the codes in the cells merely *represent* the actual character in the binary matrix. Numbered correctly, all three-item statements that relate to a particular character can be traced back to the written discussion of the character, which is, hopefully, included in the taxonomic account at some point. For example, suppose 'binary character 1' is an anatomical feature of some kind, perhaps a scapula (the shoulder blade). Of taxa A–D, it is only present in taxon C and D. Binary character 1 can also be represented as AB(CD) and can be represented by two three-item statements. If appropriately numbered, it would render the binary character equivalent to the three-item data:

Box 12.1 (*cont.*)

	B1	T1 a	T2 b
	1	a	b
A	0	0	
B	0	=	0
C	1	1	1
D	1	1	1

Thus binary character 1 = three-item data T1 (a) + T2 (b) = scapula + its implied relationship.

Later Farris noted:

> The 3ts matrix reflects only pairwise apomorphic similarities, discarding all further information on the identities and distributions of features. (Farris 2000a, p. 405)

This comment is of more significance. What Farris must have feared was that *re-interpretations* of the binary character (the apomorphies) would not now be available. By re-interpretations we mean that the binary character AB(CD) could now no longer be *distorted* to mean (AB)CD or (AB)(CD) when fitted to the optimal cladogram. That is, for Farris, the data are uninformative before analysis. *Data distortion* is dealt with in more detail in the main body of this chapter.

To be sure, as three-item matrices got larger, they did acquire numerous question marks, and the addition of question marks did raise some genuine issues already noted for analysis of conventional binary matrices (Platnick et al. 1991; Nelson & Ladiges 1993, see below). Thus, there was a necessity to discuss *minimal trees.*

Minimal trees: Some programs, at that time (in the 1990s), included in their output cladograms with nodes that have *zero-length branches*, in some cases caused by the addition of question marks. In cases of multiple cladograms of equal length, a strict consensus tree of all the output cladograms would find the minimal tree: that is, the shortest cladogram with the fewest nodes (Nelson & Ladiges 1993). It was relatively easy, if not a little time-consuming, to either manually collapse each node of every cladogram found retaining only those nodes on cladograms where *tree length did not change* (Nelson & Ladiges 1996; Siebert & Williams 1997; Kitching et al. 1998). That is, finding a tree *minimal* from the resulting array of multiple cladograms was certainly tiresome but not excessively complex.

Just to recap: The addition of an outgroup and the addition of question marks for blank cells were both merely *heuristic devices* included to *implement* the method of analysis and had no bearing on the principles behind three-item data, its representation, meaning or subsequent analysis.

Such problems are now, more or less, irrelevant (see Box 12.2).

Box 12.1 (*cont.*)

Cladogram length and three-item statements in parsimony analyses: When a three-item matrix was analysed using a parsimony program, the relationship between tree length and three-item statements is simple:

> Tree Length (L) = Accommodated Three-item Statements (ATS) + (2 x Not accommodated Three-item Statements) (NATS). (Nelson 1992, p. 357; Nelson & Ladiges 1994, p. 195)

This might be de-constructed further: 'Accommodated Three-item Statements' (ATS) = those that fit the tree exactly; 'Not accommodated Three-item Statements' (NATS) = those that do not fit the tree. Thus, tree length would be made up of a score of 1 for each ATS plus a score of 2 for each NATS. Tree length would accurately reflect the fit of all statements to every cladogram. For fractionally weighted cladograms (see below) the length is the product of the number of statements and their fractional weight, so shorter trees do not necessarily fit more statements but fit a greater total weight (Nelson 1993; Nelson & Ladiges 1994).

Information measures: Nelson and Platnick (1991, p. 358) noted that the consistency index (ci) for each statement will either be 1.00 (because each statement that fits a particular cladogram fits it exactly) or 0.50 (because each statement that does not fit a particular cladogram fits it inexactly). They rejected this as a useful overall measure of the data/tree relationship because the retention index (ri) scales each statement's fit correctly and, along with tree length, accurately measures the fit of data to tree (in spite of the superfluous objections of Farris 2000a).

Fractional weighting: For analytical purposes, there is a direct relationship between three-item data and binary data with each binary character equivalent to a number of three-item statements, given by the formula $(t - n)n(n - 1)/2$ where t = the number of taxa and n = number of taxa with the informative state. When $n > 2$, there is some redundancy among the statements as not all are independent.

The number of independent three-item statements is given by the formula $(n - 1)(t - n)$. The absolute value is the ratio of independent statements to total statements, $2/n$ (Nelson & Ladiges 1992). Some simple cases are:

	Total statements		Independent statements
ABC(DE)	3:	A(DE)	3
		B(DE)	
		C(DE)	
ABC(DE)	6:	A(CD)	4
		A(DE)	
		A(CE)	
		B(CD)	

Consider a more complex example of five taxa (ABCDE) with three (CDE) having the informative state. The total number of statements is 6 [A(CD) + A(DE) + A(CE) + B(CD) + B(DE) + B(CE)], whereas the number of independent statements is $(n - 1)(t - n) = (3 - 1)(5 - 3) = 4$. The six statements might be thought of as two sets, one corresponding

NATURAL METHODS II **293**

Box 12.1 (*cont.*)

to the relationships of A with CDE, the other corresponding to the relationships of B with CDE. When all three statements for the relationships of A are considered together, any two of the three will yield the same solution: A(DE) + A(CE) = A(CDE). The same holds for the relationships of B to CDE. That is, only four statements are required to yield the correct answer, two from each suite of three, such as:

$$A(DE) + A(CE) + B(DE) + B(CE) = AB(CDE)$$

Nevertheless, all six statements of relationship are 'true' for the actual observation hence no statement can, or should, be eliminated and a fractional weight can be applied to compensate, in this case 'weighting' by 2/3 giving the total proportional weight of 4/6 (Nelson & Ladiges 1992). As redundant statements offers spurious information, fractionally weighted data will be more precise (Nelson & Ladiges 1994).

Missing data: Missing data have long been considered a problem in the analysis of systematic data leading to empty cells in a data matrix[2]. How much missing data is artefact depends on the type of data. For example, two molecular sequences may actually be incomplete due to a poor sequence alignment or actual missing bases; in morphological data, actual parts may simply be missing due to poor preservation, or not yet available. The issue first arose with reference to fossil taxa, which will, to some extent, always have missing data because of the vagaries of preservation (Gauthier 1986, p. 8: 'The multiple trees obtained in the second run resulted from the missing data in the 10 less well known taxa', see Kearney & Clark 2003; but the general problem was addressed by Sneath & Sokal 1973, p. 180). Solutions were offered, usually by attempts to determine which taxa could be safely excluded. Solutions of this kind to one side, these empty matrix cells should be empty – *for there are no available data*. Examine the following matrix (after Platnick et al. 1991):

	1	2	3	4
A	1	0	0	0
B	1	0	0	1
C	1	1	0	0
D	1	1	1	1
E	1	1	1	0
F	1	1	1	?
G	1	1	1	?

There are no data for character 4 for taxa F and G. The matrix should really be:

	1	2	3	4
A	1	0	0	0
B	1	0	0	1

Box 12.1 *(cont.)*

C	1	1	0	0
D	1	1	1	1
E	1	1	1	0
F	1	1	1	
G	1	1	1	

But no cells in a data matrix can (as yet) be left empty and therefore question marks are used to fill the space. These question marks supposedly represent every available option. Thus character 4 becomes characters 4a–d:

	1	2	3	4	4a	4b	4c	4d
A	1	0	0	0	0	0	0	0
B	1	0	0	1	1	1	1	1
C	1	1	0	0	0	0	0	0
D	1	1	1	1	1	1	1	1
E	1	1	1	0	0	0	0	0
F	1	1	1	?	1	0	1	0
G	1	1	1	?	1	0	0	1

Rather than tackle the issue as one of *data representation*, the 'problem' was understood as a computational exercise with options such as *safe taxonomic reduction* (Wilkinson 1995), which required the 'a priori identification and removal of those terminal taxa that, by virtue of their abundance of missing entries, increase the number of MPTs but that can be removed from an analysis without any danger of affecting relationships among the remaining terminal taxa that are supported by the parsimonious interpretation of the complete data' (Wilkinson 1995, p. 502, for further studies along these lines see Wilkinson 2003 and Siu-Ting et al. 2015, and others, but see Kearney & Clark 2003, for counter arguments[3]).

Further, some systematists have explored empirical and simulation studies to monitor the effects of missing data (Wiens 2003; Wiens & Morrill 2011). Regardless, we see missing data as a *pseudoproblem* (data are always missing) and we prefer to call these various computational operations attempts to *massage the data* (or attempts to '. . . wring truth from recalcitrant data', Patterson 1994, p. 185). We repeat: the matrix 'gaps' *should be empty – for there is no available data*[4]. And we repeat again: solutions can be found with the data themselves, rather than any intricate manipulations.

Box 12.1   *(cont.)*

It would be much better if *actual* missing data was treated as if they really were missing. Consider the simple character here represented as a binary variable:

	1
A	0
B	
C	1
D	1
E	1

For this character we have no *actual* observations for taxon B: We do not know what state character 1 has (it is not available, it is an incomplete fossil, it is a missing sequence read, etc.); we have not been able to examine it. The conventional representation in a matrix is:

	1
A	0
B	?
C	1
D	1
E	1

For the same data, the three-item equivalent is A(CD) + A(CE) + A(DE) or:

	1	2	3
A	0	0	0
B			
C	1	1	
D	1		1
E		1	1

There are no relationships for B as these data are missing. Three-item statements represent exactly what has been observed: *missing data are treated as missing* and therefore present no special problem.

Box 12.1   (*cont.*)

Nelson and Platnick presented a more complex example using a matrix with numerous missing entries for 7 characters and 7 taxa (A–G):

	1	2	3	4	5	6	7
A	1	0	?	?	?	?	?
B	1	1	0	?	?	?	?
C	1	1	1	0	?	?	0
D	1	?	1	1	0	?	1
E	0	?	?	1	1	0	1
F	0	?	?	?	1	1	?
G	0	?	?	?	?	1	?

Analysis of this matrix using parsimony yielded one cladogram: A(B(C(D(E(FG). Remarkably, the data for character 1, EFG(ABCD) is not included directly but is *distorted* to mean the opposite of what was discovered: ABCD(EFG). Thus, the missing values influence the outcome being able to switch from 0→1 or 1→0 depending on what is considered the shortest tree, in this case switching character 1. This result can only be interpreted as *methodological artefact*.

The same matrix can be represented by 24 three-item statements. Analysis of those yields a different cladogram: (E(FG))(A(B(CD))). If numbers of statements found in both cladograms are considered a measure of the information content of the matrix, results are as follows:

	1	2	3	4	5	6	7	Total n	Total %
Total three-item statements per character	18	1	1	1	1	1	1	24	100
Results for raw data cladogram	0	1	1	1	1	1	1	6	25
Results for three-item cladogram	18	1	1	0	1	1	0	22	92

The three-item cladogram includes 92% of the original data. That measure may not be acceptable to some. Even so, when the raw data are treated as binary characters, it allows a massive distortion to the original observations when processed by one or another program (see also Zaragüeta i Bagils & Bourdon 2007).

Paired homologues: One further, and an altogether more serious exploration of three-item data, was the issue of paired homologues (Carine & Scotland 1999, 2000; Scotland 2000a, 2000b; Scotland & Carine 2000). Based on further developments of Patterson's approach to homology, Carine and Scotland modified the three-item

Box 12.1 (*cont.*)

approach to take into account what they referred to as 'paired homologues'. This was based on the idea that the complement relation – Patterson's presence versus absence – is but one taxic version of homology, with paired homologues – having two different states – being the more usual form of homology statement and that each pair deserved equal treatment when rendered into its three-item equivalents. The relation was depicted in their figure 2 (Carine & Scotland 1999, figure 2, reproduced here as our Figure B12.1.1) and is a representation of the standard unrooted binary character (compare with our discussion in Box 3.4). One might appreciate from this diagram the difficulty in placing the character separating the pair on the central branch, where X and X^1 sit rather uncomfortably adjacent to one another, side by side. Effectively, Carine and Scotland were coding the plesiomorphic aspect of the data *as well as* rendering it a somewhat quasi-phenetic approach ('The new method is sensitive to autapomorphies and can group by symplesiomorphy[5]. Carine and Scotland have reinvented phenetics'; Farris et al. 2001, p. 83). Nevertheless, it was an interesting exercise that inadvertently revealed yet another aspect of the irrelevance of unrooted trees ('*Paired homologues*').

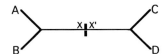

Figure B12.1.1 'Paired homologues' (redrawn from Carine & Scotland 1999, figure 2).

Optimisation: Some of the implementation issues above were discussed as if they were fundamental to the method, and the journal *Cladistics* published a number of articles as if they were matters of principle. This inevitably delayed – or more correctly obscured – discussion that related to what was, and still is, a very real problem with conventional approaches to parsimony analysis when using binary data matrices: *optimisation*. This process effectively needs to *distort* data in its search for the shortest cladogram – the shortest cladogram was, perhaps, a success for programmers but a hindrance to biologists, creating artefact in its wake (Platnick, 1989).

The critique of *optimisation* was by far the most extensive, which was no surprise, as it threatened the very basis of Wagner parsimony (and other methods to be sure). By distort, we mean the data are manipulated – and in many cases (most cases?) actually *require* manipulation – to find the shortest cladogram, the *sine qua non* of Wagner parsimony. Discussions relating to optimisation have smouldered in the relevant literature ever since. We discuss this topic in more detail via examples later in the body of this chapter.

References

Brazeau, MD., Guillerme, T. & Smith, MR. 2019. An algorithm for morphological phylogenetic analysis with inapplicable data. *Systematic Biology* 68: 619–631.

Gauthier, J. 1986. Saurischian monophyly and the origin of birds. *Memoirs of the California Academy of Sciences* 8: 1–55.

Box 12.1 *(cont.)*

Kearney, M. & Clark, JM. 2003. Problems due to missing data in phylogenetic analyses including fossils: a critical review. *Journal of Vertebrate Paleontology* 23: 263–274.

Maddison, W. 1993. Missing data versus missing characters in phylogenetic analysis. *Systematic Biology* 42: 576–581.

Nelson, G. & Ladiges, PY. 1996. Paralogy in cladistic biogeography and analysis of paralogy-free subtrees. *American Museum Novitates* 3167: 1–58.

Nelson, GJ. & Platnick, NI. 1991. Three-taxon statements: a more precise use of parsimony? *Cladistics* 7: 351–366.

Patterson, C. 1994. Null or minimal models. In: Scotland, R., Siebert, DJ. & Williams, DM. (eds.), *Models in Phylogeny Reconstruction*. Systematics Association Special Volume 52. Oxford University Press, Oxford, pp. 173–192.

Platnick, NI. 1989.Cladistics and phylogenetic analysis today. In: Fernholm, B., Bremer, K. & Jörnvall, H. (eds.), *The Hierarchy of Life*. Elsevier, Amsterdam, pp. 17–24.

Platnick, NI., Griswold, CE. & Coddington, JA. 1991. On missing entries in cladistic analysis. *Cladistics* 7: 337–343.

Siu-Ting, K., Pisani, D., Creevey, CJ. & Wilkinson, M. 2015. Concatabominations: identifying unstable taxa in morphological phylogenetics using a heuristic extension to safe taxonomic reduction. *Systematic Biology* 64: 137–143.

Sneath, PHA. & Sokal, RR. 1973. *Numerical Taxonomy: The Principles and Practice of Numerical Classification*. W. H. Freeman, San Francisco.

Wiens, JJ. 2003. Missing data, incomplete taxa, and phylogenetic accuracy. *Systematic Biology* 52: 528–538.

Wiens, JJ. & Morrill, MC. 2011. Missing data in phylogenetic analysis: reconciling results from simulations and empirical data. *Systematic Biology* 60: 719–731.

Wilkinson, M. 1995. Coping with abundant missing entries in phylogenetic inference using parsimony. *Systematic Biology* 44: 501–514.

Wilkinson, M. 2003. Missing entries and multiple trees: instability, relationships, and support in parsimony analysis. *Journal of Vertebrate Paleontology* 23: 311–323.

Zaragüeta i Bagils, R. & Bourdon, E. 2007. Three-item analysis: hierarchical representation and treatment of missing and inapplicable data. *Compte Rendu Palevol* 6: 527–534.

References: The first 10 years of three-item studies (1991–2002)

Below we have provided the reader with all the literature from the first 10 years of discussion, pro and con, for the decade 1991–2002. We have omitted the three-item biogeographic literature from this list but it is certainly relevant and interested persons will find many of them in the main bibliography of this chapter. It should go without saying that it is possible that three-item representation for character data might never have been developed had its potential in biogeography not been first appreciated.

We have divided the publications below into *Theoretical contributions* and *Practical applications*. Some listed in *Practical applications* do sometimes have a theoretical component (e.g., De Laet 1997, con; Nelson & Ladiges 1994, pro).

The *Theoretical contributions* have been listed in chronological order so those interested may follow the discussions as they developed.

Box 12.1 (cont.)

Theoretical Contributions

1991

Nelson, GJ. & Platnick, NI. 1991. Three-taxon statements: a more precise use of parsimony? *Cladistics* 7: 351–366.

1992

Harvey, AW. 1992. Three-taxon statements: more precisely, an abuse of parsimony? *Cladistics* 8: 345–354.

Nelson, GJ. 1992. Reply to Harvey. *Cladistics* 8: 355–360.

Nelson, GJ. & Ladiges, PY. 1992. Information content and fractional weight of three-taxon statements. *Systematic Biology* 41: 490–494.

1993

Kluge, AG. 1993. Three-taxon transformation in phylogenetic inference: ambiguity and distortion as regards explanatory power. *Cladistics* 9: 246–259.

Nelson, GJ. 1993. Reply. *Cladistics* 9: 261–265.

Nelson, GJ. & Ladiges, PY. 1993. Missing data and three-item analysis. *Cladistics* 9: 111–113.

Platnick, NI. 1993. Character optimization and weighting: differences between the standard and three-taxon approaches to phylogenetic inference. *Cladistics* 9: 267–272.

1994

Kluge, AG. 1994. Moving targets and shell games. *Cladistics* 10: 403–413.

Wilkinson, M. 1994. Three-taxon statements: when is a parsimony analysis also a clique analysis? *Cladistics* 10: 221–223.

1995

Farris, JS., Kallersjo, M., Albert, VA., Allard, M., Anderberg, A., Bowditch, B., Bult, C, Carpenter, JM., Crowe, TM., De Laet, J., Fitzhugh, K., Frost, D., Goloboff, P., Humphries, CJ., Jondelius, U., Judd, D., Karis, PO., Lipscomb, D., Luckow, M., Mindell, D., Muona, J., Nixon, K., Presch, W., Seberg, O., Sidall, ME., Struwe, L., Tehler, A., Wenzel, J., Wheeler, Q. & Wheeler, W. 1995. Explanation. *Cladistics* 11: 211–218.

1996

Deleporte, P. 1996. Three-taxon statements and phylogeny construction. *Cladistics* 12: 273–289.

Nelson, GJ. 1996. *Nullius in Verba*. Published by the Author. New York. and *Journal of Comparative Biology* 1: 141–152.

Platnick, NI., Humphries, CJ., Nelson, GJ. & Williams, DM. 1996. Is Farris optimization perfect? *Cladistics* 12: 243–252.

1997

Farris, JS. 1997. Cycles. *Cladistics* 13: 131–144.

Siebert, DJ. & Williams, DM. 1997. Book Review [Nullius in Verba]. *Biological Journal of the Linnean Society* 60: 145–6.

Box 12.1 *(cont.)*

1998

De Laet, J. & Smets, E. 1998. On the TTSC–FTSC formulation of standard parsimony. *Cladistics* 14: 239–248.

De Laet, J. & Smets, E. 1998. On the three-taxon approach to parsimony analysis. *Cladistics* 14: 363–381.

Farris, JS. & Kluge, AG. 1998. A/The brief history of three-taxon analysis. *Cladistics* 14: 349–362.

Kitching, IJ., Forey, PL., Humphries, CJ. & Williams, DM. 1998. *Cladistics: The Theory and Practice of Parsimony Analysis*. Oxford University Press, Oxford, pp. 168–186.

Siebert, DJ. & Williams, DM. 1998. Recycled. *Cladistics* 14: 339–347.

1999

Carine, MA. & Scotland, RW. 1999. Taxic and transformational homology: different ways of seeing. *Cladistics* 15: 121–129.

Kluge, AG. & Farris, JS. 1999. Taxic homology = Overall similarity. *Cladistics* 15: 205–212.

2000

Carine, MA. & Scotland, RW. 2000. 68 taxa and 32 characters: resolving species relationships using morphological data. In: Morton, CM., Harley, MM. & Blackmore, S. (eds), *Pollen and Spores: Morphology and Biology*. Royal Botanic Garden, Kew, pp. 365–384.

Farris, JS. 2000a. Diagnostic efficiency of three-taxon analysis. *Cladistics* 16: 403–410.

Farris, JS. 2000b. Paraphyly, outgroups, and transformations. *Cladistics* 16: 425–429.

Scotland, RW. 2000a. Homology, coding and three-taxon statement analysis. In: Scotland, RW. & Pennington, RT. (eds) *Homology and Systematics: Coding Characters for Phylogenetic Analysis*. Taylor and Francis, London, pp. 145–182.

Scotland, RW. 2000b. Taxic homology and three-taxon statement analysis. *Systematic Biology* 49: 480–500.

Scotland, RW. & Carine, MA. 2000. Classification or phylogenetic estimates. *Cladistics* 16: 411–419.

Williams, DM. & Siebert, DJ. 2000. Characters, homology and three-item analysis. In: Scotland, RW. & Pennington, RT. (eds), *Homology and Systematics: Coding Characters for Phylogenetic Analysis*. Taylor and Francis, London, pp. 183–208.

2001

Farris, JS., Kluge, AG. & De Laet, JE. 2001. Taxic revisions. *Cladistics* 17: 79–103.

2002

Williams, DM. 2002. Parsimony and precision. *Taxon* 51: 143–149.

Practical Applications

De Laet, J. 1997. A Reconsideration of Three-Item Analysis, the Use of Implied Weights in Cladistics, and a Practical Application in Gentianaceae. Ph.D. thesis, Katholieke Universiteit Leuven.

Box 12.1 *(cont.)*

Ebach, MC. & McNamara, KJ. 2002. A systematic revision of the family Harpetidae (Trilobita). *Records of the Western Australian Museum* 21: 135–167.

Ladiges, PY., McFadden, GI., Middleton, N., Orlovich, DA., Treloar, N. & Udovicic, F. 1999. Phylogeny of *Melaleuca*, *Callistemon*, and related genera of the *Beaufortia* suballiance (Myrtaceae) based on 5S and ITS-1 spacer regions of nrDNA. *Cladistics* 15: 151–172.

Nelson, GJ. & Ladiges, PY. 1994. Three-item consensus: empirical test of fractional weighting. In: Scotland, RW., Seibert, DJ. & Williams, DM. (eds), *Models in Phylogeny Reconstruction*. Systematics Association Special Volume 51. Clarendon Press, Oxford, pp. 193–209.

Patterson, C. & Johnson, GD. 1995. The intermuscular bones and ligaments of teleostean fishes. *Smithsonian Contributions in Zoology* 559: 1–78.

Udovicic, F., McFadden, GI. & Ladiges, PY. 1995. Phylogeny of *Eucalyptus* and *Angophora* based on 5s rDNA spacer sequence data. *Molecular Phylogenetics and Evolution* 4: 247–256.

Williams, DM. 1996. Fossil species of the diatom genus *Tetracyclus* (Bacillariophyta, 'ellipticus' species group): Morphology, interrelationships and the relevance of ontogeny. *Philosophical Transactions of the Royal Society, London, Series B* 351: 1759–1782.

[1] We do not deal with biogeography in this book, but the theoretical studies are relevant.

[2] Of course, one must expect missing data in any cladistic analysis – few matrices achieve n = all.

[3] This issue of the *Journal of Vertebrate Palaeontology* has several papers on the topic of missing data, see volume 23(2): 263–323, 2003.

[4] Likewise, the notion that inapplicable data (e.g., Maddison 1993) present difficulties is also a *pseudoproblem*: if data are inapplicable then they are *inapplicable*: they do not exist. *The handicap is matrix representation not the lack of data or lack of some computational fix.* We hesitate to use the word, but many modern attempts to solve these problems by computational trickery will be found ultimately futile (e.g., Brazeau et al. 2019): 'To a man with a hammer, everything looks like a nail' (Mark Twain).

[5] Note here the nature of the critique. If an unrooted tree is converted into its three-item equivalent, it is a reinvention of phenetics. If an unrooted tree is retained as a binary character in a matrix, it is not phenetic. This example, perhaps inadvertently, supports our view that binary character representation is indeed inherently phenetic. The difference here is that 'paired homologues' convert apomorphic and plesiomorphic data into something informative, while binary characters are converted into something informative only made informative by the program (e.g., reversals). Either way, we see both as phenetic.

In Box 12.1 we provide a brief summary of the issues raised, responses to those issues as well as a full bibliography, should any interested reader wish to follow the topic up for themselves, for this first critique, which lasted for roughly a decade, from 1991 until 2002.

Two further detailed and focused explorations of three-item analysis have recently emerged. The first based in the Laboratoire Informatique et Systématique (LIS), Paris, France, the second in the Florida Museum of Natural History, University of Florida, USA. We have attempted to summarise their approaches, although it should be noted that both of these research programmes are active and much more is expected (for more details see Box 12.2).

Box 12.2 Subsequent Exploration of Three-Item Analysis (2002–2018)

(a) LisBeth (www.infosyslab.fr/, Zaragüeta i Bagils R. et al. 2012)

LisBeth (Zaragüeta i Bagils et al. 2012), a computer program for the analysis of three-item data, first released under the name *Nelson09* (Cao et al. 2009), uses a parenthetical matrix, thereby avoiding all the methodological and quasi-theoretical baggage of unrooted networks and hypothetical outgroups used to construct a binary matrix, and, most importantly, does not need question marks for missing values. LisBeth's use of parentheses means that it is easier to visualise the relationships between two or more character states, for instance, in the form of (0,(1,1)), (0,(1,2)) and (0,1,(2)), as well as visualising the characters themselves. Effectively each character is a cladogram (i.e., branching diagram or mini-cladogram, see Chapter 11), which is then converted into uniformly or fractionally weighted three-item statements (see Box 12.1), which may be added together using the compatibility function in the program to find the minimal tree (see Box 12.1); or exported to any current tree-building program (e.g., PAST, TNT, PAUP, etc.) using whichever method is preferred (e.g., ML, Parsimony, compatibility, etc.). The use of parentheses, rather than a binary data matrix, avoids the use of question marks for cells that need to remain empty. For example, consider the following data matrix of eight characters (1–8) and four taxa (A–D):

	1	2	3	4	5	6	7	8
A	0	0	0	0	0	0	0	0
B	0	0	1	1	1	1	1	1
C	1	1	0	0	0	0	0	0
D	1	1	1	1	1	1	1	1

In LisBeth these data will appear as so:

Character 1	(A,(B,(C,D))
Character 2	(A,(B,(C,D))
Character 3	(A,(C,(B,D))
Character 4	(A,(C,(B,D))
Character 5	(A,(C,(B,D))
Character 6	(A,(C,(B,D))
Character 7	(A,(C,(B,D))
Character 8	(A,(C,(B,D))

Box 12.2   (*cont.*)

Set out this way, no question marks are necessary as the data is not restricted to a grid in which every cell needs a value. Moreover, the relationships are directly visible without the need for any all-zero hypothetical 'ancestor' (in cases where one is needed). The only problem is adding the three-items together to find the optimal tree. The implicit search function in LisBeth, while effective, is limited to a certain size. Exporting the data in a NEXUS or TNT format means the whole parenthesis matrix is converted back to a binary matrix with the seemingly infinite number of question marks as well as an all-zero hypothetical outgroup. We do not see this as a disadvantage of LisBeth but, rather, the highly limited nature in which systematists view their own data, namely in a strictly grid-like fashion in which all cells require a value, when there may not be one.

(b) TAXODIUM (Mavrodiev & Madorsky 2012) and FORESTER (Mavrodiev et al. 2017)

TAXODIUM, similar to the original three-item program TAX (Nelson & Ladiges 1991), converts a CSV binary/grid matrix into either a uniformly or fractionally weighted three-item binary/grid matrix in NEXUS and, like *Lisbeth*, the exported data matrix may be used with the phylogenetic software of your choice (e.g., MaxL, Mr Bayes, Mesquite etc.) to find the minimal tree. The problem with TAXODIUM is that it requires a zero outgroup and question marks for empty grid cells.

Mavrodiev et al. (2017), however, have created FORESTER, which converts a molecular or morphological ordinary binary/grid matrix into a series of cladograms, similar to the input in Lisbeth above. The benefit of the FORESTER program is that it can convert supermatrices into cladograms for further analysis. FORESTER is not in the strict sense a three-item program as it does not look for three-item statements, but it does deal with the problem of multiple question marks.

References

Cao, N., Bourdon, E., El Azawi, M. & Zaragüeta i Bagils, R. 2009. Three-item analysis and parsimony, intersection tree and strict consensus: a biogeographical example. *Bulletin de la Société Géologique de France* 180: 13–15.

Mavrodiev, EV. & Madorsky, A. 2012. TAXODIUM Version 1.0: a simple way to generate uniform and fractionally weighted three-item matrices from various kinds of biological data. *PLoS One* 7(11): e48813.

Mavrodiev, EV., Dell, C. & Schroder, L. 2017. A laid-back trip through the Hennigian Forests. *PeerJ* 5: e3578. http://doi.org/10.7717/peerj.3578

Nelson, GJ. & Ladiges, PY. 1991. Standard assumptions for biogeographic analysis. *Australian Systematic Botany* 4: 41–58.

Zaragüeta i Bagils, R., Ung, V., Grand, A., Vignes-Lebbe, R., Cao, N. & Ducasse, J. 2012. LisBeth: new cladistics for phylogenetics and biogeography. *Comptes Rendus Palevol* 11: 563–566.

Box 12.2 *(cont.)*

References: The Next 15 Years (2002–2017)

Theoretical Contributions

Bininda-Emonds, ORP. (ed.) 2004a. *Phylogenetic Supertrees: Combining Information to Reveal the Tree of Life*. Kluwer/Academic Press, Dordrecht.

Cao, N., Zaragüeta i Bagils, R. & Vignes-Lebbe, R. 2007. Hierarchical representation of hypotheses of homology. *Geodiversitas* 29: 5–15.

Cao, N., Bourdon, E., El Azawi, M. & Zaragüeta i Bagils, R. 2009. Three-item analysis and parsimony, intersection tree and strict consensus: a biogeographical example. *Bulletin de la Société Géologique de France* 180: 13–15.

Ebach, MC. & Williams, DM. 2012 [2013]. E quindi uscimmo a riveder le stelle. *Cladistics* 29: 227.

Grand, A. 2013. Représentation sémantique des phénotypes: métamodèle et ontologies pour les caractères taxonomiques et phylogénétiques. Ph.D. thesis, MNHN Paris.

Mavrodiev, EV. 2015. Three-taxon analysis can always successfully recognize groups based on putative reversals. *PeerJ PrePrints* 3: e1206.

Mavrodiev, EV. 2016. Dealing with propositions, not with the characters: the ability of three-taxon statement analysis to recognise groups based solely on 'reversals', under the maximum-likelihood criteria. *Australian Systematic Botany* 29: 119–125.

Mavrodiev, EV. & Madorsky, A. 2012. TAXODIUM Version 1.0: a simple way to generate uniform and fractionally weighted three-item matrices from various kinds of biological data. *PLoS One* 7 (11): e48813.

Mavrodiev, EV. & Yurtseva, OV. 2017. "A character does not make a genus, but the genus makes the character": three-taxon statement analysis and intuitive taxonomy. *European Journal of Taxonomy* 377: 1–7. https://doi.org/10.5852/ejt.2017.377

Mavrodiev, EV., Dell, C. & Schroder, L. 2017. A laid-back trip through the Hennigian Forests. *PeerJ* 5: e3578. https://doi.org/10.7717/peerj.3578

Mavrodiev, EV., Williams, DM. & Ebach, MC. 2019. On the typology of relations. *Evolutionary Biology* 46: 71–89.

Mooi, RD., Williams, DM. & Gill, A.C. 2011. Numerical cladistics, an unintentional refuge for phenetics – a reply to Wiley et al. *Zootaxa* 2946: 17–28.

Nelson, GJ., Williams, DM. & Ebach, MC. 2003. A question of conflict: three item and standard parsimony compared. *Systematics & Biodiversity* 1: 145–149.

Prin, S. 2012. Structure mathématique des hypothèses cladistiques et conséquences pour la phylogénie et l'évolution – Avec une perspective sur l'analyse cladistique. Ph.D. thesis, MNHN Paris.

Rineau, V., Grand, A., Zaragüeta i Bagils, R. & Laurin, M. 2015. Experimental systematics: sensitivity of cladistic methods to polarization and character ordering schemes. *Contributions to Zoology* 84: 129–148.

Rineau, V., Zaragüeta i Bagils, R. & Laurin, M. 2018. Impact of errors on cladistic inference: simulation-based comparison between parsimony and three-taxon analysis. *Contributions to Zoology* 87: 25–40.

Sevillya, G., Frenkel, Z. & Snir, S., 2016. TripletMaxCut: a new toolkit for rooted supertree. *Methods in Ecology and Evolution* 2016: 1359–1365.

Williams, DM. 2004. Homology and homologues, cladistics and phenetics: 150 years of progress. In: Williams, DM. & Forey, PL. (eds), *Milestones in Systematics*. Taylor & Francis, London, pp. 191–224.

Box 12.2 (cont.)

Williams, DM. & Ebach, MC. 2005. Drowning by numbers: a new view on numerical phylogenetics. *Botanical Reviews* 71: 355–387.

Williams, DM. & Ebach, MC. 2006. The data matrix. *Geodiversitas* 28: 409–420.

Williams, DM. & Ebach, MC. 2008. *The Foundations of Systematics & Biogeography*. Springer-Verlag New York Inc., New York.

Williams, DM. & Ebach, MC. 2012. Confusing homologs as homologies: a reply to "On homology". *Cladistics* 28: 223–224.

Williams, DM. & Ebach, MC. 2012. "Phenetics" and its application. *Cladistics* 28: 229–230.

Williams, DM. & Ebach, MC. 2017. What is intuitive taxonomic practice? *Systematic Biology* 66: 637–643.

Williams, DM. & Humphries, CJ. 2003. Component coding, three-item coding and consensus methods. *Systematic Biology* 52: 255–259.

Williams, DM., Schmitt, M. & Wheeler, Q. (eds) 2016. *The Future of Phylogenetic Systematics: The Legacy of Willi Hennig*. Cambridge University Press, Cambridge, UK.

Zaragüeta i Bagils, R. & Bourdon, E. 2007. Three-item analysis: hierarchical representation and treatment of missing and inapplicable data. *Compte Rendu Palevol* 6: 527–534.

Zaragüeta i Bagils, R. & Pécaud, S. 2016. Why cladograms should be dichotomous? In: Williams, DM., Schmitt, M. & Wheeler, Q. (eds) *The Future of Phylogenetic Systematics – The Legacy of Willi Hennig*. Cambridge University Press, Cambridge, UK, pp. 230–257,

Zaragüeta i Bagils, R., Ung, V., Grand, A., Vignes-Lebbe, R., Cao, N. & Ducasse, J. 2012. LisBeth: new cladistics for phylogenetics and biogeography. *Comptes Rendus Palevol* 11: 563–566.

Practical Applications

Cao, N. 2008. Analyse à trois elements et anatomie du bois des FAGALES Engl. Ph.D. dissertation, Muséum national d'Histoire naturelle, Paris.

Corvez, A. 2012. L'origine de la Mégaphylle chez les Monilophytes. Ph.D. dissertation, Muséum national d'Histoire naturelle, Paris, France.

Gill, AC. & Leis, JM. 2019. Phylogenetic position of the fish genera *Lobotes, Datnioides and Hapalogenys*, with a reappraisal of acanthuriform composition and relationships based on adult and larval morphology. *Zootaxa* 4680: 1–81.

Grand, A., Zaragüeta i Bagils, R., Vélezi, LM. & Ung, V. 2014. A cladistic re-analysis of the Gadiformes (Teleostei, Paracanthopterygii) using three-item analysis. *Zootaxa* 3889: 525–552.

Mavrodiev, EV., Martinez-Azorin, M., Dranishnikov, P. & Crespo, MB. 2014. At least 23 genera instead of one: the case of *Iris* L. s.l. (Iridaceae). *PLoS One* 9 (8): e106459.

Rineau, V. 2017. Un nouveau regard cladistique sur l'anatomie comparée, la systématique, la phylogénie et la paléoécologie des rudistes (Bivalvia, Hippuritidia). Ph.D. thesis, MNHN Paris.

Williams, DM. & Reid, G. 2006. *Amphorotia* nov. gen., a new genus in the family Eunotiaceae (Bacillariophyceae), based on *Eunotia clevei* Grunow in Cleve et Grunow. *Diatom Monographs* 6.

Yurtseva, OV., Severova, EE. & Mavrodiev, EV. 2017. *Persepolium* (Polygoneae): a new genus in Polygonaceae based on conventional Maximum Parsimony and Three-taxon statement analyses of a comprehensive morphological dataset. *Phytotaxa* 314: 151–194.

Zaragüeta i Bagils, R. 2002. Tests morphologique et moléculaire des hypothèses de phylogénie des Clupeomorpha (Teleostei). Ph.D. thesis, MNHN Paris.

As noted above, here we treat the subject from the perspective of a few generalities. While three-item analysis was initially developed with biogeography in mind, we deal here only with character data (for useful accounts of three-item analysis in biogeography see Parenti & Ebach 2009 and Cecca & Zaragüeta i Bagils 2015).

If we cast back to some previous chapters, we noted that in the last few decades the creation of numerous methods of data analysis, metrics and measures have, and still are, increasing exponentially, creating, in its wake, a niche within the discipline of *bioinformatics*, leaving taxonomists with a wealth of programs but precious little guidance. It is time, we believe, to step back and reconsider some basic issues all over again.

Initially, three-item analysis was seen, mistakenly, as simply yet another *method of analysis* when it was primarily intended as an alternative way to *view and represent systematic data (evidence)*. Of course, the first analyses using three-item data did happen to expose certain shortcomings, of Wagner parsimony in particular (Nelson & Platnick, 1991), but representing data in this fashion was part of a general quest for obtaining *method-free* results (Nelson & Platnick, 1981), if such is possible (Chapters 7–10). To emphasise here: *method-free* is not the same as *theory-free*. We see *method-free* approaches to systematics as a continuation of the mid-nineteenth century pursuit of *natural classification*[4]. As we noted elsewhere, without some measure of 'truth', there is simply no way to be certain that the relationships discovered by whatever method, even with a three-item approach, are indeed real rather than artefactual (Chapter 8). Below we first deal with a few basic issues and then deal with some aspects of what we have referred to as Data distortion.

How Are Systematic Data Best Represented?

With general acceptance of the cladistic approach, it is now commonplace for systematists to publish a matrix, with only a brief list of characters, and to concentrate on the results of manipulating that matrix with parsimony or other programs. In other words, the emphasis has shifted from observation, the source of the matrix, to whatever message may be extracted from the matrix … This change of emphasis replaces our pernicious old black box, evolutionary systematics, with a new one, the matrix.

(Patterson & Johnson 1997a, pp. 361–362)

One might argue about the origins of the data matrix itself, but it is generally understood as an essential component of any numerical analysis of systematic

---

[4] We are aware that the search for a natural system of classification began earlier, perhaps beginning with Linnaeus, John Ray even.

data. The data matrix has become central to any paper on systematics or taxonomy, even if not compulsory, it is more than commonplace to include one. Patterson and Johnson (above) were referring to the *quality* of the data rather than the *form* in which it is presented, but the latter is of significance as well. The *quality* is part of the initial study and, of course, the authors' responsibility to explore the organism's characters – elsewhere we discussed some kinds of characters that are not really structural, such as measurements, etc., which we consider to be *false homologues (false characters)* in terms of natural classification[5] (see Chapter 7). Here, we concentrate on the *form* of the data: how best to represent it.

Consider this simple example, one explored earlier in Chapter 8 with respect to conventional methods of analysis. A single binary character distributed among four taxa, A–D, is represented thus:

A   0

B   0

C   1

D   1

Stated differently, this represents evidence that supports the relationship AB (CD): in words, C and D are more closely related to each other than either are to A or B. C and D have the apomorphic value, with 'apomorphic', 'evidence' and 'defining feature' all considered equivalent.

The style of representation above is the classic *'taxon × character' matrix* with rows for taxa, columns for characters. Matrix representation is now an almost universally standard style for tabulating data. We refer to this form of representation as a *binary matrix*. This includes all matrices where the coding of data is a series of zeros and ones (and higher numbers if necessary) for the coding of characters or cladograms (see below).

The matrix entry above and the statement(s) of relationship derived from it are one and the same, or at least should be.

The addition of an *outgroup* to the matrix renders the data more straightforward *for analysis.* The outgroup is a *simple heuristic device* to allow the 'pleisomorphic' and 'apomorphic' values to be recognised properly *prior to analysis* – to allow what we have observed to remain true:

---

[5] Less we be misunderstood, these kinds of characters can be useful for *identification*.

OUT	0
A	0
B	0
C	1
D	1

This might be summarised as follows:

			Out	0				
A	0		A	0				
B	0	=	B	0	=	AB(CD)	(AB)CD?	(AB)(CD)?
C	1		C	1				
D	1		D	1				

For some, this outcome is not the only one possible, at least where some parsimony analyses are concerned. As noted in Chapter 8 and in Box 12.1 of this chapter, two further relationships are considered possible: (AB)CD and (AB)(CD).

Consider, then, the three-item equivalent of the same binary character AB(CD). It is important to emphasise that three-item *data* deals with *relationships* rather than *groups – that is what is meant by relational.* And we emphasise again that *three-item analysis primarily concerns how we view data not how we undertake subsequent analysis.* That three-item data exposed some shortcomings of methods of analysis is one of its virtues, which we return to later.

The character AB(CD) can be represented by two three-item statements of relationship: A(CD) and B(CD). If these are now considered to be the raw data, instead of the binary representation, then the *only* outcome of their combination or analysis is AB(CD).

This might be summarised as follows:

Out	0		0	0			0				
A	0		0	?	A	0		A(CD)			
B	0	=	?	0	B	0	=	+	=	AB(CD)
C	1		1	1	C	1		B(CD)			
D	1		1	1	D	1					

Compatibility analysis might be thought of as dealing with binary data as having only one option, AB(CD), rather than the three possible for parsimony analysis.

While this is an improvement over the simple binary variable, it is still constrained by what we call *the maximal statement of relationship*, AB(CD), rather than the *minimal statements of relationships*, A(CD) + B(CD) and still deals with *groups* rather than *relationships* (see section in Chapter 11: Cladistic Analysis and Synthesis: Cladogram Structure). To recap:

1. The binary matrix entry AB(CD) for parsimony analysis might be understood to mean either AB(CD), (AB)(CD) or (AB)CD after analysis;

2. The binary matrix entry AB(CD) for compatibility analysis might be understood to mean AB(CD) only;

3. The three-item data for AB(CD) and its subsequent analysis is only understood to mean A(CD) + B(CD) and, as a consequence, AB(CD).

Consider another simple example, related to that above, using a character that is more likely to be encountered in molecular analyses but does occur in morphological studies. Data distributed among five taxa A–E such that B, CD and EF all have the same character but of a different state. This might be represented in a matrix thus[6]:

A	0
B	0
C	1
D	1
E	2
F	2

Here we might consider two further points. Again, this matrix has no specific outgroup. One can be added, having the same value as A. Alternatively, A itself could function as the outgroup:

OUT	0
A	0
B	0
C	1
D	1
E	2
F	2

---

[6] There are other ways, but we deal with this only to illustrate our point.

Under the assumption that we do not know the *order* of the three states 0, 1 and 2, the character is equivalent to the relationship AB(CD)(EF).

In summary:

A	0		A	0		
B	0		B	0		
C	1	=	C	1	=	AB(CD)(EF)
D	1		D	1		
E	2		E	2		
F	2		F	2		

This kind of character is usually called a *multistate character* and can be represented either as one character or as a suite of binary characters. Rendering multistate characters as series of binary characters (as in *additive binary coding*), includes redundant information and may not allow the binary and multistate to be direct equivalents. A multistate character should be more informative (as states are assumed dependent) than a pair of binary characters (which are assumed independent).

From the perspective of three-item statements a multistate character is equivalent to a suite of unique three-item statements with no statement appearing more than once (Nelson & Ladiges, 1992b). In these cases the characters have clear relationships among their states.

Our example is of an *unordered multistate character*. To be sure, the unordered multistate character AB(CD)(EF) is not particularly informative: it is equivalent to only four three-item statements, A(CD), B(CD), A(EF) and B(EF). A summary is:

A	0		A	0		A(CD)		
B	0		B	0		A(EF)		
C	1	=	C	1	=	+	=	AB(CD)(EF)
D	1		D	1		B(CD)		
E	2		E	2		B(EF)		
F	2		F	2				

What does parsimony analysis, for example, yield in the case of this multistate character? Depending on the program used (and the outgroup option), either three or four cladograms will result[7]: AB(CD(EF)), AB(EF(CD)), AB(CD)(EF) and

---

[7] There are other outcomes as well, but this is again to illustrate our point.

Table 12.1 Three possible optimisations of the character AB(CD)(EF)

Parsimony solution	Transformation (coded)	Transformation (character)
AB(CD(EF))	00 → 11→ 22	AB → (CD → (EF))
AB(EF(CD))	00 → 22 → 11	AB → (EF → (CD))
AB(CD)(EF)	00 → 11 00 → 22	AB → (CD) AB → (EF)

AB((CD)(EF)). For our purposes, the last of these four can be ignored as it results from an additional unsupported basal node added due to the inclusion of a zero-length branch[8]. An overall summary can be obtained by the strict consensus of the remaining three, which yields an unresolved bush (ABCDEF). This should cause concern. Let us recap: the original datum is AB(CD)(EF) – this is based on observations of some kind, either bases pairs or some point(s) of morphology. When represented as a multistate character, the solution to its analysis is (ABC-DEF). How so? Surely to conclude that AB(CD)(EF) = (ABCDEF) is intuitively incorrect.

The problem here is, of course, optimisation. For parsimony programs, to minimise tree length all possible shortest trees have to be considered, of which there are three (Table 12.1). Yet, overall, this suggests that there is no optimal solution as all three permutations are equally possible – this contradicts knowledge we already have from the original data. Once again, it appears that the three possible solutions are arrived at by applying *a method* – parsimony in this case – using conventional binary data matrices, and is a product *of the method and its assumptions rather than the data themselves* (Table 12.1). This is not a good situation. But at least it allows the identification of what could be considered artefact.

Whatever method of analysis is applied to the suite of four three-item statements – whether they are simply added together, processed via a parsimony program, processed via a compatibility program, etc. – the solution is always the same: AB(CD)(EF). From any point of view, as far as we understand the matter, the only acceptable answer is that AB(CD)(EF) = AB(CD)(EF). These two simple examples highlight the effect optimisation has on a single character. And highlights what we mean by artefactual resolution – method-based resolution rather than data-based resolution. What of more complex situations when conflict occurs?

---

[8] It is unlikely that any current programs output this last cladogram.

Character Conflict: Representation and Data-Distortion

In nearly all conventional data analyses, some characters will conflict. That is, different characters suggest alternative relationships that are incompatible with one another. With conventional results, the conflict remains as part of the result but is explained away as homoplasies, reversals and so on. Take another simple data matrix of four taxa, A–D, and eight characters (0 = plesiomorphic value; 1 = apomorphic value):

	1	2	3	4	5	6	7	8
A	0	0	0	0	0	0	0	0
B	0	0	1	1	1	1	1	1
C	1	1	0	0	0	0	0	0
D	1	1	1	1	1	1	1	1

Characters 1 and 2 support the relationship AB(CD), characters 3–8 support the alternative relationship AC(BD). The weight of evidence suggests a preference for AC(BD) (6 characters, 3–8) against that of AB(CD) (2 characters, 1–2). What of characters 1 and 2? They can be made to fit any particular cladogram by adding them to the optimal solution (Figure 12.1a). The conflicts are explained away by invoking *homoplasy*: characters 1 and 2 are not *really* the same; they may look alike, but have really originated separately in taxon C and D.

If we continue our studies and find 10 more characters, our matrix now looks different:

	1	2	3	4	5	6	7	8	9	10	11	12	13	14	15	16	17	18
A	0	0	0	0	0	0	0	0	0	0	0	0	0	0	0	0	0	0
B	0	0	1	1	1	1	1	1	0	0	0	0	0	0	0	0	0	0
C	1	1	0	0	0	0	0	0	1	1	1	1	1	1	1	1	1	1
D	1	1	1	1	1	1	1	1	1	1	1	1	1	1	1	1	1	1

Characters 1 and 2, and now characters 9–18, support the relationship AB(CD), while characters 3–8 support the relationship AC(BD). The weight of evidence now suggests a preference for AB(CD) (10 characters, 1, 2 and 9–18) instead of AC(BD) (6 characters, 3–8). We might conclude that we have learnt something about characters 1 and 2: that they are *not* really homoplasious after all but *homologous*, and that in reality characters 3–8 were the homoplasious ones (Figure 12.1b). This

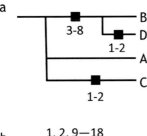

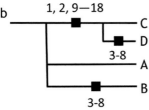

Figure 12.1 (a) Cladogram derived from matrix of eight characters and four taxa, characters 1 and 2 support the relationship AB(CD), characters 3–8 support the alternative relationship AC(BD). (b) Cladogram derived from matrix of eight characters and four taxa, characters 1 and 2, and now characters 9–18, support the relationship AB(CD), while characters 3–8 support the relationship AC(BD).

switching back and forth might constitute an endless process, apparently resolved only by acquiring more and more data until the weight of numbers definitively suggests one or another solution – if that point is ever reached:

> The hope, nevertheless, is that the historical signal will overwhelm any other forces shaping characters and their distribution if only enough data have been collected. But when is 'enough'? (Rieppel 2007, p. 143)

When indeed. Homoplasy merely acts as an *explanation* for characters that do not exactly fit the cladogram of relationships favoured by the preponderance of other characters – it does not help resolve the very real issue of character conflict. Of course, we might use another method, compatibility, for example, and see if the conflict remains. But the obvious question must be: Why do some characters suggest alternative relationships?

Our example above is perhaps rather trivial. In some cases the notion of *rampant homoplasy* is invoked for what appears to be simply a massive amount of ill-fitting data:

> ... we examined 19 diagnostic characters, and analysed these employing both maximum-parsimony and maximum-likelihood approaches. Our results revealed that 16 characters exhibited multiple state changes within the family, with ten exhibiting >eight changes and three exhibiting between 28 and 40. (Wu et al. 2015)

Rather than wishful thinking, or collecting more and more data, or resorting to more and more explanations, another solution might be to return to the specimens. Did we figure out the homologues correctly and in sufficient detail? Did we make a mistake in our studies of the parts of the organisms we are interested in? Are we really dealing with homologues or are our characters merely superficial? Are we dealing with 'false' homologues, measurements and so on (see Chapter 7)? These are all worthy questions and in many cases (if not all cases?) a return to the specimens – or getting more specimens, or better specimens, or perhaps more importantly, having a better understanding of existing specimens, all part of Patterson and Johnson's quest for quality – resolves the issues (see Chapter 7).

Here we are interested in the idea that to find any particular summary diagram (any fundamental cladogram, Chapter 7) of a group of organisms, some data may have to be made to fit that diagram; it has to be *distorted* to make sense. Below we outline a few simple cases of what we call *data-distortion* to illuminate the process of analysis.

In spite of the fact we suggest the classic data matrix is more a hindrance than a help, here we start with the assumption that all *character × taxon matrices* may have a 'true' solution. By that we mean every data matrix has a 'true' summary of the data contained therein. The summary can be represented by a cladogram or a classification.

We are more than aware that many would suggest that, no, a data matrix does not have any 'true' solution and results are only as good as the method applied. If the results of our analyses are only as good as our methods, then it is the *method* rather than the *data* that creates, as it were, some of our results – and some subscribe to the view that *all* methods (and their various *implementations*) are useful *under certain circumstances*. Such a view would deny that natural classification exists or is even possible and that all classifications are based on *imposition* of some kind or other (which again is another aspect of phenetics). We do not subscribe to that view: natural classification is possible, as evidenced by the accumulating results over many decades, regardless of how philosophies have changed (Chapter 3). *That natural classifications can be discovered is one of the major reasons for writing this book.*

How, then, can the 'true solution' of any data matrix be explored? To be clear, the classic data matrix in conventional understanding is a summary of *binary variables*. Below in our examples we make comparisons with results found from Wagner parsimony. As noted above, we could change the method of analysis and use another approach if we wished. Although we have been, and remain, critical of Wagner parsimony as a general method (as we are of all other methods), one of its virtues is that it deals directly with character data and hence offers direct exposure to its faults.

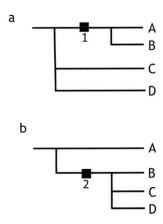

Figure 12.2 (a,b) Two cladograms representing character 1, (AB)CD, and character 2, A(BCD) (redrawn from Nelson et al. 2003).

Character Conflict #1: A Simple Case

Take two simple binary characters that conflict: Character 1 is (AB)CD, character 2 is A(BCD) (example after Nelson et al. 2003). How might we combine these characters to arrive at an optimal solution? We might ask: Is there an optimal solution? Surely, one or another method might find one or another answer and it is therefore not necessarily the *solution* that matters as we should prefer the *method* that we can best justify? We take a different view here, one that suggests for any dataset there is an optimal solution.

The data above can be represented in a binary data matrix (Matrix 1) as well as two one-node cladograms (Figure 12.2a, b):

	Matrix 1	
	Character 1 **(Figure 12.2a)**	**Character 2** **(Figure 12.2b)**
Outgroup	0	0
A	1	0
B	1	1
C	0	1
D	0	1

Table 12.2 The eight parsimony solutions to a matrix of two characters that conflict: (AB)CD and A(BCD); these solutions can be understood as two sets of four

Solutions	Cladogram 1 (= Character 1)	Cladogram 2 (= Character 2)
1	(AB)CD	
2		A(BCD)
3	(AB)(CD)	
4		A(B(CD))
5	C(D(AB))	
6	D(C(AB))	
7		A(C(BD))
8		A(D(BC))

When these data are analysed using Wagner parsimony, the number of cladograms found depends on the program used and the options applied, but the maximum number of most parsimonious solutions is eight (Table 12.2). These can be understood as two sets of four, each set relating to the original two characters.

For Cladogram 1 (= Character 1), Cladograms 3, 5 and 6 are possible resolutions that also have the shortest length but with additional nodes that have no direct support from the data (Table 12.2).

For Cladogram 2 (= Character 2), there are also three possible resolved cladograms, 4, 7 and 8, with additional nodes having no potential support from the data (Table 12.2).

The 'best' solutions, then, are (AB)CD and A(BCD), which cannot be combined without loss of all resolution. These are the original two characters.

For three-item representation, the two characters have five three-item statements (Matrix 2):

Matrix 2					
	Character 1		Character 2		
	a	b	a	b	c
Outgroup	0	0	0	0	0
A	1	1	0	0	0
B	1	1	1	1	?
C	0	?	1	?	1
D	?	0	?	1	1
Weight	3	3	2	2	2

This might be better represented as a table of five statements, none of which overlap:

(AB)CD:	C(AB)	D(AB)			
A(BCD):			A(BC)	A(BD)	A(CD)
Weight	3	3	2	2	2

In this example, fractional weighting (see Box 12.1 for an explanation) is applied giving a weight of 3 to each statement of Character 1a = (AB)C and Character 1b = (AB)D; and a weight of 2 to each statement from Character 2a = A(BC), Character 2b = A(BD) and Character 2c + A(CD).

Analysis of this matrix yields just one cladogram: Cladogram 3 from the eight listed in Table 12.2: (AB)(CD). The remaining Cladograms 1–2, 4, and 7–8 (Table 12.2) are all suboptimal for the three-item data. How can these results be understood? Is Cladogram 3 better in some sense than Cladograms 1 or 2?

Parsimony solution

Cladogram 1 & 2 (Table 12.2): Cladograms 1 and 2 are identical to Characters 1 and 2. The parsimony solution suggests no optimal solution is possible from these two characters, no combination is possible.

Cladogram 3 (Table 12.2): Cladogram 3 is one of three possible resolutions derived from Cladogram 1. Thus, Character 1, (AB)CD, forms one node of Cladogram 3, and the second node gains support by changing Character 2 to Character 2[a], such that:

Character 2		Character 2[a]
A(BCD)	=	AB(CD)

The data are now:

Character 1		Character 2[a]		Tree 3
(AB)CD	+	AB(CD)	=	(AB)(CD)

The change of Character 2 to Character 2[a] is made by fitting the original Character 2 to Cladogram 3. The change suggests that the 1-entry for Taxon B and the 1-entries for Taxa C and D are not homologous. The change can be represented by new Matrix 1A, where Character 2 of Matrix 1 is transformed (changed) into two novel characters (x and y) that sum to the relationship of Character 2[a]. These data can also be represented by its additive multistate equivalent, Character MS 2[a]. An additional implication is that the original data of Matrix 1 is incorrect for Character 2 and Character 2[a], instead, is correct. Unsurprisingly, analysis of the data in Matrix 1A yields only Cladogram 3:

	Matrix 1A			
	Character 1	**Character 2[a]**		**Character MS 2[a] (multistate)**
		x	y	
Outgroup	0	0	0	1
A	1	0	0	1
B	1	0	1	0
C	0	1	0	2
D	0	1	0	2

Cladogram 4 (Table 12.2): Cladogram 4, A(B(CD)), is one of three resolutions possible for Cladogram 2, A(BCD). Character 2, A(BCD), forms one node of Cladogram 4, the second node can be supported by a change in Character 1, (AB)CD, to Character 1[a], AB(CD) – a change often referred to as a 'reversal' (see Chapter 7 and below):

Character 1	Character 1[a]		Character 2		Tree 3
(AB)CD	AB(CD)	+	A(BCD)	=	A(B(CD))

The change in Character 1 to Character 1[a] is made by fitting the original Character 1 to Cladogram 4. The change suggests that the 0-entries for Taxa CD and the 0-entry for the Outgroup are not homologous. The change is equivalent to altered Matrix 1B, in which Character 1 of Matrix 1 is transformed (changed) into two novel characters (x and y) that sum to the relationship of Character 1[a]. These data can also be represented by its additive multistate equivalent, Character MS 1[a].

An additional implication is that the original data of Matrix 1 is incorrect for Character 1 and Character 1ª, instead, is correct. Unsurprisingly, in this instance as well, analysis of Matrix 1B yields only Cladogram 4:

	Matrix 1B			
	Character 1ª		Character MS 1ª (multistate)	Character 2
	x	y		
Outgroup	0	0	0	0
A	1	0	1	0
B	1	0	1	1
C	1	1	2	1
D	1	1	2	1

Three-Item Solution

Tree 3 (Table 12.2): As in Matrix 2, Character 2 is represented by three statements (2a-c) with only Character 2c fitting Tree 3. Thus relevant data are:

Character 1		Character 2^c		Tree 3
(AB)CD	+	A(CD)	=	(AB)(CD)

Character 1 and Character 2c form Cladogram 3 and statements from Character 2a and 2b are not included (Matrix 2A):

	Matrix 2A		
	Character 1		Character 2
	a	b	c
Outgroup	0	0	0
A	1	1	0
B	1	1	?
C	0	?	1
D	?	0	1

Thus, Cladogram 3 is a real sum of the original two conflicting characters, even though some statements are not included.

The three statements of Character 2 are logically related: if any two of them are (assumed) true, then the third is as well – but if any two are false, the third need not be so. This means also that if any one is true and any second false, then the third, too, is false. And also if any one is false, then the other two cannot both be true; at most, only one. Finally, if any one is true, then the other two must be either both true or both false. Relevant here is the possibility that one statement is true and the others not.

The one statement that fits Cladogram 3 is treated as possibly true with the other two treated as possibly false. In short, A(CD) is the only one of the three statements that does not conflict with Character 1 and, therefore, can combine with it.

If statements 2a and 2b are false, this means that in those statements the 1-entries for Taxon B are not homologous with the 1-entries for Taxa CD. This possibility is reflected in altered Matrix 1C where the 1-entry for Taxon B is replaced by a question mark. The implication is that the data of Matrix 1 are not correct for Character 2 but that data of Matrix 1C are correct as far as they go. Analysis of Matrix 1C yields only Cladogram 3:

	Matrix 1C	
	Character 1	Character 2
Outgroup	0	0
A	1	0
B	1	?
C	0	1
D	0	1

Some Generalities

All of the corrected versions of Matrix 1 can be compared:

Matrix		1		1A		1B		1C	
Character		**1**	**2**	**1**	**2**	**1**	**2**	**1**	**2**
	Outgroup	0	0	0	0	0	0	0	0
	A	1	0	0	0	1	0	1	0
	B	1	1	0	1	1	0	1	?
	C	0	1	1	1	0	1	0	1
	D	0	1	1	1	0	1	0	1

The four matrices (1, 1A–1C), as well as Cladogram 3, can be represented by their implied 3-item statements:

	1	1A	1B	1C	Cladogram 3
(AB)C	+	−	+	+	+
(AB)D	+	−	+	+	+
A(BC)	+	+	−	−	−
A(BD)	+	+	−	−	−
A(CD)	+	++	+	+	+
Total	5	3	3	3	3
B(CD)		+	+	−	+
Total	5	4	4	3	4

Of the five statements in Matrix 1, three are implied by each 'corrected' version: 1A implies the same statement, A(CD) twice (++); Versions 1A and 1B imply one other statement, B(CD), not among the five for Matrix 1. In implying no spurious statements, version 1C is the most accurate representation of the original data in Matrix 1 (* = duplicate item included in count):

Matrix	Total statements	Included in Matrix 1	Not included in Matrix 1
1A	5	3	2*
1B	4	3	1
1C	3	3	0
Cladogram 3	4	3	1

It is of interest that Matrix 1C yields Cladogram 3 as its unique result even though the matrix lacks one statement, B(CD), implied by the cladogram.

Derived from this, it will be true that:

(AB)CD	+	A(BCD)	=	(AB)(CD)
(AB)CDE	+	A(BCDE)	=	(AB)(CDE)
(AB)CDEF	+	A(BCDEF)	=	(AB)(CDEF)
(AB)CDEFG	+	A(BCDEFG)	=	(AB)(CDEFG)

In general, however, results arrived at using binary characters almost always involve some character *alteration* or *distortion* to make *all* the data fit the final solution. That is, fitting data to a cladogram by distortion means, to a certain extent, *all solutions found contain methodological artefacts.*

Character Conflict #2: The 'Nullius' Matrices

The 'Nullius' matrices are so named because they first appeared in the pamphlet *Nullius in Verba* (Nelson 1996a[9]). The simplest 'Nullius' matrix is of four taxa ABCD with three characters that conflict in all possible combinations for two of three taxa:

	1	2	3
Out	0	0	0
A	0	0	0
B	1	1	0
C	1	0	1
D	0	1	1

A parsimony analysis of this matrix yields a maximum of six trees, three with a single node and three with two nodes. Each single-node tree has an equivalent two-node tree, resolved without data supporting the extra node (Table 12.3). The three two-node trees appeared because tree length is still minimal even though the node is unsupported by data.[10]

The strict consensus of all six trees is uninformative, as is the strict consensus of the three one-node trees. One explanation for the lack of common resolution is there is no *unequivocal synapomorphy* for the group BCD[11], even though the BCD group is common to all six (or three) solutions, albeit with partial support.

---

[9] Also published in Nelson (1996b) and as an appendix to Williams & Ebach (2005). Goloboff explored similar matrices for his notion of 'data decisiveness' (Goloboff 1991) and something similar was discussed by Lyons-Weiler et al. (1996).

[10] Most parsimony programs today yield only the one-node trees.

[11] This was an objection raised by Farris: 'If matrix 1 supported (BCD), then the stem-species of (BCD) would need to have at least one apomorphy among the characters of that matrix' (Farris 2012, p. 552).

Table 12.3 Parsimony analysis of the above matrix yielding
six trees: three with a single node (column 1), three with two
nodes (column 2). Each single-node tree has an equivalent
two-node tree (rows 1–3). The three two-node trees have
minimal tree length with the extra node unsupported by data

One-node tree	Two-node tree
AB(CD)	A(B(CD))
AC(BD)	A(C(BD))
AD(BC)	A(D(BC))

Unequivocal synapomorphies to one side, Farris objected to the three-item
result because

> One could just as well say that character 1 excludes terminal D . . ., 2 excludes C,
> and 3 excludes B, so emptying the group. In fact the characters are simply
> incongruent. (Farris 1997, p. 134)

Examination of *all* three-item statements puts these comments into proper per-
spective. Each character in the matrix above yields two three-item statements.
They form two sets. Each set might be added together. One set is: D(BC) + C(BD) +
B(CD), which does indeed result in no resolution. But that is only part of the data.
The remaining set of three is A(BC) + A(BD) + A(CD), which sum to A(BCD). Thus,
statements D(BC), C(BD) and B(CD) are irrelevant (not included) to the solution
(Table 12.4; see also Siebert & Williams 1998, p. 342, table 2).

If matrices with all conflicting characters increase in size, by taxa and characters,
the results are revealing (Table 12.5).

Of the 57 analyses in Table 12.5, only 16 lack resolution (28%), the remaining 41
(72%) find the relevant basal group: A(B–E), A(B–F), A(B–G) and A(B–H), respectively (Table 12.5, column 6). Regardless of missing *unequivocal synapomorphies*,
nearly three-quarters of these kinds of matrices find the basal group[12]. This is
enough to suggest that parsimony fails to be consistent – consistent in finding the
basal group for which there is evidence in each case. Recall that in each Nullius
matrix taxon A has no apomorphic value at all and so there is no a priori evidence
as to its relationships with *anything* in the larger group. But these results should
not necessarily be seen as just a failure of the parsimony algorithm – it fails not *just*
because of the idiosyncratic nature of optimisation, but because of the way

---

[12] The total number of matrices analysed by Nelson was 120 – the B–I results are not included in
our Table 12.6. With these included, 95% of the matrices find their relevant basal group (Nelson
1996a, 1996b).

Table 12.4 Each character in the matrix above yields two three-item statements that form two sets. Set 1 has no resolution; Set 2 = A(BCD). Set 1 is irrelevant to the solution as none is included in the answer (see also Siebert & Williams 1998, p. 342, table 2)

		Set 1		Set 2
AD(BC)	=	D(BC)	+	A(BC)
AC(BD)	=	C(BD)	+	A(BD)
AB(CD)	=	B(CD)	+	A(CD)
		(BCD)	+	A(BCD)

Table 12.5 Parsimony results for B–D (four), B–E (five), B–F (six), B–G (seven) and B–H (eight)

B–D	2	None	**B–H**	None
B–E	2	None	2	None
	3	A(B–E)	3	A(B–H)
	2–3	None	4	A(B–H)
B–F	2	None	5	A(B–H)
	3	None	6	A(B–H)
	4	A(B–F)	2+3	None
	2+3	None	2+4	None
	2+4	A(B–F)	2+5	A(B–H)
	3+4	A(B–F)	2+6	A(B–H)
	2–4	A(B–F)	3+4	None
B–G	2	None	3+5	A(B–H)
	3	None	3+6	A(B–H)
	4	A(B–G)	4+5	A(B–H)
	5	A(B–G)	4+6	A(B–H)
	2+3	None	5+6	A(B–H)
	2+4	A(B–G)	2+3+4	None
	2+5	A(B–G)	2+3+5	A(B–H)

Table 12.5 (*cont.*)

B–D	2	None	**B–H**	None
B–E	2	None	2	None
	3+4	A(B–G)	2+3+6	A(B–H)
	3+5	A(B–G)	2+4+5	A(B–H)
	4+5	A(B–G)	2+4+6	A(B–H)
	2+3+4	None	2+5+6	A(B–H)
	2+3+5	A(B–G)	3+4+5	A(B–H)
	2+4+5	A(B–G)	3+4+6	A(B–H)
	3+4+5	A(B–G)	3+5+6	A(B–H)
	2–5	A(B–G)	4+5+6	A(B–H)
			2+3+4+5	A(B–H)
			2+3+4+6	A(B–H)
			2+3+5+6	A(B–H)
			3+4+5+6	A(B–H)
			2–6	A(B–H)

characters are represented. Representation of data is a wider issue, relevant to *all* methods of analysis no matter how sophisticated the algorithm is for any subsequent analysis. Again, the results indicate that *methodological artefacts abound* with binary data.

Of course, the anomalous results in Table 12.5 would not be possible if optimisation did not work the way it does – the search for the shortest cladogram. For the same matrices in Table 12.5 (A–D, A–E, A–F, A–G and A–H) when represented by their three-item equivalents, all relevant basal groups are found every time. This we consider to be a contribution to a *method-free approach*: nothing is added via the complexities (= sophistication) of the algorithm to the data that can affect the results. *The results are a product of the data alone.*

Nelson noted that 'Of more practical concern is the evidence for a group in the absence of unequivocal synapomorphy' (Nelson 1996a, 1996b). Yet it may be true that certain taxa do not have, at any particular moment, an unequivocal synapomorphy defining it (Williams & Ebach, 2017; Mavrodiev & Yurtseva, 2017 and Chapter 5).

Character Conflict #3: 'Reversals' and Optimisation

Some kinds of *data-distortion* are explained away as 'character reversals' (for a critique see Mavrodiev 2016; Mooi & Gill 2016). Here we treat the subject from another point of view (see also Mavrodiev 2016).

Example 1 (after Williams 2002): In this series the first matrix is composed of six congruent Characters (1–6) that describe the cladogram A(B(C(D(E(F(GHIJK)))))) (Table 12.6). The second matrix is identical to the first except now taxon K1 has the plesiomorphic value for Character 1 and is substituted for K so the matrix is composed of taxa A–J + K1 (Table 12.6). The third matrix is identical to the first except now taxon K2 has the plesiomorphic value for Characters 1 *and* 2 and is substituted for K so the matrix is composed of taxa A–J + K2 (Table 12.6). Substitutions of further characters are made until the final sixth matrix has taxon K6 instead of taxon K with all Characters (1–6) having the plesiomorphic value (Table 12.6).

This results in six matrices in all (Table 12.7).

Each matrix was analysed using Wagner parsimony. The results are in Table 12.8.

The cladograms for matrices K, K1 and K2 are identical. Matrix K3 yields two cladograms, one retaining taxon K in the terminal group G–J, while the second places taxon K in a basal group with taxon A. Matrices K4–K6 have a single cladogram with taxon K at the base with A (Table 12.8).

The change in the position of taxon K occurs abruptly in matrix K3 where three of the characters are optimised as plesiomorphic (Characters 1–3, fitting with two steps) and three are optimised as apomorphic (Characters 4–6, fitting with one step) for cladogram AK(B(C(D(E(F(GHIJ))))))). The opposite happens in the second cladogram A(B(C(D(E(F(GHIJK))))))), where Characters 4–6 are optimised as plesiomorphic (fitting with two steps) and Characters 1–3 are optimised as apomorphic (fitting with one step, see Table 12.9).

For the first cladogram (Table 12.9, upper cladogram), where K is in a terminal group with G–J, the plesiomorphic states of Characters 1–3 would be called derived 'reversals'. For the second cladogram (Table 12.9, lower cladogram), where K is in a basal position with A, the apomorphic states of Characters 4–6 would be explained as having unique origins. Explanations to one side, the alterations (distortions) are a result of optimising binary characters.

When matrices of this kind are analysed using Wagner parsimony, if less than c. 50% of the entries for one taxon are plesiomorphic then it is grouped terminally, and if above c. 50%, it is grouped apically. With investigation, a precise numerical relationship exists between matrix entries, regardless of how the data are distributed among the organisms. This relationship is 'caused' by the optimisation procedure and the character 'alterations' required to achieve an optimal most

Table 12.6 Binary matrix of Characters 1–6 and 11 taxa A–K. For each analysis taxon K is reduced by one apomorphic entry at a time creating six matrices: taxa A–K, A–J + K1, A–J + K2, A–J + K3, A–J + K4, A–J + K5 and A–J + K6

	1	2	3	4	5	6
OG	0	0	0	0	0	0
A	0	0	0	0	0	0
B	1	0	0	0	0	0
C	1	1	0	0	0	0
D	1	1	1	0	0	0
E	1	1	1	1	0	0
F	1	1	1	1	1	0
G	1	1	1	1	1	1
H	1	1	1	1	1	1
I	1	1	1	1	1	1
J	1	1	1	1	1	1
K	1	1	1	1	1	1
K1	0	1	1	1	1	1
K2	0	0	1	1	1	1
K3	0	0	0	1	1	1
K4	0	0	0	0	1	1
K5	0	0	0	0	0	1
K6	0	0	0	0	0	0

parsimonious cladogram. For fear of repeating ourselves: *these results reflect the method not the data.*

For all the matrices in the series K-K6 (Tables 12.6 and 12.7), when represented by its equivalent three-item matrix, the analyses produced different results. As taxon K becomes more plesiomorphic it moves *gradually*, rather than abruptly, from the terminal position to the basal position – which, by any standard, appears a more reasonable summary of the data (Table 12.10) – and in our view, *results are not affected by any method as there is none.*

Table 12.7 Matrices in the series are constructed from taxa A-K, A-J + K1, A-J + K2, A-J + K3, etc. producing seven matrices in total

Matrix	Taxa (see Table 12.12)
K	A–K
K1	A–J + K1
K2	A–J + K2
K3	A–J + K3
K4	A–J + K4
K5	A–J + K5
K6	A–J + K6

Table 12.8 Results of the analysis of matrices K–K6 using Wagner parsimony

Binary matrix (Table 12.13)	Cladogram
K	A(B(C(D(E(F(GHIJK))))))
K1	A(B(C(D(E(F(GHIJK))))))
K2	A(B(C(D(E(F(GHIJK))))))
K3	A(B(C(D(E(F(GHIJK)))))) AK(B(C(D(E(F(GHIJ))))))
K4	AK(B(C(D(E(F(GHIJ))))))
K5	AK(B(C(D(E(F(GHIJ))))))
K6	AK(B(C(D(E(F(GHIJ))))))

This holds for all examples of this kind (summarised in Table 12.11).

Example 2: A number of 'character × taxon' matrices with plesiomorphic values nested as a block of apomorphic values have also been used to demonstrate the usefulness of 'reversals' for the conventional approach. For example, a matrix of nine taxa (A–I) and six Characters (1–6) might be thought of as including some *obvious* reversals: Character 1 in taxon H and I and Character 2 in taxon I, identified by their zero value might describe a group HI (Table 12.12).

Table 12.9 The two cladograms resulting from matrix K3; the fit (length) of Characters 1–6 is given along with the total length of the cladogram (L)

Cladograms from Matrix K3	1	2	3	4	5	6	L
A(B(C(D(E(F(GHIJK))))))	2	2	2	1	1	1	9
AK(B(C(D(E(F(GHIJ))))))	1	1	1	2	2	2	9

Table 12.10 Results from three-item analysis of the matrices in Table 12.6

Three-item Matrix (Table 12.14)	Cladogram
K	A(B(C(D(E(F(GHIJK))))))
K1	A(B(C(D(E(F(K(GHIJ)))))))
K2	A(B(C(D(E(K(F(GHIJ)))))))
K3	A(B(C(D(K(E(F(GHIJ)))))))
K4	A(B(K(C(D(E(F(GHIJ)))))))
K5	A(K(B(C(D(E(F(GHIJ)))))))
K6	AK(B(C(D(E(F(GHIJ)))))))

The idea is that as Character 1 in taxon H and I and Character 2 in taxon I share the zero value with the outgroup (i.e., plesiomorphic value) and because parsimony analysis recovers the cladogram A(B(C(D(E(F(FG(HI)), these characters actually represent a 'reversal' from 0 →1→0 in H and I and thereby define the group (HI). But are they?

Inspecting the matrix, however, only shows that Character 1 in taxon H and I, and Character 2 in taxon I simply have the plesiomorphic value. A more useful interpretation of this situation can be appreciated from viewing data such as these in some context. By context, we mean how the data behave in a suite of similar examples, rather than just a single isolated result (as in Example 1 above).

To begin, consider a matrix of nine taxa (A–I) with a perfectly congruent suite of six characters (Table 12.13, Matrix 1, Table 12.14, Matrix 1). Each character in the matrix specifies a group: Character 1 groups A–I, Character 2 groups B–I, Character 3, groups C–I, Character 4 groups D–I, Character 5 groups E–I and Character 6 group F–I (Table 12.13, Matrix 1, Table 12.14, Matrix 1). Analysis of these data yields just one cladogram: (A(B(C(D(E(FGHI).

Table 12.11 Results of series of matrices with increasing plesiomorphic values for one taxon. The size of each series of matrices is increased by one taxon starting with four taxa (matrices D to D1–3) through to nine taxa (matrices J to J1–9). In each case the last taxon (in bold) has the maximum level of plesiomorphy. Farris and Kluge presented a similar example (Farris and Kluge, 1998, p. 353, fig. 3, after Kluge, 1994, table 3) and is equivalent to matrix G12 (their three-item result differs from that in our table as they did not implement fractional weighting in their analysis)

Matrix	Cladogram	Three-item
D	A(B(**CD**))	A(B(**CD**))
D1	A(B(**CD**))	A(B(**CD**))
D12	**D**(A(BC))	**D**(A(BC))
D1–3	**D**(A(BC))	**D**(A(BC))
E	A(B(C(**DE**)))	A(B(C(**DE**)))
E1	A(B(C(**DE**)))	A(B(C(**DE**)))
E12	A(B(C(**DE**))) **E**(A(B(CD))	A(B(**E**(CD))) A(**E**(B(CD)))
E1–3	**E**(A(B(CD))	**E**(A(B(CD))
E1–4	**E**(A(B(CD))	**E**(A(B(CD))
F	A(B(C(D(**EF**))))	A(B(C(D(**EF**))))
F1	A(B(C(D(**EF**))))	A(B(C(D(**EF**))))
F12	A(B(C(D(**EF**))))	A(B(C(D(**EF**))))
F1–3	**F**(A(B(C(DE))))	A(**F**(B(C(DE))))
F1–4	**F**(A(B(C(DE))))	**F**(A(B(C(DE))))
F1–5	**F**(A(B(C(DE))))	**F**(A(B(C(DE))))
G	A(B(C(D(E(F**G**)))))	A(B(C(D(E(F**G**)))))
G1	A(B(C(D(E(F**G**)))))	A(B(C(D(E(F**G**)))))
G12	A(B(C(D(E(F**G**)))))	A(B(C(D(E(F**G**)))))
G1–3	A(B(C(D(E(F**G**))))) **G**(A(B(C(D(EF)))))	A(B(C(**G**(D(EF)))))
G1–4	**G**(A(B(C(D(EF)))))	A(**G**((B(C(D(EF))))))
G1–5	**G**(A(B(C(D(EF)))))	**G**(A(B(C(D(EF)))))
G1–6	**G**(A(B(C(D(EF)))))	**G**(A(B(C(D(EF)))))

Table 12.11 (*cont.*)

Matrix	Cladogram	Three-item
H	A(B(C(D(E(F(G**H**))))))	A(B(C(D(E(F(G**H**))))))
H1	A(B(C(D(E(F(G**H**))))))	A(B(C(D(E(F(G**H**))))))
H12	A(B(C(D(E(F(G**H**))))))	A(B(C(D(E(F(G**H**))))))
H1–3	A(B(C(D(E(F(G**H**))))))	A(B(C(D(E(**H**(FG))))))
H1–4	**H**(A(B(C(D(E(FG))))))	A(B(**H**(C(D(E(FG))))))
H1–5	**H**(A(B(C(D(E(FG))))))	A(**H**(B(C(D(E(FG))))))
H1–6	**H**(A(B(C(D(E(FG))))))	**H**(A(B(C(D(E(FG))))))
H1–7	**H**(A(B(C(D(E(FG))))))	**H**(A(B(C(D(E(FG))))))
I	A(B(C(D(E(F(G(H**I**)))))))	A(B(C(D(E(F(G(H**I**)))))))
I1	A(B(C(D(E(F(G(H**I**)))))))	A(B(C(D(E(F(G(H**I**)))))))
I12	A(B(C(D(E(F(G(H**I**)))))))	A(B(C(D(E(F(G(H**I**)))))))
I1–3	A(B(C(D(E(F(G(H**I**)))))))	A(B(C(D(E(F(G(H**I**)))))))
I1–4	A(B(C(D(E(F(G(H**I**))))))) **I**(A(B(C(D(E(F(GH)))))))	A(B(C(D(E(**I**(F(GH)))))))
I1–5	**I**(A(B(C(D(E(F(GH)))))))	A(B(**I**(C(D(E(F(GH)))))))
I1–6	**I**(A(B(C(D(E(F(GH)))))))	A(**I**(B(C(D(E(F(GH)))))))
I1–7	**I**(A(B(C(D(E(F(GH)))))))	**I**(A(B(C(D(E(F(GH)))))))
I1–8	**I**(A(B(C(D(E(F(GH)))))))	**I**(A(B(C(D(E(F(GH)))))))

If matrix 1 is altered slightly and Character 1 for taxon I is changed from a 1 to a zero, from an apomorphic value to a plesiomorphic value, then a slightly different dataset results (Table 12.13, Matrix 2, Table 12.14, Matrix 2) with slightly different groups possible (Table 12.13, Matrix 2, Table 12.14, Matrix 2). Parsimony analysis of matrix 2 yields the same perfectly congruent tree: (A(B(C(D(E(FGHI).

Matrix 2 can be further modified so that Character 1 for taxon I and H, and Character 2 for taxon I are changed from a 1 to a zero, so that they now have the plesiomorphic value for those characters, and another different dataset results (Table 12.13, Matrix 3, Table 12.14, Matrix 3) with different groups possible (Table 12.13, Matrix 3, Table 12.14, Matrix 3). Somewhat oddly, a parsimony analysis yields one cladogram that is *more* resolved than the previous two: (A(B

Table 12.12 Matrix of nine taxa (A–I) with six characters, with Character 1 in taxon H and I and Character 2 in taxon I assumed to be reversals, identified by their zero value, identical to that in the outgroup as the cladogram from a parsimony analysis is (A(B(C(D(E(FG(HI) (this matrix is the same as Matrix K2 in Table 12.6)

	1	2	3	4	5	6
O	0	0	0	0	0	0
A	1	0	0	0	0	0
B	1	1	0	0	0	0
C	1	1	1	0	0	0
D	1	1	1	1	0	0
E	1	1	1	1	1	0
F	1	1	1	1	1	1
G	1	1	1	1	1	1
H	0	1	1	1	1	1
I	0	0	1	1	1	1

(C(D(E(FG(HI). The zeros in H and I for Character 1 are acting as 'reversals', allowing the lack of evidence (plesiomorphic values) to *become* evidence to satisfy the programs criteria for finding the shortest tree. What was once considered to be the lack of evidence is now converted into evidence (see Chapter 7).

Matrix 3 can be modified further so that Character 1 for taxon I, H and G, Character 2 for taxon H and I, and Character 3 for taxon I now all have the plesiomorphic value for those characters, and a different dataset results (Table 12.13, Matrix 4, Table 12.14, Matrix 4), with slightly different groups possible (Table 12.13, Matrix 4, Table 12.14, Matrix 4). Parsimony analysis yields one cladogram that is even *more* resolved than that of Matrix 3: (A(B(C(D(E(F(G(HI), the additional groups all based on non-evidence (plesiomorphy).

The last matrix to be considered is modified such that Character 1 for taxon I, H, G and F, Character 2 for taxon I, H and G, Character 3 for taxon I and H, and Character 4 for taxon I all have plesiomorphic values (Table 12.13, Matrix 5, Table 12.14, Matrix 5). Here the results begin to dismiss the notion that plesiomorphy = apomorphy. Two cladograms are recovered, the resolved 'reversal' cladogram, (A(B(C(D(E(F(G(HI) and an alternative that recognises two separate groups of A–E and F–I: (A(B(C(DE))))) ((I(H(FG))). Subsequent matrix alterations retain these two groups.

Table 12.13 Five matrices of nine taxa (A–I) with six characters; for Characters 1–4 there are increasing numbers of plesiomorphic values in taxa F–I (bold) in each subsequent matrix, 1 through 5

	Matrix 1						Matrix 2						Matrix 3						Matrix 4						Matrix 5					
	1	2	3	4	5	6	1	2	3	4	5	6	1	2	3	4	5	6	1	2	3	4	5	6	1	2	3	4	5	6
O	0	0	0	0	0	0	0	0	0	0	0	0	0	0	0	0	0	0	0	0	0	0	0	0	0	0	0	0	0	0
A	1	0	0	0	0	0	1	0	0	0	0	0	1	0	0	0	0	0	1	0	0	0	0	0	1	0	0	0	0	0
B	1	1	0	0	0	0	1	1	0	0	0	0	1	1	0	0	0	0	1	1	0	0	0	0	1	1	0	0	0	0
C	1	1	1	0	0	0	1	1	1	0	0	0	1	1	1	0	0	0	1	1	1	0	0	0	1	1	1	0	0	0
D	1	1	1	1	0	0	1	1	1	1	0	0	1	1	1	1	0	0	1	1	1	1	0	0	1	1	1	1	0	0
E	1	1	1	1	1	0	1	1	1	1	1	0	1	1	1	1	1	0	1	1	1	1	1	0	1	1	1	1	1	0
F	1	1	1	1	1	1	1	1	1	1	1	1	1	1	1	1	1	1	1	1	1	1	1	1	**0**	1	1	1	1	1
G	1	1	1	1	1	1	1	1	1	1	1	1	1	1	1	1	1	1	**0**	1	1	1	1	1	**0**	**0**	1	1	1	1
H	1	1	1	1	1	1	1	1	1	1	1	1	**0**	1	1	1	1	1	**0**	**0**	1	1	1	1	**0**	**0**	**0**	1	1	1
I	1	1	1	1	1	1	**0**	1	1	1	1	1	**0**	**0**	1	1	1	1	**0**	**0**	**0**	1	1	1	**0**	**0**	**0**	**0**	1	1

Table 12.14 The five matrices of Table 12.13 with Characters 1–6 as 'groups' rather than binary variables

	Matrix 1	Matrix 2	Matrix 3	Matrix 4	Matrix 5
Character 1	(A–I)	(A–H)	(A–G)	(A–F)	(A–E)
Character 2	(B–I)	(B–I)	(B–H)	(B–G)	(B–F)
Character 3	(C–I)	(C–I)	(C–I)	(C–H)	(C–G)
Character 4	(D–I)	(D–I)	(D–I)	(D–I)	(D–H)
Character 5	(E–I)	(E–I)	(E–I)	(E–I)	(E–I)
Character 6	(F–I)	(F–I)	(F–I)	(F–I)	(F–I)

As data are successively changed from apomorphic values (evidence) to plesiomorphic values (lack of evidence), parsimony analysis will treat the zero entries as if they really were evidence, yielding the same resolved cladogram until a critical quantity is achieved and the alternative evidence-supported arrangement prevails. In other words, matrices 1–4 (Tables 12.13 and 12.14) assume the data to be getting better when, in fact, only more plesiomorphy is being added.

Two conclusions are possible:

1. Parsimony treats all the data as equal, regardless of whether they are apomorphic or plesiomorphic values (hence our equating it with phenetics, see Chapter 7);

2. The resulting cladograms include artefactual nodes created by the method rather than the data (the nodes including G, H and I).

Three-item representation of the data in matrices 1–5 yields different cladograms upon analysis (Table 12.15).

The cladogram from matrix 1 is identical to that of the parsimony analysis, as one would expect given there is no conflict in the data. Matrix 2 yields the cladogram (A(B(C(D(E(I(FGH) with seven inter-nested groups:

ABCDEFGHI

BCDEFGHI

CDEFGHI

DEFGHI

EFGHI

FGHI

FGH

Table 12.15 Cladograms found after parsimony analysis of binary matrices 1–5 (from Tables 12.13 and 12.14) and their three-item representation

Matrix	Parsimony cladogram(s)	Three-item cladogram(s)
1	(A(B(C(D(E(FGHI))))))	(A(B(C(D(E(FGHI))))))
2	(A(B(C(D(E(FGHI))))))	(A(B(C(D(E(I(FGH)))))))
3	(A(B(C(D(E(FG(HI)))))))	(A(B(C(D(E(I(H(FG)))))))))
4	(A(B(C(D(E(F(G(HI)))))))))	(A(B((CD)(I(E(H(FG))))))) (A(B((CD)(I(H(E(FG))))))) = (A(B((CD)(IEH(FG)))))
5	(A(B(C(D(E(F(G(HI))))))))) (A(B(C(DE))))) ((I(H(FG))))	(A((B(C(DE)))(I(H(FG)))))

Table 12.16 Matrix of nine taxa (A–I) with six characters, with Character 1 in taxon H and I and Character 2 in taxon I as reversals, identified by their 2 value

	1	2	3	4	5	6
O	0	0	0	0	0	0
A	1	0	0	0	0	0
B	1	1	0	0	0	0
C	1	1	1	0	0	0
D	1	1	1	1	0	0
E	1	1	1	1	1	0
F	1	1	1	1	1	1
G	1	1	1	1	1	1
H	2	1	1	1	1	1
I	2	2	1	1	1	1

Differences rest with the terminal group FGHI, where the three-item result includes (I(FGH)) and the conventional result (F(GHI). *Here the latter is an invention of the method, not a property of the data.*

From the perspective of parsimony and the constraints imposed by the necessity to fit the data to whatever might be the shortest cladograms, reversals become a

Table 12.17 Results from series of matrices designed to show 'obvious' reversals reanalysed by Mavrodiev (2015, 2016)

Data source	Cladogram(s): Binary data, upper cladogram; three-item data, lower cladogram
Kluge (1994, table 2); Mavrodiev (2015: modified table 2)	(X(A(B(C(D(E(FGH(I(JK))))))))) (X(A(B(C(D(E(FGH(I(JK)))))))))
Kluge (1994: table 3) and Mavrodiev (2015: modified table 3)	(X(A(B(C(D(E(FG))))))) (X(A(B(C(D(E(FG)))))))
Farris (1997: fig. 4) and Mavrodiev (2015: modified figure 4)	(O((A(B(C(D(E(FG))))))(H(I(J(K(L(MN))))))))) (O((A(B(C(D(E(FG))))))(H(I(J(K(L(MN)))))))))
Farris and Kluge (1998: Fig. 5) and Mavrodiev (2015: figure 5)	(O(A(B(C(D(E(F(G(H(I(J(K(L(M(N(P(Q(R(S(T(U(V(W(X(YZ))))))))))))))))))))))))) (O(A(B(C(D(E(F(G(H(I(J(K(L(M(N(P(Q(R(S(T(U(V(W(X(YZ)))))))))))))))))))))))))

useful explanation for this data-distortion. Of course, if one might be certain that 'reversals' have occurred, they can be coded as such (Table 12.17, after Mavrodiev 2016).

In Table 12.16, rather than code the 'reversals' as plesiomorphic, they can be coded as apomorphic (see Box 12.3). This, then, is a different matrix and yields results the same as the parsimony results (Mavrodiev 2016, summarised in our Table 12.17).

Box 12.3   'Special Similarity' and Reversals (Farris 1977)

In 1977 Farris suggested that taxa might best be grouped ('clustered') using a quantity or measure he called *special similarity* (Farris 1977) understood as a proxy for synapomorphy: 'I studied a measure of resemblance, special similarity, that can be used to quantify synapomorphy – similarity in derived features alone' (Farris 1979a, p. 200). The concept was found useful in demonstrating certain shortcomings in the more usual phenetic 'overall similarity' approaches to data analysis. One example went like so. The data in the matrix in Table B12.3.1 if 'clustered' using some measure of overall similarity, yielded no optimal cladogram as clustering occurred on the 0s (zeros) as well as the 1s forcing the characters to conflict in an absolute way (Figure B12.3.1a). Clustering by special similarity – effectively clustering on just the 1s – yielded a fully resolved cladogram (Figure B12.3.1b).

   Later Farris expanded on the idea, and it is worth quoting some of it in full:

> It is not generally possible for each state of a character to distinguish a natural taxon. For example, while the presence of hair distinguishes the natural taxon

Box 12.3 *(cont.)*

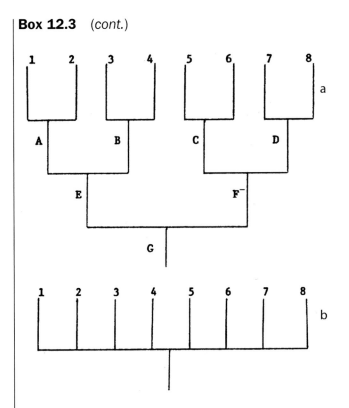

Figure B12.3.1 (a) Result when data from matrix in Table 12.8 is 'clustered' using some measure of overall similarity; (b) result when clustering by special similarity.

Mammalia, the complementary condition, absence of hair, distinguishes the set of non-mammals, which is neither a natural group nor a taxon. Consequently there must be character states, the joint possession of (similarity in) which is un-informative – or even misleading – as to whether organisms should be united in a natural taxon. It is therefore possible that overall similarity – similarity computed from matches in all states may likewise sometimes be uninformative or misleading on the formation of natural taxa. (Farris 1978, p. 236)

The reasoning [above] suggests that in order to identify natural taxa it may be desirable to use some form of special similarity, rather than overall similarity. This may be accomplished by omitting from consideration similarity in states that are uninformative in the sense of [the above]. I showed that this approach leads to a method which properly constructs natural classifications in all the hypothetical cases of [the above, and in Figure B12.3.1a and b], and which therefore lacks some of the shortcomings of overall similarity. (Farris 1978, p. 236)

He further wrote:

This means that the process of clustering by special similarity behaves as if no reversal occurred in evolution, as under the model of Camin and Sokal (1965).

Box 12.3 *(cont.)*

Table B12.3.1 Taxa are listed in the first row (1–8), the characters (Ch.) are listed in the first column (1–14); the final column includes the weight (factor, ×5 or ×1) of each character

	Taxa								
	1	**2**	**3**	**4**	**5**	**6**	**7**	**8**	**Factor**
Ch.									
1	1	0	0	0	0	0	0	0	5
2	0	1	0	0	0	0	0	0	1
3	0	0	1	0	0	0	0	0	5
4	0	0	0	1	0	0	0	0	1
5	0	0	0	0	1	0	0	0	5
6	0	0	0	0	0	1	0	0	1
7	0	0	0	0	0	0	1	0	5
8	0	0	0	0	0	0	0	1	1
9	1	1	0	0	0	0	0	0	3
10	0	0	1	1	0	0	0	0	1
11	0	0	0	0	1	1	0	0	3
12	0	0	0	0	0	0	1	1	1
13	1	1	1	1	0	0	0	0	1
14	0	0	0	0	1	1	1	1	1

Data after Farris (1977, p. 837, table 1).

> This should hardly be surprising, since the special similarity measure was originally devised by Farris, Kluge and Eckardt for application to the Camin–Sokal model. (Farris 1979b, p. 494)

He continued:

> The assumption of irreversibility of evolution is not usually realistic, so that from the standpoint of phylogenetic inference it would undoubtedly be better to do without it. ... Grouping by special similarity does not directly recognize synapomorphies either, since it operates only on a similarity matrix, but it does group by synapomorphy nonetheless. Further, we have arrived at the more general methods simply by removing from the special similarity method the *a priori* assumption of irreversibility. (Farris 1979b, p. 494)

Box 12.3 *(cont.)*

It is clear from these passages that Farris was exploring various models of evolution rather than being concerned with the data and how they behaved. In fact, a year later he writes:

> But all this discussion of grouping by special similarity is to a degree obsolete. In a more recent paper (Farris, 1979a) I showed that retrieving raw similarities from an ultrametric special similarity corresponds to fitting a path-length tree distance, one in which the tree-derived distances are determined as sums of branch lengths. The technique of using ultra-metric special similarities, it turns out, amount to fitting branch lengths subject to the restriction that evolution proceed irreversibly away from the reference point. Better fits can be obtained by dropping the restriction, and I showed that for the real data sets from my 1979 paper the improvement can be substantial. In the improved method, the reference point has no special role, and the degree of fit obtained is independent of how the tree is rooted. ... The utility of clustering by special similarity, insofar as it figures in my criticism of phenetics, is simply that that method provides a more effective way of retrieving phenetic distance information than do phenetic clustering techniques. Even better results can be obtained by dropping the irreversibility restriction. (Farris 1980, p. 364)

'Better fits' and 'better results' here simply meant shorter trees, not more realistic summaries of the data to hand – and to a certain degree, analysis of 'special similarity' appears to be a version of compatibility, or at least the way binary data might have been understood using that method.

As Mooi and Gill noted of an earlier Farris paper, 'The Farris et al. (1970) reformulation was a step towards the general opinion that character evidence resides in the tree through parsimony rather than in the characters (observations) themselves' (Mooi & Gill 2016, p. 263). This was its undoing.

References

Farris, JS. 1976. On the phenetic approach to vertebrate classification. In: Hecht, MK., Goody, P. & Hecht, BM. (eds), *Major Patterns in Vertebrate Evolution*. Plenum, New York, pp. 823–850.

Farris, JS. 1977. On the phenetic approach to vertebrate classification. In: Hecht, MK., Goody, P. & Hecht, BM. (eds), *Major Patterns in Vertebrate Evolution*. NATO Advanced Study Institute Series, no. 14. Plenum Press, New York, pp. 823–850.

Farris, JS. 1978. The 11th Annual Numerical Taxonomy Conference – and part of the 10th. *Systematic Zoology* 27: 229–238.

Farris, JS. 1979a. On the naturalness of phylogenetic classification. *Systematic Zoology* 28: 200–214.

Farris, JS. 1979b. The information content of the phylogenetic system. *Systematic Zoology* 28: 483–519.

Farris, JS. 1980. The efficient diagnoses of the phylogenetic system. *Systematic Zoology* 29: 386–401.

Box 12.3 *(cont.)*

Farris, JS., Kluge, AG. & Eckardt, MJ. 1970. A numerical approach to phylogenetic systematics. *Systematic Zoology* 19: 172–189.

Mooi, R. & Gill, A. 2016. Hennig's auxiliary principle and reciprocal illumination revisited. In: Williams, DM., Schmitt, M. & Wheeler, Q. (eds), *The Future of Phylogenetic Systematics: The Legacy of Willi Hennig*. Systematics Association Special Volume Series. Cambridge University Press, Cambridge, UK, pp. 258–285.

Character Conflict #4: The 'Consensus–Total Evidence' Debate Revisited

Chippindale and Wiens (1994) offered an example of the differences between 'consensus and total evidence', with the first dataset of 225 characters, the second with 200 characters (the binary matrix is given in Chapter 10).

Dataset 1 yields one cladogram: C(AB)(DE); dataset 2 yields a different cladogram: D(AB)(CE); the strict consensus of the two is (AB)CDE, and the Adams consensus yields further resolution of taxon E, such that (AB)E)CD) is the result. The results are notable as they all retain the (AB) relationship. The combined data (dataset 1 + dataset 2), however, yields a further result, A(B(C(DE). The results are summarised below:

	Dataset 1	Dataset 2
Separate analyses	C(AB)(DE)	D(AB)(CE)
Combined analysis	A(B(C(DE)	
Consensus – strict	(AB)CDE	
Consensus – Adams	(AB)E)CD)	

The equivalent in components for these data as follows:

Dataset 1		Dataset 2	
1 ABC(DE)	50	6 ABD(CE)	25
2 C(ABDE)	75	7 AB(CDE)	25
3 (AB)CDE	125	8 (AB)CDE	25
4 A(BCDE)	50	9 A(BCDE)	50
5 AB(CDE)	25	10 D(ABCE)	75

With these data represented as a suite of three-item statements, dataset 1 yields two cladograms C(A(B(DE))) and C((AB)(DE)) and dataset 2 yields two cladograms D(A(B(CE))) and D((AB)(CE)). The combined data (dataset 1 + dataset 2) yields just one, (AB)(C(DE)). The results of binary and three-item analyses are summarised below:

	Parsimony	Three-Item
Dataset 1	C((AB)(DE))	C(A(B(DE))) C((AB)(DE))
Dataset 2	D((AB)(CE))	D((AB)(CE)) D((AB)(CE))
All data (1 + 2)	A(B(C(DE)))	(AB)(C(DE))

Results from the combined data, when represented as binary characters, yields a cladogram with three components: BCDE, CDE and DE, in spite of the conflict in the full character set between a DE and CE component.

Huelsenbeck et al. (1994, p. 290) noted the 'unexplained fact' that '... their [Chippindale & Wiens'] dataset 1 would contain 50 uncontradicted characters supporting a D + E grouping when dataset 2 contains 25 uncontradicted characters supporting the incompatible C + E grouping. Moreover, fully one half (100/200) of the characters in the second dataset are incompatible with a D + E grouping, yet this grouping prevails in the analysis'. This 'unexplained fact' is explained via more *data-distortion*.

The desire to equate more observations with more evidence (e.g., Wiens & Chippindale, 1994, p. 566, 'most characters ... should share a common history'; 'You stop getting different answers or different resolutions when you add new data', McLennan & Brooks 2001, p. 26) does share some similarities with the phenetic approach (Felsenstein 1982; de Queiroz & Good 1997). Missing from the early commentary on combining different datasets was consideration of a more general approach to branching diagrams (cladogram) in biogeography, in particular cladistic biogeography. Thus, there is a certain synonymy between the cladistic approach to biogeography and the cladistic approach to characters, inasmuch as they can both be expressed as specific sets of relationships (see Box 12.4 for more recent critiques).

Box 12.4   J.S. Farris: Further Critiques (2011–2015)

In a series of Letters to the Editor (11 in all) along with three lengthy book reviews, all published in the journal *Cladistics* over a period of four years (2011–2015), Farris returned to his critique of three-item analysis and other related matters. Quite what inspired this volley of bad-tempered contributions remains unknown, but it would try the patience of our readers for us to review all of these contributions as they mostly repeat what had already been noted earlier and ignores most developments made by other interested groups during the period between 2002 and 2011 (see our Boxes 12.1 and 12.2 and some references in this chapter). In many ways, these contributions are best quietly passed over. But we are compelled to make one or two comments.

Many of Farris's Letters include copious quotations. Appropriate quotations often help convey the core idea of any particular contribution. The danger, of course, is that readers are tempted to absorb the message in any given quotation as if it really is a fair representation of whatever point was being made by the quoted author. In many cases, a simple glance at the original text is illuminating but it would be a bit much to expect every reader to devote that much time to verifying every single quotation they read.

One simple example might assist. In a short piece entitled 'Pattern Poses' (Farris 2014b), Farris adds a plethora of quotations (roughly a third of the piece), one taken from Colin Patterson (1982) and reproduced as so:

> [Beatty's claim is] Odd, for I still hold the view (Patterson, 1978: 221) that systematists are mistaken if they take falsification seriously. My own change of heart to pattern or transformed cladistics had nothing to do with Popper. Instead, it came mainly from the realisation that the interminable argument about method and philosophy in systematics had only one [source], evolutionary theory.

The missing word 'source' drew this passage to our attention[1]. The actual quotation (Patterson 1982, p. 284, emphasis ours, and highlighting the missing words) reads:

> Odd, for I still hold the view (Patterson, 1978:221) that systematists are mistaken if they take falsification seriously.
>
> My own change of heart to pattern or transformed cladistics had nothing to do with Popper. Instead, it came mainly from the realisation that the interminable argument about method and philosophy in systematics had only one source, evolutionary theory, *specifically the belief that there is a necessary connection between phylogeny and systematics, and that knowledge of phylogeny exists and should influence our systematics.*

The missing part is reasonably significant when attempting to understand Patterson's point of view. With sufficient energy, and heaps of enthusiasm, one might read through all of Farris's contributions and search for something that tells us how 'knowledge of phylogeny exists and [how it] should influence our systematics'. Our search was in vain. Presumably that is why the crucial section of Patterson's quote was missing.

Box 12.4 *(cont.)*

More trivial puzzles emerge in some of the quotations from David Hull. Just one will suffice (from Hull 1988, p. 237):

> Although both Patterson and Platnick are supposed to be pattern cladists, Patterson rejects a falsificationist view of science, while Platnick is partial to it. Conversely, Brooks is a falsificationist but not a pattern cladist. Nelson supposedly maintains that the ontogenetic criterion for plesiomorphy cannot be misleading, while Patterson agrees with Brooks and Wiley that outgroup analysis can refute inferences from ontology . . . Parsimony is used by practically all cladists, including such antipattern cladists as Brooks and Wiley, while Patterson doubts its value. Similar discordances exist, according to Farris (1985), with respect to the dispute over species being classes or individuals.

Possibly another error in transcription (it would be too early for wayward spell-checkers) but surely *ontogeny* is meant rather than *ontology* in the second sentence?[2]

Some may still be tempted to read through these contributions to gain enlightenment about cladistics and its development (Farris 2011, 2012a–2012f, 2013a, 2013b, 2014a–2014d, 2015, 2018). If so, the reader is advised to proceed with care.

Maybe these contributions will be better appreciated by future historians and social scientists.

References

Farris, JS. 2011. Systemic foundering [Book review]. *Cladistics* 27: 207–221.

Farris, JS. 2012a. Fudged "phenetics". *Cladistics* 28: 231–233.

Farris, JS. 2012b. Counterfeit cladistics. *Cladistics* 28: 227–228.

Farris, JS. 2012c. 3ta Sleeps with the fishes [Book review]. *Cladistics* 28: 422–436.

Farris, JS. 2012d. Early Wagner trees and "the cladistic redux". *Cladistics* 28: 545–547.

Farris, JS. 2012e. Nelson's arrested development. *Cladistics* 28: 551–553.

Farris, JS. 2012f. Homology and historiography. *Cladistics* 28: 554–559.

Farris, JS. 2013a. Symplesiomorphies and explanation. *Cladistics* 29: 13–14.

Farris, JS. 2013b. Pattern taxonomy. *Cladistics* 29: 228–229.

Farris, JS. 2014a. "Taxic homology" is neither. *Cladistics* 30: 113–115.

Farris, JS. 2014b. Pattern poses. *Cladistics* 30: 116–119.

Farris, JS. 2014c. "Pattern cladistics" really means paraphyly. *Cladistics* 30: 236–239.

Farris, JS. 2014d. Homology and misdirection. *Cladistics* 30: 555–561.

Farris, JS. 2015. Histories on thin IISE [Book review]. *Cladistics* 31: 330–336.

Hull, DL. 1988. *Science as Process: An Evolutionary Account of the Social and Conceptual Development of Science*. University of Chicago Press, Chicago.

Patterson, C. 1982. Classes and cladists or individuals and evolution. *Systematic Zoology* 31: 284–286.

[1] We assume the omission of the word 'source' was simply an error in transcription.

[2] Ontogeny is, of course, the correct word.

Quo Vadis

We have discussed a few examples of *data-distortion* found with conflicting characters. But one might ask: Are there any *genuine* cases of conflict, or are they all simply reflections of an absence of knowledge rather than real phenomena? What of this massive amount of apparently ill-fitting data, this avalanche of *rampant homoplasy*?

> As it grows, the matrix of total evidence tends toward the biggest, the matrix with no real homologies at all, with all its characters homoplastic, when all of them change by optimization on the best fitting tree of an inexact solution, when the least amount of that better fit distorts each character, at least a little. (Nelson 2004, p. 139)

In some respects early explorations of DNA data did just that – but now we rarely learn what characters (bases) contribute to the nodes found, and like every other analysis, it either ends up in 'the blender of optimization' (Mooi & Gill 2010; Williams & Ebach 2010) or the fog of various statistical manipulations (e.g., Assis 2018; Brower 2018). With respect to these distortions, some have gone as far as suggesting that 'Homoplasy *Increases* Phylogenetic Structure' (Källersjö et al. 1999, their emphasis), when in fact 'the claims of Källersjö et al. (1999) can themselves be reinterpreted; homoplasy does not increase phylogenetic structure, it merely increases structure' (Mooi & Gill 2010, p. 31). And that structure includes *false nodes*, those we refer to as artefact.

But if even there are genuine cases of conflict, is invoking homoplasy an entirely satisfactory explanation? Maybe the only place genuine conflict does occur is with DNA data. With respect to morphology, we suggest that conflict will for the most part disappear as more is understood about the organisms and their parts rather than by merely accumulating more and more data – our classifications will progress from *polythetic* definitions of taxa towards *monothetic* definitions (Chapter 5), from Interpretation 2 options to Interpretation 1 options (Chapter 11). Quite what will become of DNA data in this comparative framework is, at present, anyone's guess.

We suggest that attention is best directed towards the *nature* of evidence (characters), a better understanding of the cladistic parameter and how that understanding relates to the data matrices constructed (even if a matrix is required at all), rather than the bald assumption that the more data we have the better our analyses will be (= phenetics), or the better our 'evolutionary models' we have the better our cladograms will be (= gradistics) (Chapter 7). The core of the matter was expressed some time ago as one of data quality rather than quantity, here noted by Rudolf Meier:

> It is probably safe to assume that most mistakes in phylogenetic investigations occur prior to cladistic analysis and that with the advent of cladistic computer

programs the focus of at least the theoretical discussion has unduly shifted from character analysis to finding parsimonious trees. (Meier 1995, p. 164[13])

We hope it is obvious that from a certain perspective our discussion above resolves the 'total evidence-consensus' debate, which has become a somewhat sterile discussion of methods – as did the phenetics of a previous era.

The key appears to reside in a general understanding of relationships, expressed so: A(BC). Thus, we offer a way forward, a protocol for taxonomic studies:

1. Study the organisms, understand the parts (even if the parts are nucleic acids), determine the homologues;

2. Determine the homology statements derived from those homologues, parse A + B (the homologues) into C(AB) (the relationship);

3. Determine the simplest form of statement of relationship from those homology statements;

4. Derive an optimal cladogram from those data, one that might be used for the classification.

Rest assured: this approach does not depend upon any program (algorithm, model) to assess whether the parts (homologues) studied are 'correct' (homology) or not (homoplasy). Only further study of actual organisms can achieve that.

References

Akanni, WA., Wilkinson, M., Creevey, CJ., Foster, PG. & Pisani, D. 2015. Implementing and testing Bayesian and maximum-likelihood supertree methods in phylogenetics. *Royal Society Open Science* 2(8): 140436. doi: 10.1098/rsos.140436. eCollection 2015 Aug.

Assis, LCS. 2018. Revisiting the Darwinian shortfall in biodiversity conservation. *Biodiversity and Conservation* 27: 2859–2875.

Beatty, J. 1982. Classes and cladists. *Systematic Zoology* 31: 25–34.

Brady, RH. 1982. Theoretical issues and "pattern cladists". *Systematic Zoology* 31: 286–291.

Brooks, DR. & Wiley. EO. 1985. Theories and methods in different approaches to phylogenetic systematics. *Cladistics* 1: 1–11.

Brower, AV. 2018. Statistical consistency and phylogenetic inference: a brief review. *Cladistics* 34: 562–567.

Bryant, D. 2003. A classification of consensus methods for phylogenetics. In: Janowitz, MF., Lapointe, F-J.,

---

[13] A more detailed examination, and one that draws attention to the preoccupation with matrices, can be found in Patterson & Johnson (1997a, b).

McMorris, FR., Mirkin, B. & Roberts, FS. (eds) *Bioconsensus, DIMACS Working Group Meetings on Bioconsensus.* DIMACS Series in Discrete Mathematics and Theoretical Computer Science, Volume 61. American Mathematical Society, Providence, pp. 163–184.

Bryant, D., Francis, A. & Steel, M. 2017. Can we "future-proof" consensus trees? *Systematic Biology* 66: 611–619.

Cecca, F. & Zaragüeta i Bagils, R. 2015. *Paléobiogéographie.* EDP Sciences, Paris.

Charig, AJ. 1982. Systematics in biology: a fundamental comparison of some major schools of thought. In: Joysey, KA. & Friday, AE. (eds), *Problems of Phylogenetic Reconstruction.* Academic Press, New York, pp. 363–440.

Chippindale, PT. & Wiens, JJ. 1994. Weighting, partitioning, and combining characters in phylogenetic analysis. *Systematic Biology* 43: 278–287.

Dawkins, R. 1986. *The Blind Watchmaker.* W. W. Norton & Company, Inc., New York [2nd edition 1996; numerous other editions exist, including a 25th anniversary edition and it has its own Wikipedia page, https://en.wikipedia .org/wiki/The_Blind_ Watchmaker#Reception].

de Queiroz, K. & Good, DA. 1997. Phenetic clustering in biology: a critique. *The Quarterly Review of Biology* 72: 3–30.

Ebach, MC. & Williams, DM. 2020. Ronald Brady and the cladists. *Cladistics* 36: 218–226

Farris, JS. 1997. Cycles. *Cladistics* 13: 131–144.

Farris, JS. 2012. Nelson's arrested development. *Cladistics* 28: 551–553.

Farris, JS. & Kluge, AG. 1998. A/The brief history of three-taxon analysis. *Cladistics* 14: 349–362.

Felsenstein, J. 1982. Numerical methods for inferring evolutionary trees. *Quarterly Review of Biology* 57: 379–404.

Goloboff, PA. 1991. Homoplasy and the choice among cladograms. *Cladistics* 7: 215–232

Harvey, AW. 1992. Three-taxon statements: more precisely, an abuse of parsimony? *Cladistics* 8: 345–354.

Huelsenbeck, JP., Swofford, DL., Cunningham, CW., Bull., JJ. & Waddell, PJ. 1994. Is character weighting a panacea for the problem of data heterogeneity in phylogenetic analysis? *Systematic Biology* 43: 288–291.

Hull, DL. 1988. S*cience as Process: An Evolutionary Account of the Social and Conceptual Development of Science.* University of Chicago Press, Chicago.

Jansson, J., Nguyen, NB. & Sung, W-K. 2006. Algorithms for combining rooted triplets into a galled phylogenetic network. *Siam Journal of Computing* 35 (5): 1098–1121.

Källersjö, M., Albert, VA. & Farris, JS. 1999. Homoplasy *increases* phylogenetic structure. *Cladistics* 15: 91–93.

Kluge, AG. 1993. Three-taxon transformation in phylogenetic inference: ambiguity and distortion as regards explanatory power. *Cladistics* 9: 246–259.

Kluge, AG. 1994. Moving targets and shell games. *Cladistics* 10: 403–413.

Lin, HT., Burleigh, JG. & Eulenstein, O. 2009. Triplet supertree heuristics for the tree of life. *BMC Bioinformatics* 10 (Suppl. 1): S8.

Lyons-Weiler, J., Hoelzerand, GA. & Tausch, RJ. 1996. Relative apparent synapomorphy analysis (RASA) I: the statistical measurement of phylogenetic signal. *Molecular Biology & Evolution* 13: 749–757.

Mavrodiev, EV. 2015. Three-taxon analysis can always successfully recognize groups based on putative reversals. *PeerJ PrePrints* 3: e1206.

Mavrodiev, EV. 2016. Dealing with propositions, not with the characters: the ability of three-taxon statement analysis to recognise groups based solely on 'reversals', under the maximum-likelihood criteria. *Australian Systematic Botany* 29: 119–125.

Mavrodiev, EV. & Madorsky, A. 2012. TAXODIUM Version 1.0: a simple way to generate uniform and fractionally weighted three-item matrices from various kinds of biological data. *PLoS One* 7(11): e48813.

Mavrodiev, EV. & Yurtseva, OV. 2017. "A character does not make a genus, but the genus makes the character": three-taxon statement analysis and intuitive taxonomy. *European Journal of Taxonomy* 377: 1–7.

McLennan, DA. & Brooks, DR. 2001. Phylogenetic systematics: five steps to enlightenment. In: Adrain, JM., Edgecombe, GD. & Lieberman, BS. (eds), *Fossils, Phylogeny, and Form: An Analytical Approach*. Kluwer Academic/ Plenum, New York, pp. 7–28.

Meier, R. 1995. Advantages and disadvantages of computerized phylogenetic analyses. *Zoologische Beiträge N. F.* 36(2): 141–167.

Mooi, R. & Gill, A. 2010. Phylogenies without synapomorphies – a crisis in fish systematic: time to show some character. *Zootaxa* 2540: 26–40.

Nelson, G. 1996a. *Nullius in verba.* Published by the author, New York and Melbourne.

Nelson, G. 1996b. *Nullius in verba. Journal of Comparative Biology* 1: 141–152.

Nelson, GJ. 2004. Cladistics: its arrested development. In: Williams, DM. & Forey, PL. (eds), *Milestones in Systematics*. CRC Press, Boca Raton, FL, pp. 127–147.

Nelson, GJ. & Ladiges, PY. 1991. Standard assumptions for biogeographic analysis. *Australian Systematic Botany* 4: 41–58.

Nelson, GJ. & Ladiges, PY. 1992a. Three-area statements: standard assumptions for biogeographic analysis. *Systematic Zoology* 40: 470–485.

Nelson, GJ. & Ladiges, PY. 1992b. Information content and fractional weight of three-taxon statements. *Systematic Biology* 41: 490–494.

Nelson, GJ. & Platnick, NI. 1981. *Systematics and Biogeography: Cladistics and Vicariance*. Columbia University Press, New York.

Nelson, GJ. & Platnick, NI. 1991. Three-taxon statements: a more precise use of parsimony? *Cladistics* 7: 351–366.

Nelson, GJ., Williams, DM. & Ebach, MC. 2003. A question of conflict: three item and standard parsimony compared. *Systematics & Biodiversity* 1: 145–149.

Parenti, LR. & Ebach, MC. 2009. *Comparative Biogeography: Discovering and Classifying Biogeographical Patterns of a Dynamic Earth*. University of California Press, Berkeley.

Patterson, C. & Johnson, GD. 1997a. The data, the matrix, and the message: Comments on Begle's "Relationships of the Osmeroid Fishes". *Systematic Biology* 46: 358–365.

Patterson, C. & Johnson, GD. 1997b. Comments on Begle's "Monophyly and relationships of Argentinoid fishes". *Copeia* 1997: 401–409.

Ridley, M. 1986. *Evolution and Classification: The Reformation of Cladism*. Longman, London.

Rieppel, O. 2007. Parsimony, likelihood, and instrumentalism in systematics. *Biology and Philosophy* 22: 141–144.

Siebert, DJ. & Williams, DM. 1998. Recycled. *Cladistics* 14: 339–347.

Steel, M. & Rodrigo, A. 2008. Maximum likelihood supertrees. *Systematic Biology* 57: 243–250.

Tazehkand, SJ., Hashemi, SN. & Poormohammadi, H. 2013. New heuristics on rooted triplet consistency. *Algorithms* 6: 396–406.

Thu-Hien, T. & Habib, M. 2009. Level-k phylogenetic network can be constructed from a dense triplet set in polynomial time. *arXiv*: 0901.1657v1.

Wiens, JJ. & Chippindale, PT. 1994. Combining and weighting characters and the prior agreement approach revisited. *Systematic Biology* 43: 564–566.

Williams, DM. 2002. Parsimony and precision. *Taxon* 51: 143–149.

Williams, DM. & Ebach, MC. 2005. Drowning by numbers: a new view on numerical phylogenetics. *Botanical Reviews* 71: 355–387.

Williams, DM. & Ebach, MC. 2008. *Foundations of Systematics and Biogeography*. Springer, New York.

Williams, DM. & Ebach, MC. 2010. Molecular systematics and the 'blender of optimization': is there a crisis in systematics? *Systematics and Biodiversity* 8: 481–484.

Williams, DM. & Ebach, MC. 2017. What is intuitive taxonomic practice? *Systematic Biology* 66: 637–643.

Wu, BY. 2004. Constructing the maximum consensus tree from rooted triples. *Journal of Combinatorial Optimization* 8: 29–39.

Wu, Z-Y., Milne, RI., Chen, C-J., Liu, J., Wang, H. & Li, D-Z. 2015. Ancestral state reconstruction reveals rampant homoplasy of diagnostic morphological characters in Urticaceae, conflicting with current classification schemes. *PLoS ONE* 10(11): e0141821.

Zaragüeta i Bagils, R., Ung, V., Grand, A., Vignes-Lebbe, R., Cao, N. & Ducasse, J. 2012. LisBeth: new cladistics for phylogenetics and biogeography. *Comptes Rendus Palevol* 11: 563–566.

Further Reading

A number of items we list below deal with the gene tree/species tree resolution (e.g., Doyle 1992, 1996, Page & Charleston 1997, Doyle & Davis 1998). We understand these as attempts to reconcile different and conflicting cladograms rather than just exercises in molecular systematics. We simply ask that readers approach these contributions from that perspective so that cladogram combination can be understood in this most general form. This, as it happens, is the essence of cladistic analysis as we understand it.

Doyle, JJ. 1992. Gene trees and species trees: molecular systematics as one-character taxonomy. *Systematic Botany* 17: 144–163.

Doyle, JJ. 1996. Homoplasy connections and disconnections: genes and species, molecules and morphology. In: Sanderson, MJ. & Hufford, L. (eds), *Homoplasy: The Recurrence of Similarity in Evolution*. Academic Press, San Diego, pp. 37–66.

Doyle, JJ. & Davis, JI. 1998. Homology in molecular phylogenetics: a parsimony perspective. In: Soltis, DE., Soltis, PS. & Doyle, JJ. (eds), *Molecular Systematics of Plants II*. Springer, Boston, MA, pp. 101–131.

Mooi, R. & Gill, A. 2010. Phylogenies without synapomorphies –a crisis in fish systematic: time to show some character. *Zootaxa* 2540: 26–40.

Mooi, R. & Gill, A. 2016. Hennig's auxiliary principle and reciprocal illumination revisited. In Williams, D., Schmitt, M. & Wheeler, Q. (eds.), *The Future of Phylogenetic Systematics: The Legacy of Willi Hennig.* Systematics Association Special Volume Series. Cambridge University Press, Cambridge, pp. 258–285.

Nelson, GJ. 1996. *Nullius in verba.* Published by the author, New York and Melbourne.

Nelson, GJ. 2004. Cladistics: its arrested development. In: Williams, DM. & Forey, PL. (eds), *Milestones in Systematics.* CRC Press, Boca Raton, FL, pp. 127–147.

Nelson, GJ. & Ladiges, PY. 1991. Standard assumptions for biogeographic analysis. *Australian Systematic Botany* 4: 41–58.

Page, RDM. 2000. Extracting species trees from complex gene trees: reconciled trees and vertebrate phylogeny. *Molecular Phylogenetics and Evolution* 14: 89–106.

Page, RDM. & Charleston, MA. 1997. From gene to organismal phylogeny: reconciled trees and the gene tree/species tree problem. *Molecular Phylogenetics and Evolution* 7: 231–240.

Simmons, MP. & Freudenstein, JV. 2002. Uninode coding vs gene tree parsimony for phylogenetic reconstruction using duplicate genes. *Molecular Phylogenetics and Evolution* 23: 481–498.

Wheeler, WC. 2012. *Systematics: A Course of Lectures.* Wiley-Blackwell, Oxford.

Williams, DM. & Ebach, MC. 2005. Drowning by Numbers: A new view on numerical phylogenetics. *Botanical Reviews* 71:355–387.

Williams, DM. & Ebach, MC. 2008. *Foundations of Systematics and Biogeography.* Springer, New York.

Williams, DM. & Ebach, MC. 2010. Molecular systematics and the 'blender of optimization': is there a crisis in systematics? *Systematics and Biodiversity* 8: 481–484.

Williams, DM. & Ebach, MC. 2017. What is intuitive taxonomic practice? *Systematic Biology* 66: 637–643.

PART V

Beyond Classification

Beyond classification, and the cladistic parameter, lie explanations: 'What does homology and monophyly mean? What can it tell us of the world?' and so on. The explanatory side of historical biology has little to do with observation (i.e., comparative biology) and understanding evidence (i.e., synapomorphy), but more to do with reasoning and modelling hypotheses and theories based on labyrinthine data. Phylogenetics, and its recent molecular incarnations, such as phylogenomics (see Chapter 10) are usually understood through complex models that attempt to create versions of the natural world with which to explain the swathes of data derived from whole genomes. No one claims to know what, exactly, these data mean, hence the use of complex models. We have not avoided the complexities of the molecular approach to systematics (taxonomy) and refer the reader to Parts I–IV in this book, which, in our view, applies to all kinds of data: all data, whether just a few homologues found in a fragmented fossil or those found in entire genomes, is relational and therefore homology and monophyly apply within the context of the cladistic parameter.

In this section we will address some issues that have arisen in recent years, not necessarily directly linked to a classificatory outlook to comparative biology but of some significance for the general views expressed herein. In Chapter 13 we briefly investigate some aspects of phylogeny. In Chapter 14 we comment on problems using phylogenetic inferences for classification (e.g., phylogenetic classifications such as the *PhyloCode*). In Chapter 15 we cover 'more myths and misunderstandings' with respect to cladistics (as opposed to those 'myths and misunderstandings' in taxonomy in general, topics covered in Part II), such as its portrayal in the popular literature; its portrayal in the unfortunate '#parsimonygate' episode; its portrayal by some philosophers of biology; and some other (relatively) minor issues (the taxonomic impediment, the 'end user' and counting species).

13

Beyond Classification:
How to Study Phylogeny

When I was very young, I was suitably impressed to learn that, appearances notwith-standing, the whale is not a fish. Nowadays these questions of classification move me less; and it does not worry me unduly when I am assured that history is not a science.

(Carr 1961, What is History?*)*

Threefold Parallelism: Form, Time and Space

Haeckel's genealogical project began in 1866 with his monumental two-volume *Generelle Morphologie der Organismen* (Haeckel 1866), written partly under the influence of Darwin's *Origin of Species* (Darwin 1859), terminating some 30 years later with another equally exhaustive survey – this time in three volumes: *Systematische Phylogenie: Entwurf eines natürlichen Systems der Organismen auf Grund ihrer Stammesgeschichte* (Haeckel 1894–1896[1]).

Towards the end of his life, Haeckel commented on his efforts:

> The first rough drafts of pedigrees that were published in the 'Generelle Morpho-logie' have been improved time after time in the ten editions of my 'Naturaliche Schopfungsgeschichte' (1868-1902) … A sounder basis for my phyletic hypoth-eses, derived from a discriminating combination of the three great records–morphology, ontogeny, and palaeontology–was provided in the three volumes of my 'Systematische Phylogenie' … (Haeckel 1909)

Today the 'three great records' have been modified and are often referred to as *Form, Time and Space* (sensu Nelson and Platnick 1981), another threefold parallelism[2]. The

---

[1] 1. *Systematische Phylogenie der Protisten und Pflanzen*; 2. *Systematische Phylogenie der wirbellosen Thiere (Invertebrata)*; 3. *Systematische Phylogenie der Wirbelthiere (Vertebrata)*.

[2] For a general history of the threefold parallelism, see Williams & Ebach (2008).

threefold parallelism is not equally divisible into Form, Time and Space, but rather, are the relative spatial and temporal aspects of form. For example, an individual organism of the common wombat (*Vombatus ursinus*) occurs in space and time, as well as in form. *Form* is represented by its morphology (anatomy and physiology) and its molecules (its DNA; see Ebach et al. 2006) – but also by any individual fossils that might be found, as well as its developmental sequence from birth to adulthood, its ontogeny. Time and space are different aspects of form: ontogeny is form developing through *time*, the endemic area that it is found in is form through *space* and *time*. Neither, however, is representative of a process, that is, an ontogenetic process such as allometry or paedomorphosis (see Nelson 1978). That is to say, we know ontogeny occurs, but it is the comparative aspect of form that concerns the threefold parallelism. For instance, the embryos of a wombat and kangaroo (e.g., the red kangaroo, *Macropus rufus*) can be compared to that of a dingo (*Canis familiaris*). The discovery that the embryos of a common wombat and the northern hairy-nosed wombat (*Lasiorhinus krefftii*) share different manifestations of the same form than they do to the kangaroo, is evidence for constructing a natural classification (homology, monophyly) and, derived from that, a shared history (e.g., they share a common ancestor). If we can show that two forms share the same morphology in all its ontogenetic stages, then we can be certain of a homologous relationship.

The same is true for the *space* aspect of form, namely geographical distribution. We know that form occupies space and organisms move, thereby creating a *distribution*. Add in the temporal aspect and we have *form moving in space and through time*, from which we infer an area of endemism. Taxa and areas of endemism (or biotic areas) are abstract in the sense that we cannot actually see one, only parts of it. Moreover, we may not be certain, until monophyly is discovered, whether our taxa are actually natural. While the cladistic parameter ensures that homology and monophyly are used as the theoretical and practical elements of a natural classification, the threefold parallelism ensures that the evidence for our natural classification is congruent across space and time. For instance, a non-homologous and analogous characteristic may not share the same ontogeny (temporal aspect of form) or the same distribution (spatial aspect of form). For example, both tenrecs (Tenrecidae) and echidnas (Tachyglossidae) have spines, but unlike tenrecs, the spines in echindas precede the hairs in ontogeny (Alibardi & Rogers 2015). Moreover, echidnas are endemic to Australia and southern New Guinea (Weisbecker & Beck 2015), while tenrecs are endemic to Madagascar (Olson 2013). If form differs in its spatial and temporal aspects, then it is highly likely that the characteristic is non-homologous and analogous (functional). In practice, the threefold parallelism is used prior to any analysis – it is part of a taxonomist's training not to confuse characteristics that share no temporal or spatial aspects in their form. Yet, well-known examples slip by as in the case of the Australian Jurassic fossil 'Agathis jurassica[3]'.

The temporal aspect of form also applies to fossils. Certainly Lazarus taxa[4], such as the coelacanth (Latimeria chalumnae, Smith 1939), may 'reappear' as representatives of thought-to-be extinct orders (Coelacanthiformes); but, in general, when fossil forms of living taxa appear in the fossil record, much caution is required when assigning them to an extant taxon[5]. The 170-million-year-old 'Agathis jurassica' was assigned to a living genus belonging to the coniferous Araucaria family, which has modern relatives, such as A. robusta, that are also found along the east coast of Australia and in New Guinea in a similar temperate environment. Given the morphological similarity of both species (no DNA is preserved in the fossil specimens) as well as ontogenetic similarities (both have similar fruiting bodies), and the fact they are found in the same area, we may assume that 'jurassica' is likely to be part of Agathis. Yet, little is known of the ontogeny of 'jurassica', even though the morphological data is striking similar to living Araucaria genera such as Agathis and Wollemia. While the morphology and spatial aspects of 'jurassica' are similar to living Araucaria, it is the temporal aspects (namely ontogeny and age), which throws its taxonomy into doubt. In order to discover which taxon 'jurassica' belongs, a cladistic analysis is required in order to determine whether it is, in fact, part of Agathis or another member of the Araucariaceae altogether. We know, for example, that Agathis is monophyletic (Escapa & Catalano 2013, p. 1158, fig. 4 and Escapa et al. 2018, p. 1362, fig. 12, see our Figure 13.1a) and is known to have fossil taxa, such as A. zamunerae, which is found in the early Eocene in Patagonia (Wilf et al. 2014).

---

[3] 'Agathis jurassica' White (1981) is not a considered to be a valid name and was originally identified as Podozamites lanceolatus by Walkom (1921). Turner et al. (2009), however, do not believe A. jurassica to be a 'close match' to any 'living extant araucariacean genera' (Turner et al. 2009, pp. 58–59; Vajda & McLoughlin, 2005). Since White (1981), 'A. jurassica' has had a taxonomic rollercoaster ride: Ruth Stockey first designated 'A. jurassica' as incertas sedis but noted that it 'may prove to be a podocarp' (Stockey 1990, p. 188), while later pondering the 'close similarities [of "A. jurassica"] to those of Araucaria Section Eutacta' and reassigning it to Podozamites (Stockey 1994, p. 497). Turner et al. (2009, fig. 3) accepted the reassignment to Podozamites (an araucariacean conifer). Clearly a cladistic analysis is necessary to clear up the taxonomic affinity of 'A. jurassica' both at the familial level as well as at the generic level: 'no phylogenetic analysis of the characters used to place ["A. jurassica"] in the genus Agathis has been made, and they may merely represent shared plesiomorphies' (Gilmore & Hill 1997, p. 277). Due to the aphyly of 'A. jurassica', we use the generic name in inverted commas to signify its uncertain taxonomic status.

[4] These are extinct taxa that are said to have temporarily disappeared from the fossil record, only to re-appear again later. First noted by Flessa and Jablonski (1983), the idea has been much discussed and occasionally doubted (see Rickards & Wright 2002 and for a brief review see Fara 2001).

[5] There are exceptions, diatoms being one such group of organisms (see Williams 2007).

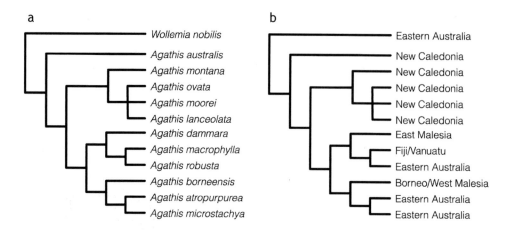

Figure 13.1 (a) *Agathis* cladogram (redrawn from Escapa & Catalano 2013, fig. 4); (b) *Agathis* areagram in which the names of taxa have been replaced by the names of the areas in which they occur.

The fact that the threefold parallelism works in *conjunction* with the cladistic parameter speaks volumes for the *hierarchical nature of classification.*

For a further example, consider wombats (*Vombatus*) again. Part of its classification is:

Mammalia

Marsupialia

Diprotodontia

Vombatus

Common wombats (*Vombatus ursinus*) are known from the Bassian subregion, which is part of an area classification, namely,

Austral Kingdom

Australian region

Bassian subregion

The common wombat is known only from the Holocene, which is also part of a temporal classification, namely,

Phanerozoic

Cenozoic

Quaternary

Holocene

All of the classifications above are equally abstract in existence (i.e., taxon, biotic area, geological period) but only the latter temporal classification is artificial (but see Novacek & Wheeler 1992, Norell 1993 and Zaragüeta i Bagils et al. 2004, for further commentary). Given this we have morphological/molecular homologues and homologies as well as *area homologues* and *homologies* (Parenti & Ebach 2009). This would mean that morphological/molecular homologies form character cladograms and area homologies form areagrams. The relationships between areas are derived from the relationships between taxa and their characteristics. Any temporal information, such as the age and distribution of the taxa, are derived from the taxa themselves. In effect, the threefold parallelism (*Form, Time, Space*) and the cladistic parameter (*homology and monophyly*) are *the* basis for comparative biology. The cladogram, then, is the basis for communicating natural classification. Phylogeny, on the other hand, is the act of deriving a scenario *from* natural classifications (and more often than not, from artificial classifications).

Beyond the Cladogram: Inferring Phylogenies

As we pointed out in the above chapters (and many others have done so before us), cladograms *are not* phylogenetic trees (see Chapter 7). Cladograms are classifications, whereas phylogenies are historical inferences derived from cladograms. Hence one can only derive phylogenies from classifications, not the other way around. With this in mind, the first step in any phylogenetic analysis is to obtain a classification in the form of a cladogram and convert it into a phylogenetic tree (see Chapter 7). The conversion is mostly interpretive. Generally, the phylogenetic tree still *looks* like a cladogram, but how we *interpret* the tree differs. The interpretations, like the *Agathis* example above, may lead to some highly questionable assumptions (e.g., '*A. jurassica*' is the ancestor to all other species of *Agathis*). How, then, do we curb such speculative (and largely non-scientific) interpretations? The answer lies in discovering *patterns*.

Unlike statistical correlations, such as trends (e.g., diversification of flowering plants versus the 'demise of dinosaurs'), topographical patterns gain context within the threefold parallelism – that is, they are tied to notions of homology, ontogeny, etc. For example, in the fossil Brachymetopid trilobite genus *Cordania* there are seven species: *C. becraftensis, C. buicki, C. cyclurus, C. falcata, C. gasepiou, C. macrobius* and *C. wessmanii.* On the basis of 11 characters, Ebach and Edgecombe found one cladogram (Ebach & Edgecombe 1999, p. 432, figure 2, reproduced here as Figure 13.2). The characters were not randomly chosen. Rather, they were compared across species of *Cordania* based on morphological homologues and knowledge of trilobite ontogeny. The characters chosen form *character cladograms*. For example, character 6 is 'Presence of pits on anterior area of anterior cranidial border', which occurs in two states: absent and present (Ebach & Edgecombe 1999, p. 435). Naturally, the present state is considered derived and

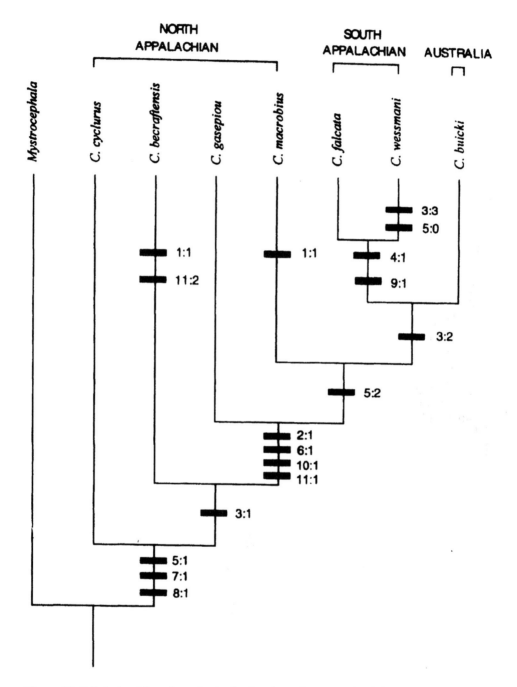

Figure 13.2 Relationships of species in the fossil Brachymetopid trilobite genus *Cordania* (seven species: *C. becraftensis, C. buicki, C. cyclurus, C. falcata, C. gasepiou, C. macrobius* and *C. wessmanii*) (after Ebach & Edgecombe 1999, p. 432, fig. 2, with permission).

homologous to the presence of other pits on the anterior cranidial border, and together results in the character cladogram:

(Mystrocephala, C. becraftensis, C. cyclurus, (C. buicki, C. falcata, C. gasepiou, C. macrobius and C. wessmanii))

The notion of homology, also from the threefold parallelism, assumes that the character cladograms are all part of the same organism, based on the sameness of their homologues and ontogeny, meaning they can be added together to form a general cladogram (see Chapter 2 for further examples). When all 11 character cladograms are combined, the following general cladogram results:

(Mystrocephala, (C. cyclurus, (C. becraftensis, (C. gasepiou, (C. macrobius, (C. buicki, (C. wessmanii, C. falcata)))))))

This general cladogram is based on notions of ontogeny and morphology, reflecting the threefold parallelism. What the cladogram above tells us is that *Cordania* is monophyletic and part of a natural classification – the derived phylogenetic tree, however, tells us that all members of *Cordania* share a common ancestor. The threefold parallelism ensures that we are combining meaningful data in order to derive solid assumptions (e.g., common ancestry) rather than speculations (e.g., *C. buicki* is ancestral to either *C. wessmanii* and/or *C. falcata*). Moreover, the threefold parallelism avoids combining two possibly unrelated statistical trends (i.e., phenetics), which may result in a statistically significant, but cladistically irrelevant, branching diagram.

Cladograms and Distributional Data: Comparative Biogeography

In comparative biogeography (e.g., Parenti & Ebach 2009), cladograms can be combined with distributional information in order to form a biogeographical classification called an *areagram*. By simply replacing the names of the taxa with the areas that they occur in, our *Agathis* cladogram can be converted into an areagram (Figure 13.1b).

An areagram is not unlike a character cladogram – the relationships between taxa now apply to the areas in which they are found – all areas are found at the terminal branches and, like taxa, only one area is represented at any given time. Areagrams (from different cladograms) may be added together (if they overlap geographically) to form a *general areagram*, and like cladograms, these are based on the threefold parallelism, ensuring that we arrive at a solid assumption supported by data, in this case common ancestry of a biota.

In the *Agathis* example above, we come across a problem commonly encountered when dealing with areagrams, namely *geographical paralogy* (i.e., area duplication, see Nelson & Ladiges 1996). As some characters are not ideal for use in a cladistic analysis (e.g., apomorphic characters), so too are certain taxa in areagrams. In the case of *Agathis*, Eastern Australia is duplicated, when in fact only a single occurrence is possible (there is, after all, only one Eastern Australia). A way to resolve *paralogy* is to work out to which other area Eastern Australia is most closely related. At first we find that Eastern Australia is more closely related to Borneo/Western Malesia than it is to anything else. In the sister clade we also find another instance of Eastern Australia, this time it is more closely related to Fiji/Vanuatu than it is to East Malesia. Clearly we have two unique areagrams containing one instance of Eastern Australia:

1. (((Eastern Australia, Borneo/Western Malesia), New Caledonia), New Zealand)
2. ((((Eastern Australia, Fiji/Vanuatu), Eastern Malesia), New Caledonia), New Zealand)

These two unique areagrams are called *paralogy-free subtrees*, as they contain no duplicated areas (New Caledonia occurs four times in one clade and therefore can be reduced to a single terminal branch). They do, however, contain *multiple areas on a single terminal branch* (MAST), namely Fiji/Vanuatu and Borneo/Western Malesia, meaning the two original areagrams above can be broken down into four, each representing a unique relationship:

1. (((Eastern Australia, Borneo), New Caledonia), New Zealand)
2. (((Eastern Australia, Western Malesia), New Caledonia), New Zealand)
3. ((((Eastern Australia, Fiji), Eastern Malesia), New Caledonia), New Zealand)
4. ((((Eastern Australia, Vanuatu), Eastern Malesia), New Caledonia), New Zealand)

When added together the general areagram is:

(((((Eastern Australia, Fiji, Vanuatu, Borneo, Western Malesia), Eastern Malesia), New Caledonia), New Zealand)

A single areagram contained four unique area relationships, all based on data and equally valid. The problem, in the *Agathis* example, is that we do not know what the relationships are between Eastern Australia, Fiji, Vanuatu, Borneo and Western Malesia. In order to resolve this, we simply add more areagrams. Remember, areagrams behave in the same way as character cladograms in that they are supported by the threefold parallelism. In comparative biogeography, the threefold parallelism is represented by geographical homologues (i.e., the areas), which ensures that Eastern Australia in one *Agathis* subtree is the same as in another.

While it is possible to add areagrams together to form a general areagram, there are certain problems. Due to various reasons (e.g., extinction, lack of data, etc.) not all taxa share the exact same distributions, meaning that we need to add several new areagrams until the geographical overlap covers Eastern Australia, Fiji, Vanuatu, Borneo and Western Malesia. Finding more area relationship may be problematic for some regions due to the lack of available areagrams. Once we do find more areagrams and construct the general areagram, we may make several inferences that are supported by the data, namely:

1. Areas may form monophyletic regions because they share a common ancestral biota;

2. The main mode of diversification in the taxa under study is vicariance, as dispersal is either ad hoc or unique to a particular taxon and therefore does not form patterns;

3. If the patterns of external factors (e.g., climate, tectonics, geography, etc.) match the general areagram, a correlation may be inferred, but it is not supported by the threefold parallelism.

Any other assumptions are not supported by the data, such as *dispersalism* and the identification of *ancestral areas* – general areagrams *may* contain ancestral areas, but they lie unidentified alongside their descendants in the terminal branches.

Phylogenetic Networks: Bounded by Branching Diagrams

The threefold parallelism ensures that our inferences from phylogenies and general areagrams are supported by data. Based on this, inferences such as common ancestry are supported by the data to a certain degree. We, of course, cannot identify the *actual* common ancestor, nor can we ascertain whether any of our taxa are ancestors. If there are any ancestors, they will lie undetected in the terminal branches of the phylogenetic tree along with all their descendants. But what if we suppose that we 'know' that two closely related 'ancestors' have produced a hybrid, namely a taxon that is more closely related to two other taxa than it is to another taxon?

Consider Hybrid C, which shares an equal relationship to taxa B and D, within a set of relationships including A:

$$(D(BC) + (A(CD) + (C(AB) + (D(AB)$$

Figure 13.3 depicts a 'galled phylogenetic network' (Jansson et al. 2006, fig. 1, see also Jansson & Sung 2016), in which taxon C is equally related (connected) to B and D. From the perspective of a cladogram or tree, the galled network does not

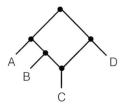

Figure 13.3 A 'galled phylogenetic network', in which taxon C is equally related (connected) to B and D (redrawn from Jansson et al. 2006, fig. 1).

reveal what the relationship is between A, B and D. If we were to add all four statements together (uniformly weighted), we would find:

$$(D(BC) + (A(CD) + (C(AB) + (D(AB) = (ABCD)$$

Note that the only conflicting statement is (D(BC). If we count the number of pairings, we find that AB occurs twice and BC and CD once each, whereas AC, AD, BC and BD are absent. If we assign weights to AB, BC and CD (with AB having the greater weight), the final result is ((AB)(CD)).

The same is true for the 'hybrid' taxon C. We find that C *conflicts,* yet we find a solution if we weight the relationships. Why, then, do we assume that C is a hybrid, when the solution shows that it is clearly more closely related to D than it is to B? The hybrid example is the same as those for HGT (horizontal gene transfer, see Chapter 3), examples of *assuming a specific process before discovering a pattern –* more precisely, assuming a process has occurred (i.e., hybridisation and HGT) when *no pattern reveals itself.* Why chose conflict over a specific process?

Systematics is about making sense of the world based on patterns, whereas phylogenetics tends to focus on why patterns tend to be missing (e.g., character conflict). What is merely a case of simple conflict has turned into a highly speculative inference. Taxon C is assumed to be a hybrid (i.e., as having undergone a *specific* process), when many equally plausible (or frankly far more plausible) explanations exist, such as poor characters or typographical errors in the data matrix. Regardless, phylogenetists have designed galled networks and reticulograms in order to show and explain the conflict (i.e., non-pattern) as natural processes (e.g., hybrids, HGT, etc.) rather than search for patterns and remove conflict (as the case above demonstrates).

Networks are part of a larger movement towards a model-based approach in systematics and phylogenetics. Rather than search for patterns in nature, models that assume certain processes are correlated against data to see if they match. This process is erroneously known as 'testing' to see if the model matches the data. The error here is assuming that correlation (fitting a model to the data) is somehow testing. Most models are mathematical representations of opinions –

they may be incredibly well-formed and well-argued opinions – nonetheless, all opinions in phylogenetics are based on inferences of data and not on empirical observations or patterns (e.g., molecular clocks, branch lengths etc.). These model-based approaches are usually disguised as 'hypothesis testing', which is to say, the hypothesis is the model and the 'test' is a way to correlate the model with the data. In the case of molecular clocks, cladograms are correlated with a marker (e.g., fossil, geological event, etc.). If the date changes, so does the correlation, meaning that each node is given a new date. Correlations have no predictive ability and do little to help us understand the natural world. Rather, models simply enforce our own scientific opinions on data that is silent about process. The threefold parallelism helps us distinguish between statistically significant and cladistically irrelevant speculations from robust inferences based on data.

Rules, Models and Relationships: Why Comparative Biology is the Science of Patterns

For every *rule* in comparative biology there is an exception, and every exception makes a rule null and void. Why, then, insist on rules?

Rules are important in defining models. Even though a model is nothing more than a mathematical representation of an opinion, the model relies on a set of rules to work. For example, the notion that every taxon originated in a single area (*monogenesis*) is an assumption that many take for a rule. After all, what evidence is there for *polygenesis*? We may pose an alternative question: 'What evidence is there for monogenesis?'

Models rely on a set of assumptions, and when those assumptions are considered true, based solely on popular perception rather than evidence, then those assumptions become 'rules'. A case in point is the *Progression Rule* in biogeography (*sensu* Brundin 1966). This makes two assumptions:

1. The most basal terminal taxon is ancestral;
2. The area which the taxon inhabits is most likely the ancestral area.

The second assumption is based on the idea that descendants move away from ancestral areas, meaning that ancestral taxa inhabit ancestral areas. In contrast, William Diller Matthew (1871–1930) believed that areas undergoing significant climate change would result in descendants adapting to the new conditions, meaning that the ancestors would move away from the ancestral area (Matthew 1915). There is growing evidence to support Matthew's assumption. In the Great Barrier Reef, for example, new metapopulations of coral (*Acropora millepora*) are adapting to the warmer conditions, meaning older populations used to cooler waters have been replaced (Matz et al. 2018). In this case there is an exception to the Progression Rule, meaning that it does not universally apply to all taxa, meaning that it is not a functional rule. What consequences does this have for other biogeographical models?

If biogeographical models are not universal, then we would have to create assumptions for individual taxa and areas, meaning that large-scale studies over longer time frames become impossible in historical biogeography. One way to overcome this is simply to abandon models for practical reasons (i.e., there are no universal assumptions of laws in comparative biology) and adopt a more universal approach, namely *relationships* (Chapter 7).

Relationships rely on the notion that all things are somehow related. Certainly, we may infer that homology is evidence for evolution and a common evolutionary history, but nineteenth century anatomists, such as Louis Agassiz (1807–1873), saw homology as evidence of a 'divine plan' (i.e., archetype or *bauplan*). Relationships contain no rules or laws. If two things share the same characteristic than they do to a third, they also share a relationship. Whether that relationship is understood to be natural is based on *patterns* (i.e., homology and monophyly) and not *rules*. Here we once again return to the cladistic parameter, namely,

$$\text{Homology} \asymp \text{Monophyly} \asymp \text{Natural Classification}$$

Homology is a relationship, which, compared to other relationships, forms a pattern, namely, monophyly. By combining character cladograms (i.e., relationships) we discover general cladograms (i.e., patterns). By combining areagrams (i.e., relationships) we discover general areagrams (i.e., patterns). In other words:

$$\text{Relationship (of homologues of a character)} \asymp \text{Relationship (of characters)} \asymp$$
$$\text{Relationship (of taxa)}$$

Patterns are discoveries based on evidence, whereas processes are inferred within our models. Comparative biology, then, is the science of pattern rather than the science of process. How then does this impact on phylogenetics?

Phylogenetics is about inferring processes from our patterns through our models. How far one wishes to take this is up to the individual – but beware: our models are only as good as our inferences, that is, our well-informed opinions[6]. We may claim that monophyly is evidence for common descent, but in most cases that is as far as it can go. Once we start to argue for explicit processes based on tree

---

[6] In her book *Weapons of Math Destruction*, Cathy O'Neil defined models as 'nothing more than an abstract representation of some process' (O'Neil 2016, p. 18). In fact, in the historical sciences where many of these processes are unobservable and immeasurable, one could say that 'models are opinions embedded in mathematics' (O'Neil 2016, p. 21).

topology or paralogy, then we find ourselves arguing for universal laws, such as the Progression Rule. In other cases, phylogenetics has sought to infer processes when patterns are missing, which is rather confounding. Finding meaning (e.g., a process) when there are no patterns (i.e., no evidence or 'noise') is bordering on what can only be described as science fiction.

Inferring from Noise: How Not to Do Phylogenetics

Phylogenetic studies use a number of terms in addition to homology, such as *homoplasy, paraphyly, polyphyly, reversals* and so on[7]. Definitions of these words or phrases vary considerably. None of these terms appear in our Chapter 1 (which includes a brief summary of taxonomic terms) or Chapter 7 (which has a more detailed discussion of the taxonomic terms listed in Chapter 2[8]); paraphyly is discussed briefly in Chapter 6. All these terms come into play after any character (or data) analysis has been undertaken to explain what has not been found by direct evidence. Thus, homoplasy does not, and cannot, exist without homology; paraphyly and polyphyly do not, and cannot, exist without monophyly, etc. Oddly, then, each of these terms – homoplasy, paraphyly, polyphyly, reversals – depend on *negative evidence*: those characters *not* homologous are thought to be homoplasious; those groups *not* monophyletic are thought to be paraphyletic. But as we stated earlier, *evidence of monophyly is not evidence of paraphyly* and likewise *evidence of homology is not evidence of homoplasy*. Both are simply terms given to *noise* as opposed to *signal*. As Colin Patterson wrote in an unpublished talk:

> As for homoplasy, as I see it, [it] is not of much interest to systematists. Certainly it's our job to discover homoplasy, to sort the correspondences into two piles, one of possible homologies and one of probable homoplasies, but once that's done, the systematist has no further interest in homoplasies – they are just the chaff left over from the thrashing. Homoplasy is interesting not to systematists but to evolutionists, who like to find explanations for it. (Patterson, unpublished data, 1990)

Each term ends up representing a process that lacks any evidence. The usual process inferred to explain homoplasy is evolutionary convergence, namely that two unrelated taxa have undergone similar evolution due to some external influence, such as an adaptation to climate. Another, simpler, explanation is that the character formulation is at fault, either because it was poorly described or coded incorrectly. In the case of

---

[7] There are many other terms, *hemiplasy* for example (Avise & Robinson 2008). A suite of new terms have recently been introduced (Ochoterena et al. 2019) and there are numerous new terms relating to homology in molecular systematics (e.g., Mindell & Meyer 2001, see their figure 2 and our Chapter 7, Box 7.5).

[8] Homoplasy appears in passing.

molecular data, it could be due to any number of errors in collecting or processing. The same is true for paraphyly and polyphyly (i.e., aphyly), which are phylogenetic narratives given to the taxa left over after monophyly is discovered. These narratives often involve unknown and unobservable ancestors for which there is no evidence either. Conversely, a more reasonable and rational explanation, such as incomplete or unknown taxonomy, would make paraphyly and polyphyly *systematic problems* rather than *phylogenetic curiosities*, perhaps explaining why phylogenetists still use these terms. In this case, the former – *systematic problems* – invokes a science, the latter – *phylogenetic curiosities* – a 'solution' found in explanation (e.g., Powell 2007; Currie 2012).

Reversals and branch lengths also invoke phylogenetic inferences, when more practical systematic explanations will do. Reversals are similar to homoplasy, when the same character state supports two synapomorphies in the same clade. Branch lengths are character states that are mapped onto the internal and terminal branches of cladograms. In molecular cladograms these character states are inferred to be accumulated *substitutions* and are used to measure rates of evolution. From a systematic perspective, they are simply autapomorphies, that is, character states unique to a single terminal taxon which contain no information about shared characteristics (i.e., synapomorphies) – autapomorphies provide no evidence or information about relationships.

Summary

Comparative biology is the science of patterns, whereas phylogenetics is a way to model our inferences based on those patterns. Phylogenetics can make highly speculative assumptions that are not based on any evidence whatsoever, or, if used properly, may make insightful narratives based on systematic and biogeographical patterns. The threefold parallelism, namely the evidence to support the cladistic parameter, is a good way to curb unobservable assumptions and narratives as well as non-homologous-analogous data from being used.

References

Alibardi, L. & Rogers. G. 2015. Observations on fur development in echidna (Monotremata, Mammalia) indicate that spines precede hairs in ontogeny. *Anatomical Record* 298: 61–70.

Avise, JC. & Robinson, TJ. 2008. Hemiplasy: a new term in the lexicon of phylogenetics. *Systematic Biology* 57: 503–507.

Brundin, L. 1966. Transantarctic relationships and their significance as evidenced by midges. *Kungliga Svenska Vetenskapsakademiens Handlinger* 11 (Series 4):, 1–472.

Carr, EH. 1961, *What is History?* University of Cambridge Press, Cambridge, UK.

Currie, A. 2012. Convergence as evidence. *The British Journal for the Philosophy of Science* 64: 763–786.

Darwin, C. 1859. *On the Origin of Species by Means of Natural Selection, or, the Preservation of Favoured Races in the Struggle for Life*. John Murray, London.

Ebach, MC. & Edgecombe, GD. 1999. The Devonian trilobite *Cordania* from Australia. *Journal of Paleontology* 73: 431–436.

Ebach, MC., Morrone, JJ. & Williams, DM. 2006. Getting rid of origins. *Rivista di Biologia* 99: 360-365.

Escapa, IH. & Catalano, SA. 2013. Phylogenetic analysis of Araucariaceae: integrating molecules, morphology, and fossils. *International Journal of Plant Sciences* 174: 1153–1170.

Escapa, IH., Iglesias, A., Wilf, P., Catalano, SA., Caraballo-Ortiz, MA. & Rubén Cúneo, N. 2018. *Agathis* trees of Patagonia's Cretaceous-Paleogene death landscapes and their evolutionary significance. *American Journal of Botany* 105: 1345–1368.

Fara, E. 2001. What are Lazarus taxa? *Geological Journal* 36: 291–303.

Flessa, KW. & Jablonski, D. 1983. Extinction is here to stay. *Paleobiology* 9: 315–321.

Gilmore, S. & Hill, KD. 1997. Relationships of the Wollemi pine (*Wollemia nobilis*) and a molecular phylogeny of the Araucariaceae. *Telopea* 7: 275e291.

Haeckel, EHPA. 1866. *Generelle Morphologie der Organismen: allgemeine Grundzüge der organischen Formen-Wissenschaft, mechanisch begründet durch die von Charles Darwin reformirte Descendenz-Theorie*. G. Reimer, Berlin.

Haeckel, EHPA. 1894–1896. *Systematische Phylogenie: Entwurf eines natürlichen Systems der Organismen auf Grund ihrer Stammesgeschichte*. G. Reimer, Berlin.

Haeckel, EHPA. 1909. Charles Darwin as an anthropologist. In: Seward, AC (ed.), *Darwin and Modern Science*. Cambridge University Press, Cambridge, pp. 137–151.

Jansson, J. & Sung, WK. 2016. Algorithms for combining rooted triplets into a galled phylogenetic network. In: Kao, MY. (eds) *Encyclopedia of Algorithms*, 2nd ed. Springer, New York, NY, pp. 48–52.

Jansson, J., Nguyen, NB. & Sung, W-K. 2006. Algorithms for combining rooted triplets into a galled phylogenetic network. *SIAM Journal on Computing* 35(5): 1098–1121.

Matthew, WD. 1915. Climate and evolution. *Annals of the New York Academy of Sciences* 24: 171–318.

Matz, MV., Treml, EA., Aglyamova, GV. & Bay, LK. 2018. Potential and limits for rapid genetic adaptation to warming in a Great Barrier Reef coral. *PLOS Genetics* 14(4): e1007220.

Mindell, DP. & Meyer, A. 2001. Homology evolving. *Trends in Ecology & Evolution* 16: 434–440.

Nelson, GJ. 1978. Ontogeny, phylogeny, paleontology and the biogenetic law. *Systematic Zoology* 27: 324–345.

Nelson, GJ. & Ladiges, PY. 1996. Paralogy in cladistic biogeography and analysis of paralogy-free subtrees. *American Museum Novitates* 3167.

Nelson, GJ. & Platnick, NI. 1981. *Systematics and Biogeography: Cladistics and Vicariance*. Columbia University Press, New York.

Norell, MA. 1993. Tree-based approaches to understanding history; comments on ranks, rules and the quality of the fossil record. *American Journal of Science* 293-A: 407–417.

Novacek, MJ. & Wheeler, QD. 1992. *Extinction and Phylogeny*. Columbia University Press, New York.

Ochoterena, H., Vrijdaghs, A., Smets, E. & Claβen-Bockhoffs, R. 2019. The search for common origin: homology revisited. *Systematic Biology* 68: 767–780.

Olson, LE. 2013. Tenrecs. *Current Biology* 23: R5–R8.

O'Neil, C. 2016. *Weapons of Math Destruction: How Big Data Increases Inequality and Threatens Democracy*. Crown, New York.

Parenti, LR. & Ebach, MC. 2009. *Comparative Biogeography: Discovering and Classifying Biogeographical Patterns of a Dynamic Earth*. University of California Press, Berkeley, CA.

Powell, R. 2007. Is convergence more than an analogy? Homoplasy and its implications for macroevolutionary predictability. *Biology and Philosophy* 22: 565-578.

Rickards, RB. & Wright, AJ. 2002. Lazarus taxa, refugia and relict faunas: evidence from graptolites. *Journal of the Geological Society* 159(1): 1-4.

Smith, JLB. 1939. A living fish of Mesozoic type. *Nature* 143: 455–456.

Stockey, RA. 1990. Antarctic and Gondwanan conifers. In: Taylor, TN. & Taylor, EL. (eds) *Antarctic Palaeobiology*. Springer, New York, pp. 179–191.

Stockey, RA. 1994. Mesozoic Araucariaceae: morphology and systematic relationships. *Journal of Plant Research* 107: 493–502.

Turner, S., Bean, LB., Dettmann, M., McKellar, JL., McLoughlin, S. & Thulborn, T. 2009. Australian Jurassic sedimentary and fossil successions: current work and future prospects for marine and non-marine correlation. *GFF* 131: 49–70.

Vajda, V. & McLoughlin, ST. 2005. A new Maastrichtian-Paleocene *Azolla* species from of Bolivia, with a comparison of the global record of coeval *Azolla* microfossils. *Alcheringa* 29: 305–329.

Walkom, AB. 1921. Mesozoic Floras of New South Wales Part 1: fossil plants from Cockabutta Mountain and Talbragar. *Memoirs of the Geological Survey of New South Wales*. Memoir Palaeontology 12. W.A. Gullick, Government Printer, Sydney.

Weisbecker, V. & Beck, RMD. 2015. Marsupial and monotreme evolution and biogeography. In: Klieve, A., Hogan, L., Johnston, S. & Murray, P. (eds), *Marsupials and Monotremes*. Nova Science Publishers, Inc., New York, pp. 1–31.

White, ME. 1981. Revision of the Talbragar Fish Bed Flora (Jurassic) of New South Wales. *Records of the Australian Museum* 33: 695–721.

Wilf, P., Escapa, IH., Cúneo, NR., Kooyman, RM., Johnson, KR. & Iglesias, A. 2014. First South American *Agathis* (Araucariaceae), Eocene of Patagonia. *American Journal of Botany* 101: 156–179.

Williams, DM. 2007. Classification and diatom systematics: the past, the present and the future. In: Brodie, J. & Lewis, J. (eds), *Unravelling the Algae*. CRC Press, Boca Raton, FL, pp. 57–91.

Williams, DM. & Ebach, MC. 2008. *Foundations of Systematics and Biogeography*. Springer-Verlag New York Inc., New York.

Zaragüeta i Bagils, R., Lelièvre, H. & Tassy, P. 2004. Temporal paralogy, cladograms, and the quality of the fossil record. *Geodiversitas* 26: 381–389.

14

The Separation of Classification and Phylogenetics

... ye may always fully and uninterruptedly enjoy every civil and religious right; and be, in your turn, the means of securing it to others; but that the example which ye have unwisely set, of mingling religion with politics, may be disavowed and reprobated by every inhabitant of America.

(Thomas Paine, Common Sense, 1987 [1776], p. 115)

Natural systematics is generally identified today with phylogenetic systematics. This is quite wrong. Natural systematics is actually far older. Its foundations were laid even before the time of Darwin ... It is therefore important to show what natural systematics was before phylogeny ... He will at least admit that the establishment of a natural system is a major task of 'comparative' morphology, in any case a product of it.

(Naef 1972[1921, p. 6], p. 12)

The trouble with evolutionary classification (or cladistics), so far as I understand it, seems to me to be that when there are gaps in the evidence you have to fill them up with fiction and storytelling (Furbank, P.N. 2005, *New York Review of Books*, 22 September 2005).

*You gotta keep 'em separated (*The Offspring – *Come out and play)*

Ernst Haeckel (1834–1919) and Karl Gegenbaur (1826–1903) first combined systematics with phylogenetics in *Systematische Phylogenie* (1894–1896, see the earlier and better known Haeckel, 1866) and the second edition of *Grundzüge der vergleichenden Anatomie* (Gegenbaur, 1870), the latter more concerned with developing the new field of *evolutionary morphology* (for a review of Gegenbaur see Hoßfeld et al. 2003); Adolf Naef (1883–1949) separated them in his *Systematic Morphology* (Naef 1919; for a review see Rieppel et al. 2013); Willi Hennig (1913–1976) brought them back together again with his *Phylogenetic Systematics* (Hennig 1950, 1966, see Rieppel et al. 2013 and Williams et al. 2016); Nelson and Platnick separated them once again in their *Systematics and Biogeography* (Nelson & Platnick 1981); and so it goes. In any case, Naef's statement that natural [systematics] classification is the major task of comparative biology and 'its foundations were laid even before the time of Darwin' (Naef 1972, p. 12, translated from Naef 1921, p. 6[1]) remains accurate and appealing (Naef 1919, 1921–1923). Given this understanding, why are systematists and the 'consumers' of natural classifications (i.e., 'end users', such as phylogeneticists, ecologists, etc., see Chapter 15) so often *confused* about this dichotomy? And why do their approaches *fuse* classification with phylogenetics?

Preamble: The Problem of Genealogy

Naef (1921, pp. 6–7, 1972, pp. 11–12) posed a series of questions concerning a natural system of classification:

1. Is a natural system possible in its accepted form, i.e., as a genealogical tree branching into a variety of species …? This is the main problem from which other questions follow:

2. How is it possible that the ontogenetic stages … correspond to their systematic stages …? (Parallelism of the successive stages of individual development and the 'animal series' of natural systematics …)

3. How is it possible that the geographical distribution depends on the 'systematic relationships'?

4. How is it possible that the sequence of systematic stages corresponds in general to the sequence of paleontological appearance (insofar as this can be established)?

The problem is that a natural classification provides evidence for a phylogeny only via a branching diagram (i.e., a *fundamental cladogram*) or classification (i.e., a *derivative cladogram*) (Chapter 7). How do these diagrams relate to a given *genealogy*, or series of ontogenetic stages, or geographical distribution, or stratigraphical (palaeontological) sequences? We can understand some aspects of genealogy as in certain cases we can provide *direct* evidence for a particular genealogical sequence in a breeding population (e.g., in *Drosophila* or, with appropriate records, any particular human family). We can also provide *direct* evidence for a natural classification (i.e., synapomorphy and monophyly). Yet we

---

[1] 'Man identifiziert heute durchweg die natürliche Systematik mit der phylogenetischen, doch geschieht dies durchaus mit Unrecht. Nicht nur geht die erstere der letzteren liistorisch lange voraus und hat bereits in der vor-Darwin'schen Zeit ihre klassischen Vertreter gefunden, sondern sie hat auch die Voraussetzungen für die Begründung der Abstammungslehre, nämlich obige Fragenstellungen, erst hervorgebracht' (Naef 1921, p. 6).

are unable to provide *direct* evidence for a phylogeny: that is, *a direct series of ancestor-descendant relationships* – we can only speculate by *inferring* common ancestry (see Chapter 7).

The problem of fitting genealogies into natural classifications has been addressed a number of times over the last 100 years. A notable example provided by George Gaylord Simpson (1902–1984), who posed the following question: 'Is a man more closely related to his father, son, or brother?' (Simpson 1961, p. 129[2]). The problem, later called *Simpson's Engima* (Nelson 2016), is based largely on the fact that genealogy is expressed through *hereditary*, meaning that individuals are hybrids of their parents. The cladistic solution to the above statement 'Is a man [m] more closely related to his father [f], son [s], or brother [b]?' is (b, (m, f, s)) as a man would share 50% of the same DNA with his father *and* son, and <50% with his brother (unless they are twins). This classification excludes the mother [mm] and daughter [d] who would also have the same genetic mix as the father and brother respectively, leading to a larger but equally unresolved classification: (b, d, (m, mm, f, s)). A similar problem occurs with a hypothetical cladogenetic ancestor and its descendant lineage. If species A gives rise to species B and C, then the classification would be (A, B, C), as A and B are equally related to C. In order to resolve the cladistic relationship, a hypothetical ancestor, species X, will need to be invoked, producing (A, (B, C, X). In an anagenetic lineage, where species A gives rise to species B and, species B gives rise to species C, then we discover some resolution, namely (A, (B, C)).

The problems above derive from what Adolf Naef cautions us not to do: confuse classification with phylogeny (or genealogy). Even if we did know a genealogical or an ancestor-descendant relationship, as in the above example, a natural classification would not be possible. This is because a natural classification, that is, *the cladistic parameter*, comes first and phylogeny follows (Chapter 7). We could say that following the cladistic parameter is where phylogenetics starts, meaning that phylogenetists are dependent on natural classifications (i.e., monophyly). From the point of view of a systematist, monophyly needs to be found prior to making any formal pronouncements about descent (e.g., ancestors, plesiomorphy, reversals, etc.).

---

[2] Simpson had outlined the problem a little earlier in his monograph on mammals: 'The existence of groups that are ancestral to two or more ultimately quite different phyla and the implication in classification that members of one group are more nearly related to each other than to members of other groups of the same rank give rise to the most difficult problems of classification of fossils. When the ancestral group is known, how is it to be classified? Can it be more nearly related to one than to the other of its descendent lines? In a sequence, is a group more nearly related to its ancestors, its descendants, or its contemporaries of like origin; in the human family analogy, is a man more nearly related to his father, son, or brother?' (Simpson 1945, p. 17).

Systematics and Phylogenetics Fused: Phylogenetic Classification (the *PhyloCode*)

> Fortunately, most practising phylogenetic systematists are either tolerant or self-conscious enough to believe that their cause, if just, will ultimately prevail without the band-aid of an intransigent set of rules tailored to their needs. The sooner the *PhyloCode* is buried the better for biology.
>
> *(Greuter 2004, p. 24)*

The *PhyloCode* is an attempt to fuse classification with phylogeny (i.e., to create a phylogenetic classification) via a new system of taxonomic nomenclature (hence the '-code' in *PhyloCode*). Its origins can be found in a paper written over 30 years ago by Kevin de Queiroz (1988), where he noted:

> If a truly evolutionary systematics is to flourish, it must take the concept of evolution as an axiom rather than a superficial interpretation. This will necessitate a reevaluation of systematic concepts and the methods used to determine systematic relationships as well as the taxonomies derived from them. The beginnings of an evolutionary approach have already led to suggestions to redefine various taxonomic terms, to reject certain systematic methods, to eliminate many currently recognized taxa, and to recognize many currently unrecognized taxa. Perhaps this approach will ultimately lead to the replacement of the Linnaean taxonomic system …. In short, embracing a truly evolutionary systematics will result in drastic taxonomic changes; this is why we talk about a Darwinian Revolution. (de Queiroz 1988, p. 257[3])

Critiquing the *PhyloCode,* or any other version of phylogenetic classification for that matter (e.g., Papavero et al. 2001; Kluge 2005; Béthoux 2007a–c, 2010a, b[4]; Avise & Liu 2011; Naomi, 2014; Carter et al. 2015[5]), is difficult for practising taxonomists: where does one start (for some general commentary, see Dubois 2007)? Here we briefly discuss what we see as the core issue addressed by the *PhyloCode* (de Queiroz 2006): that of a hierarchical classificatory system and associated taxonomic ranks, and types as actual specimens.

---

[3] 'But if the Darwinian Revolution is ever to occur in biological taxonomy ... , then the role of the principle of descent must change. It must change from an after-the-fact interpretation to a central tenet from which the principles and methods of taxonomy are deduced. The principle of phylogenetic definitions exemplifies this change. Previously, taxa were considered to be defined by characters and only interpreted after-the-fact as products of evolution. Under the principle of phylogenetic definitions, evolutionary considerations enter directly into the definitions of their names' (de Queiroz & Gauthier 1990, p. 320).

[4] A response to Dubois (2010).

[5] Carter et al. (2015) is unusual in that it adopts the phylogenetic nomenclature of the *PhyloCode* but specifically includes paraphyletic taxa.

First we offer brief comments on biological nomenclature and outline how the *PhyloCode* differs.

Biological Nomenclature

Biological nomenclature is a formal system of rules for naming taxa. It is not scientific, nor, in many cases, entirely logical – it could almost be interpreted as a legal document; it is not always consistent across taxa and needs constant adjustment and refinement – it is a work-in-progress[6]. For example, with the emergence of electronic (online) publications, all nomenclatural codes had to adjust their rules in order to keep pace with that particular advance in the technology of publishing: originally names were only formally recognised if, and only if, they were published in a printed form that was made generally available (e.g., see Nicolson et al. 2017 and Minelli 2013). Online publications and journals are just one form of 'new technology' that requires naming rules to adapt. But importantly, nomenclatural codes (Botanical and Zoological codes – and the *BioCode*, if it emerges, see McNeil 1996 and the Draft *BioCode*, https://archive.bgbm.org/IAPT/biocode/biocod97.pdf, but possibly not for Prokaryotes, see Parker et al. 2015) do not dictate how taxa *are to be, or should be,* discovered. Taxonomy and nomenclature are separate practices: the former, as we note above (and the core topic of this book), is a scientific enterprise; the latter merely a set of rules that have nothing to do with discovery. This much was eventually recognised by the *PhyloCode* proponents; however, the division was not seen as entirely separate:

> Taxonomy is concerned with the representation (and, in a broader sense, also the analysis) of relationships, including (under the common convention of representing relationships using groups) what kinds of groups are to be recognized as taxa. In contrast, nomenclature is concerned with naming taxa as well as with the application of existing taxon names in the context of subsequently proposed taxonomies. In short, taxonomy is concerned with taxa; nomenclature with their names. *The two activities, and their corresponding disciplines, are closely related*. (de Queiroz 2006, p. 160, our emphasis)

---

[6] Although a little dated now, Nicolson (1991) is a good review of developments in botanical nomenclature, from 1840 to 1991. See also www.iapt-taxon.org/historic/history.htm. Botanical nomenclature is reviewed and revised every five years with a new code published to include alterations, amendments and additions. The most recent version was published in 2018 (www .iapt-taxon.org/nomen/main.php). The most recent Zoological Code was published in 1999 and can be found here: https://www.iczn.org/. There is no accepted period of time for revised versions of the Code to appear (Michel, pers. comm.) and revisions and name changes (*opinions*) are published in the *Bulletin of Zoological Nomenclature (BZN)* in a similar fashion to *Taxon* for botanical names. The Bacteriological Code has recently been renamed as *The International Code of Nomenclature of Prokaryotes (ICSP)* and can be found here: http://ijs.microbiologyresearch .org/content/journal/ijsem/10.1099/ijsem.0.000778. Efforts to classify all viruses are now underway (Kuhn et al. 2019).

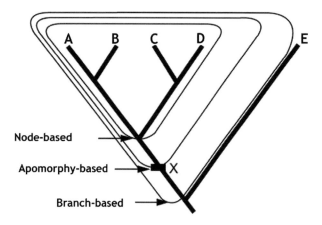

Figure 14.1 Three methods of providing taxon names when derived from any particular branching diagram (cladogram): node-based, branch-based and apomorphy-based (redrawn from Lee & Skinner 2007, fig. 1.).

Terms

Early discussions on the *PhyloCode* focused on three methods of providing taxon names when derived from any particular branching diagram (cladogram): node-based, stem-based[7] and apomorphy-based (de Queiroz & Gauthier 1990, p. 310, their fig. 1a–c; a summary diagram is reproduced as our Figure 14.1, after Lee & Skinner 2007). De Queiroz and Gauthier (1990, p. 310) defined the three options as follows (our emphasis in each):

1. By defining the name of a taxon as the clade stemming from the most recent common ancestor of two other taxa, hereafter referred to as a *node-based definition* (Figure 14.1).
2. By defining the name of a taxon as all those entities sharing a more recent common ancestor with one recognised taxon than with another, hereafter referred to as a *stem-based definition* [branch-based definition] (Figure 14.1).
3. By defining the name of a taxon as the clade stemming from the first ancestor to possess a particular synapomorphy, hereafter referred to as an *apomorphy-based definition* (Figure 14.1).

To understand the first two approaches it is necessary to distinguish between the meaning of some related terms mostly used in palaeontology: *crown-group,*

[7] Now referred to as *branch-based* (Cantino & de Queiroz 2010).

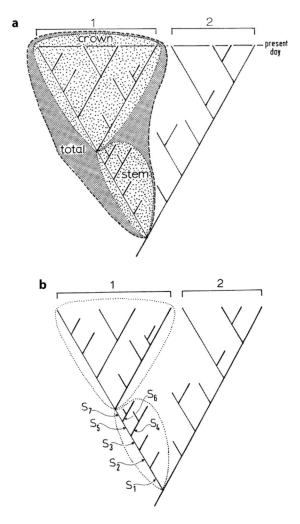

Figure 14.2 (a) *Crown-group*, *stem-group* and *total-group* (after Jefferies 1979, p. 449, fig. 5, upper, with permission). (b) Stem-groups as unnamed *plesions* S¹–S⁶ (after Jefferies 1979, p. 449, fig. 5, lower, with permission).

stem-group and *total-group*. These terms were first outlined by Hennig (1969, p. 32, Hennig 1981, p. 28 et seq.[8]), but the names crown-group[9] and stem-group were provided by Jefferies (1979, p. 449, his fig. 5 is reproduced here as our Figure 14.2a).

---

[8] The concepts were probably discussed even earlier (Willmann 2003).

[9] Hennig used the term * group (Hennig 1981, p. 29).

A *crown-group* comprises all the descendants of the presumed common ancestor of the living members of the taxon in question, while the *stem-group* comprises all the extinct species (taxa) that occur between the crown-group and its next extant taxon. Stem-groups can only be defined relative to an extant taxon (a crown group) and thus are, by definition, aphyletic (i.e., paraphyletic) with respect to that crown-group (Smith 1994, p. 95). Stem-groups correspond to *evolutionary grades* – an important concept for evolutionary taxonomists, coined by Julian Huxley who noted that 'stasigenesis results in the formation of delimitable and persistent anagenetic units, or *grades* . . .' (Huxley 1957, p. 455, original emphasis), something Huxley saw as secondary to the phylogenetic system: 'The cooperation of anagenesis and cladogenesis in all evolutionary transformations makes it necessary to supplement the customary cladal (phyletic) system with a subsidiary (gradal) terminology . . .' (Huxley 1958, p. 38, see Gascoigne, 1991). In addition, evolutionary taxonomists understood *polyphyletic* (Huxley 1957, p. 455) and *paraphyletic* taxa (Watt 1968, p. 350) as grades within a classification. Nelson (1974) referred to grade classification as *gradism*, something that was 'particularly decisive for the development of Mayr's "synthetic" theory' (Nelson 1974, p. 454[10]). De Queiroz rejected grades as 'holdovers from pre-evolutionary taxonomies based on the *Scala Naturae*, or great chain of being' (de Queiroz 1988, p. 252) but as Medawar warns us 'The methodology should therefore be measured against scientific practice to give us confidence in its worth' (Medawar 1968 [1996], p. 29) – if grades are rejected then they should not occur in any phylogenetic classification but often, when the *common* understanding of a name is revised or redefined to mean something entirely different (usually extinct fossil stem-groups, such as dinosaurs being redefined to include birds), they survive common usage (see Box 14.1). Needless to say, paraphyletic groups, stem-groups and grades are all undefinable taxa (see Chapter 6).

Box 14.1   Birds, Dinosaurs, Bird-Like Dinosaurs, Dinosaur-Like Birds

An instructive example is the classification of Aves as discussed by Patterson and Norell et al. some years ago (Patterson 1993a, 1993b; Norell et al. 1993). Although much has changed since that time, some key points remain.

In the early 1990s an account appeared of a new, extinct species of flightless vertebrate, *Mononykus olecranus*[1] (Figure B14.1.1a). It was referred to as a bird in the title ('Flightless *bird* from the Cretaceous of Mongolia') and as an 'availan theropod *dinosaur*' in the abstract (Perle et al. 1993a, our emphasis) – so: is *Mononykus olecranus* a bird, a dinosaur or both?

---

[10] One might dwell on that statement (see our Part II, especially Chapter 6).

Box 14.1 *(cont.)*

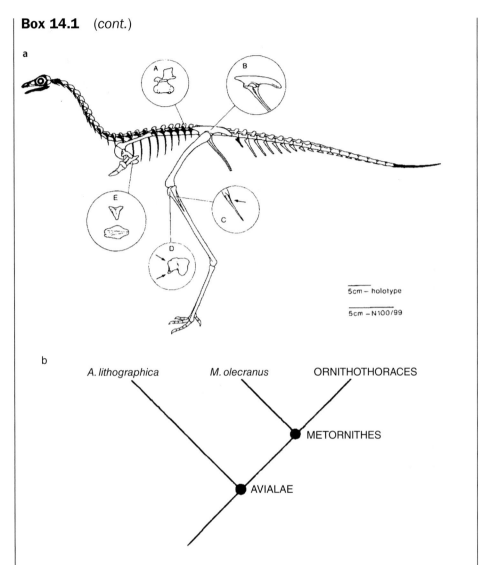

Figure B14.1.1 (a) *Mononykus olecranus* (after Perle et al. 1993a, fig. 2, with permission); (b) relationships of *Mononykus olecranus* (after Perle et al. 1993a, cladogram in the figure on p. 624, with permission).

Perle et al. adopted the crown-group approach for naming taxa (following Gauthier 1986), where all living birds comprised Aves, *Archaeopteryx* excluded. *Archaeopteryx* plus Aves comprises the Aviales with *Mononykus olecranus* belonging to a new group described by Perle et al. as the Metornithes; 'bird' referred to Aviales (Perle et al. 1993a, cladogram in the figure on p. 624, reproduced here as our Figure B14.1.1b). Patterson posed a simple question: Is *Mononychus* a bird, a dinosaur, a bird-like dinosaur or a dinosaur-like bird?

Box 14.1 *(cont.)*

Answering the question was not straightforward:

> ... Judging from their paper [Perle et al. 1993a], it is a member of both birds (a monophyletic group) and dinosaurs (a non-monophyletic group). (Patterson 1993a)

Norell's response is revealing:

> Patterson suggests that we considered *Mononychus* a member 'of birds (a monophyletic group) and dinosaurs (a non-monophyletic group)'. We did not say this; throughout the paper we consider dinosaurs to be monophyletic because birds are a member of this group. (Norell et al. 1993)

Their conclusions came with a handy phrase: 'Birds are a kind of dinosaur'. Well, what is it to be a 'kind of dinosaur'? To answer that question we need to know exactly what a dinosaur is, which is difficult as nearly everyone agrees dinosaurs, as commonly understood, are non-monophyletic, have no characters of their own and are a textbook example of a paraphyletic taxon. Birds are less problematic, at least with respect to modern birds (e.g., Prum et al. 2015, but see our comments on their tree [cladogram] in Chapter 7).

As Patterson pointed out in his response to Norell et al. (Patterson 1993b), one might salvage any particular paraphyletic group by expanding it to include the monophyletic group excluded from it (see Chapter 6). Using Patterson's examples, Reptilia could be made monophyletic by including mammals and birds, invertebrates could be made monophyletic by including vertebrates, and so on – the examples are endless. In the case of the latter, for example, could, then, vertebrates be considered 'kinds of invertebrates'? – this is clearly nonsensical. But the nonsensical becomes profound when dealing with extinct paraphyletic groups, such as dinosaurs, and one frequently encounters the idea that *birds are dinosaurs* and that *birds evolved from dinosaurs*.

For example, in an article published in *Evolution: Education and Outreach*[2] entitled 'Downsized Dinosaurs: The Evolutionary Transition to Modern Birds', the final section is entitled 'The Dinosaur in your Backyard':

> This evidence has led to the realization that the jays, finches, and hummingbirds that so peacefully frequent your backyard are indeed living dinosaurs – a surviving lineage of vicious predators that ruled the terrestrial ecosystems of the Mesozoic. (Chiappe 2009)

But jays, finches, and hummingbirds are not 'dinosaurs': they are birds.

In another article in *Evolution: Education and Outreach* we are asked to examine a 'phylogeny of tetrapods ... [that] makes it clear why phylogenetic classification leads us to view birds as dinosaurs. Since birds *evolved* from dinosaurs ...' (Thanukos 2009, p. 304). Inspection of that tree (cladogram) does not tell us these things: it tells us that birds are most closely related to 'more dinosaurs' (Thanukos 2009, p. 305, fig. 4 reproduced as our Figure B14.1.2). Finally, Gregory elaborates:

> Most of these considerations are reasonably intuitive, but many people find it surprising that phylogenetically birds are located within the 'reptiles' and represent the sister group to crocodilians. Although physical similarities would seem to suggest otherwise, *crocodiles are more closely related to birds than they are*

Box 14.1 (*cont.*)

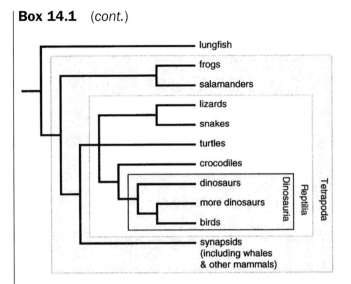

Figure B14.1.2 A 'phylogeny of tetrapods ... [that] makes it clear why phylogenetic classification leads us to view birds as dinosaurs. Since birds *evolved* from dinosaurs ...' (Thanukos 2009, p. 304); the cladogram tells us that birds are most closely related to 'more dinosaurs' (after Thanukos 2009, p. 305, fig. 4, with permission).

> to lizards. The reason for this is that the bird lineage has experienced significant modification, whereas crocodilians have remained largely unchanged for tens of millions of years. It is important to note that birds and crocodiles represent each other's closest living relatives but that birds are not descended from crocodiles nor vice versa – as taxonomic groups, crocodilians and birds both arose long after their respective lineages diverged from a common ancestor well over 200 million years ago. *Birds are, in fact, descended from a lineage of theropod dinosaurs*, making *Tyrannosaurus rex* far more similar to the last nonavian ancestor of modern birds than anything resembling a crocodile. (Gregory 2008, p. 131, emphasis ours)

Note how the text moves from observation ('crocodiles are more closely related to birds than they are to lizards') to interpretation ('Birds are, in fact, descended from a lineage of theropod dinosaurs'), confusing interpretation for data, as if it was evidence.

There might appear to be two alternative way to read trees – almost everyone agrees on how to document relationships. If one starts from the bottom (to the left of the cladogram in Figure B14.1.3), as one moves up the cladogram, *reading from bottom to top*, each group (taxon) encountered belongs to the bird stem-group, Herrerasaurideae to Deinonychosauria, prior to encountering the Aviales (Figure B14.1.3). No monophyletic groups are met with until the *entire group* is considered, that is including everything from Herrerasaurideae to Aviales. Occasionally, stem-groups, such as Herrerasaurideae to Deinonychosauria (dinosaurs), are named but '... stem groups have never existed in nature ... one might say, therefore, stem groups are just another way of expressing ... ignorance'

Box 14.1 *(cont.)*

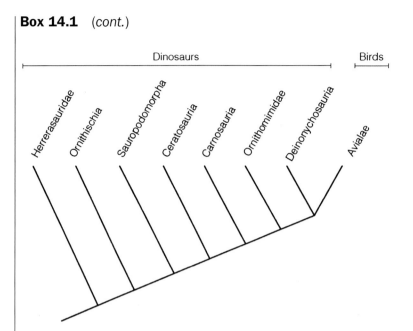

Figure B14.1.3 Cladograms of the relationships of birds and 'dinosaurs'.

(Sandvik 2008, p. 48, who give further examples of named stem-groups). It *might* then seem a moot point of nomenclature if one named the entire group (Herrerasaurideae to Aviales = the *total-group*) dinosaurs (a modified stem-group) or birds (a crown-group plus the stem-group).

Reading from *bottom to top* one encounters no specific relationships as such but simply taxon names, with the temptation to assume that the preceding ('lower') taxon *gave rise to* the one above it, and thus dinosaurs (Herrerasaurideae to Deinonychosauria, or whatever its composition might be, sometimes known as 'stem-group birds', e.g., Gee 2001), is an aphyletic (stem-group, grade, paraphyletic) group being understood as *giving rise to birds* or *birds evolved from dinosaurs*.

Yet it is the relationships that are important – and these are discovered only by reading the cladogram *from top to bottom* (or right to left in Figure B14.1.3, see also Ebach & Williams 2013). In this case we encounter several monophyletic groups: Aviales; Aviales + Deinonychosauria; Aviales + Deinonychosauria + Ornithomimoideae; and so on (Figure B14.1.3). To name each of these groups (relationships) reveals a drawback in that we may encounter a proliferation of names for each fossil added that occurs between Herrerasaurideae to Deinonychosauria, a point made by Patterson and one of the reasons the plesion concept was considered useful, as naming fossil specimens as *plesions* has no effect on the Linnaean classification of the crown-group (Patterson & Rosen 1977). To avoid this inconvenience of names, which in some cases (not all) might 'expand the classification scheme to no advantage' (Smith 1994, p. 97), a name can still be applied to the total-group (birds plus stem-group birds) but Aviales might seem more appropriate. As Patterson concluded,

Box 14.1 (*cont.*)

> Perhaps surprisingly, dinosaurs turn out to be *kinds of bird*, a theory of relation-
> ships that is most clearly and economically expressed by including them
> within Aves. (Patterson 1993b, our emphasis)

There has obviously been substantial commentary on bird relationships since 1993 and
substantial commentary on what the name Aves should be applied to with alternative
names offered for Patterson's concept of Aves, such as Avemetatarsalia (Benton 1999)
and Panaves (Gauthier & de Queiroz 2001). Neither of us are ornithologists so we have
not sought to disentangle these details (the interested reader can study Gauthier 1986,
Gauthier & de Queiroz 2001, Cau 2017 and Baron et al. 2017). The latter two contributions (Cau 2017 and Baron et al. 2017) have adopted *PhyloCode* conventions for their
classification – and the reader may judge for themselves how useful this is. Either way, the
point remains: dinosaurs are stem-group birds, isolated from their crown-group (= birds)
and they are *aphyletic* and are not *ancestors* of anything.

One final point: We noted above it would be possible to make invertebrates monophyletic by including the vertebrates. If we chose to name the group 'invertebrates + vertebrates' Vertebrata, one might be tempted to consider vertebrates as 'a kind of
invertebrate'. This, as we noted, is nonsense. Suppose we chose the alternative and
named that same group, 'invertebrates + vertebrates' Invertebrata instead and considered invertebrates as 'a kind of vertebrate' – this, too, would be nonsense. The
reasons are clear: we know what vertebrates are because, among other things, they have
a vertebral column (e.g., Benton 2004); but we cannot know what constitutes an invertebrate for they are an undefinable paraphyletic assemblage (e.g., Barnes et al. 2001): so it
is impossible to consider any vertebrate 'a kind of invertebrate' or indeed any invertebrate
'a kind of vertebrate' because of the *evidence* supporting the taxon Vertebrata, *not*
because of its relative place on the cladogram with respect to what are now colloquially
termed invertebrates, even by redefining its composition. This demonstrates that using
just the cladogram to determine which clades (monophyletic groups) to name is insufficient. These examples should offer a cautionary tale for those who name *all* clades in a
cladogram in the absence of *specific* evidence supporting the nodes being named, a
strategy adopted in many molecular studies. Without supporting evidence
(synapomorphies), we may end up with an abundance of vertebrates being considered
'kinds of invertebrates' – and with every new molecular study that yields different relationships, different 'kinds of invertebrates' emerge.

Are birds a kind of dinosaur or are dinosaurs a kind of bird? Both statements are, in
some respects, false : birds *are* birds – and birds *are most closely related to* †*Vegavis*;
birds plus †*Vegavis* are most closely related to †*Ichthyornis*; birds, †*Vegavis* plus
†*Ichthyornis* are most closely related to †Hesperornithes; birds, †*Vegavis*, †*Ichthyornis*
plus †Hesperornithes are most closely related to †*Apsaravis*; and birds + †*Vegavis* +
Ichthyornis + †Hesperornithes + †*Apsaravis* are Ornithurae (following Wang et al. 2015).
These are based on evidence.

Is *Mononykus olecranus* a bird, a dinosaur, a bird-like dinosaur or a dinosaur-like bird?
Mononykus olecranus is *Mononykus olecranus*. In this case, the relevant question is:
What is *Mononykus olecranus* related to?

Box 14.1   (*cont.*)

References

Barnes, RSK, Calow, P., Olive, PJW., Golding, DW. & Spicer, JI. 2001. *The Invertebrates: A Synthesis*. Wiley-Blackwell, Oxford.

Baron, MG., Norman, DB. & Barrett, PM. 2017. A new hypothesis of dinosaur relationships and early dinosaur evolution. *Nature* 543: 501–506.

Benton, MJ. 1999. *Scleromochlus taylori* and the origin of dinosaurs and pterosaurs. *Philosophical Transactions of the Royal Society B: Biological Sciences* 354: 1423–1446.

Benton, MJ. 2004. *Vertebrate Palaeontology*. 3rd ed. Blackwell Publishing, Oxford.

Cau, A. 2017. The assembly of the avian body plan: a 160-million-year long process. *Bollettino della Societa Paleontologica Italiana* 57(1): 1–25.

Chiappe, LM. 2009. Downsized dinosaurs: The evolutionary transition to modern birds. *Evolution, Education and Outreach* 2: 248–256.

Ebach, MC. & Williams, DM. 2013. Reading trees. *Zootaxa* 3814(2): 297–300.

Gauthier, JA. 1986. Saurischian monophyly and the origin of birds. *Memoirs of the California Academy of Sciences* 8: 1–55.

Gauthier, JA. & de Queiroz, K. 2001. Feathered dinosaurs, flying dinosaurs, crown dinosaurs, and the name 'Aves'. In: Gauthier, J. & Gall, LF. (eds), *New Perspectives on the Origin and Early Evolution of Birds: Proceedings of the International Symposium in Honor of John H. Ostrom*. Peabody Museum of Natural History, Yale University, New Haven, pp. 7–41.

Gee, H. 2001. Deuterostome phylogeny: the context for the origin and evolution of the Vertebrates. In: Ahlberg, P. (ed.), *Major Events in Early Vertebrate Evolution*. CRC Press, London, pp. 1–14.

Gregory, TR. 2008. Understanding evolutionary trees. *Evolution, Education and Outreach* 1: 121–137.

Norell, MA., Clark, JM. & Chiappe, LM. 1993. Naming names. *Nature* 366: 518.

Patterson, C. 1993a. Bird or dinosaur? (correspondence). *Nature* 365: 21–22.

Patterson, C. 1993b. Naming names (correspondence). *Nature* 366: 518.

Patterson, C. & Rosen, DE. 1977. Review of ichthyodectiform and other Mesozoic teleost fishes and the theory and practice of classifying fossils. *Bulletin of the American Natural History Museum* 158(2): 85–172.

Perle, A., Norell, MA., Chiappe, LM. & Clark, JM. 1993a. Flightless bird from the Cretaceous of Mongolia. *Nature* 362: 623–626.

Perle, A., Norell, MA., Chiappe, LM. & Clark, JM. 1993b. Correction: flightless bird from the Cretaceous of Mongolia. *Nature* 363: 188.

Prum, RO., Berv, JS., Dornburg, A., Field, DJ., Townsend, JP., Lemmon, EM. & Lemmon, AR. 2015. A comprehensive phylogeny of birds (Aves) using targeted next-generation DNA sequencing. *Nature* 526(7574): 569–573.

Sandvik, H. 2008. Tree thinking cannot be taken for granted: challenges for teaching phylogenetics. *Theory in Biosciences* 12: 45–51.

Smith, A. 1994. *Systematics and the Fossil Record*. Blackwell Scientific Publications, Oxford.

Thanukos, A. 2009. A name by any other tree. *Evolution, Education and Outreach* 2: 303–309.

Wang, M., Zheng, X., O'Connor, JK., Lloyd, GT., Wang, X., Wang, Y., Zhang, X., Zhou, Z. 2015. The oldest record of ornithuromorpha from the early Cretaceous of China. *Nature Communications* 6: 6987.

[1] See Perle et al. (1993b) for the reasons the name changed. We have retained the spelling *Mononychus* in the original quotations.

[2] This journal promotes accurate understanding and comprehensive teaching of evolutionary theory for a wide audience [...]. Targeting K-16 students, teachers and scientists alike, the journal presents articles to aid members of these communities in the teaching of evolutionary theory.

The *total-group* is the *crown-group* plus the *stem-group* and together is mono-phyletic (Figure 14.2a) and, as Smith suggested, the only groups worthy of a name (Smith 1994, p. 97, 'Crown and total groups could be given separate names . . . but this would expand the classification scheme to no advantage'[11]). Joyce et al. introduced yet another name, *panstem*:

> To accommodate fossil species that are situated outside of crowns, we decided to create new, stem-based clade names. Every crown possesses a plethora of nameable stem clades (Gauthier and de Queiroz, 2001), however, making it once again necessary to choose one among many. We will here name only the most inclusive stems that do not overlap with the stems of any other living taxon. We term this type of stem–the 'total-group' of Jefferies (1979)–a 'panstem clade' (pan = whole, entire) or simply 'panstem''. (Joyce et al. 2004, p. 993)

But as Donoghue noted, 'these authors miss the point that the total group is defined with reference to the crown group and its closest sister-crown. Thus, it is the most complete of all possible component clades' (Donoghue 2005, p. 557).

Relevant, too, is the *plesion* concept:

> We therefore propose that fossil groups or species, sequenced in a classification according to the convention that each such group is the (plesiomorph) sister group of all those, living and fossil, that succeed it, should be called 'plesions.' Plesions may be inserted anywhere (at any level) in a classification, without altering the rank or name of any other group. (Patterson & Rosen 1977, p. 160)

Craske and Jefferies later defined a plesion like so:

> A plesion, for us, comprises all those members of a stem group which, so far as can be discerned, are equally closely related to the crown group. It therefore includes all those forms which possess, or have secondarily lost, one novelty acquired in the stem lineage, but primitively lack the next recognized novelty acquired in the stem lineage. For us, therefore, *a plesion is, in principle, a paraphyletic grouping since it contains part of the stem lineage and this will have been ancestral to non-members of the plesion.* (Craske & Jefferies 1989, p. 74, emphasis ours)

All taxa (species or otherwise) that appear in stem-groups are *plesions* – but plesions are not necessarily named for any particular node (Jefferies 1979, fig. 5, reproduced as our Figure 14.2b, cf. Willmann 1987).

The three definitions of de Queiroz and Gauthier above were succinctly sum-marized, and revised, by Nixon and Carpenter who noted that:

---

[11] Of course, some stem-group taxa plus their respective crown-group, *if well-defined*, could be named.

> ... both the node- and stem-based methods are implemented in the same fashion, with the only difference being whether a name is restricted to 'crown clades' or includes taxa from the 'stem clade'. (Nixon & Carpenter 2000, p. 299)

They assigned

> ... the proper name 'Node Pointing System' (NP System) to point at the 'phylogenetic system' of de Queiroz and Gauthier, because the latter name already points to something else proposed by other authors (for instance, Hennig's 1966 *Phylogenetic Systematics*). (Nixon & Carpenter 2000, p. 298)

Platnick wrote of the NB system ('Node-Based approach', Platnick 2012a, p. 360; Platnick 2012b, p. 175, 'i.e., *PhyloCode*'), meaning much the same as Nixon and Carpenter 'Node Pointing System' (NP System). For convenience, following Platnick, the *PhyloCode* is the 'Node-Based approach', which avoids ranks.

Names and Ranks

Once taxa are discovered (by whatever means), how are they classified? For 'conventional' taxonomy it is the specimens and their characters (their parts, the homologues, see Chapter 7) that *define* taxa. For molecular data, it is (usually) the nodes of the resulting cladogram (i.e., what is referred to as the *clades*) that defines taxa. Once a clade is identified it can be (but need not be) given a name. The hierarchy is based on the structure of the cladogram (= *a derivative cladogram, a classification*, see Chapter 7 and Box 14.1).

The conventional taxonomic hierarchy is such that one node on any diagram (cladogram, one particular clade) is a subgroup of another node (clade), and of another clade, etc. The sub groups are named differently, with the structure of the name reflecting its position relative to any other subgroup (Table 14.1).

For example, if one refers to the name Harpetidae, the ending -*idae* indicates it is a Family (Tables 14.1 and 14.2); if one refers to the name Harpetida, the ending -*ida* indicates it is an Order (Tables 14.1 and 14.2). And it is generally understood that Orders contain Families. Generic names follow no specific protocol, and species names are of the usual two-part kind: the genus name and the species name (epithet), followed by the author and date of publication (the inclusion of the latter two items may vary from code to code) – the dagger (†) indicates the species is extinct, or when referring to a higher taxon, that it contains extinct species (for an example see Table 14.2: these are names of trilobites, see Ebach & McNamara 2002).

A glance at Table 14.2 is all that is required to understand the classificatory hierarchy. In phylogenetic classifications that follow the *PhyloCode* these naming conventions are (usually) dispensed with (except at the species level[12]) and taxa

---

[12] And in some cases, it has been suggested that the two-part name for species can also be dispensed with (e.g., Mishler & Wilkins 2018): 'We argue that the logical outcome of the cladistic

Table 14.1 (Some) Ranks and their endings in Botany, Zoology and Microbiology

Rank	Botany	Zoology	Microbiology
Order	*-ales*	*-ida*	*-ales*
Suborder	*-ineae*	–	*-ineae*
Family	*-aceae*	*-idae*	*-aceae*
Subfamily	*-oideae*	*-inae*	*-oideae*
Tribe	*-eae*	*-ini*	*- eae*
Subtribe	*-inae*	*-ina*	*-inae*

Table 14.2 Some names from the trilobite Order Harpetida (After Ebach & McNamara 2002)

Order	Family	Genus	Species
Harpetida			
	Harpetidae		
		Harpes	
			†*Harpes macrocephalus* Goldfuss 1839
		Dubghlasina	
			†*Dubhglasina parvula* (M'Coy 1851)

revolution in biological systematics, and the move towards rankless phylogenetic classification of nested monophyletic groups as formalized in the PhyloCode, is to eliminate the species rank along with all the others and simply name clades' (from their abstract). This is not a subject we will dwell on, suffice to say that the literature on species and the *PhyloCode* is extensive (see Dayrat et al. 2004, for a defence of uninomials as outlined by Lanham 1965). A flavour of the topic might be gained from this portion of an abstract: 'Linnaean binomial nomenclature is logically incompatible with ... phylogenetic nomenclature ... The former is based on the concept of genus, thus making this rank mandatory, while the latter is based on phylogenetic definitions and requires the abandonment of mandatory ranks. Thus, if species are to receive names under phylogenetic nomenclature, a different method must be devised to name them. Here, *13 methods for naming species in the context of phylogenetic nomenclature* are contrasted with each other and with Linnaean binomials' (Cantino et al. 1999, emphasis ours). Their conclusion is to stress the stability of any of these 13 over the Linnean binomial. Given that there are 13 to choose from, it would seem that the simple binomial might still be good for a while (see Wheeler's quote below).

require no assigned rank. Article 3 (*PhyloCode*, version 5 beta, 2014, p. 6), which concerns 'Hierarchy and Rank', reads as follows:

> *3.1.* The system of nomenclature described in this code is independent of categorical rank. Although clades are hierarchically related, and therefore intrinsically ranked in the sense that some are more inclusive than others, assignment of categorical ranks is not part of the formal naming process and has no bearing on the spelling or application of clade names.

Although they go on to say that '... This code does not prohibit, discourage, encourage, or require the use of taxonomic ranks' (*PhyloCode*, version 5 beta, 2014, p. 6, Note 3.1.1), the implication is that stability of names is greater without ranks:

> The development of the *International Code of Phylogenetic Nomenclature* (referred to here as the *PhyloCode*) grew out of the recognition that the current rank-based systems of nomenclature, as embodied in the current botanical, zoological, and bacteriological codes, are not well suited to govern the naming of clades. Clades (along with species) are the entities that make up the tree of life, and for this reason they are among the most theoretically significant biological entities above the organism level. Consequently, clear communication and efficient storage and retrieval of biological information require names that explicitly and unambiguously refer to clades and do not change over time. The current rank-based codes fail to provide such names for clades. Supraspecific names are not always associated with clades under the rank-based codes, and even when they are, they often fail to retain their associations with particular clades because the names are operationally defined in terms of ranks and types. A clade whose hypothesized composition and diagnostic characters have not changed may be given a different name under the rank-based codes based purely on considerations of rank. Such instability is particularly objectionable given the wide recognition that rank assignment is subjective and of dubious biological significance. (Preface to PhyloCode, version 4c, 2010, p. 3; there is as yet no preface to version 5 beta[13])

Simply speaking, the construction of *any* classificatory hierarchy with accepted rank endings indicates that an Order will include Families, a Family will include genera and genera will include species (Table 14.1, see our Chapters 1 and 2): the ranks are *specific guides* to the hierarchy, as has been noted in the *PhyloCode* ('Although clades are hierarchically related ...', *PhyloCode*, version 5 beta, 2014, p. 6). As such, the issue of stability with respect to ranks is a red herring, as names without any indication of their position (rank) in the hierarchy simply becomes a mélange of words:

---

[13] Since writing this piece the PhyloCode has eventually been published (de Queiroz & Cantino 2020). We have not yet been able to examine a copy.

A rank-free system of naming requires an annotated reference tree for even the simplest exchanges of information. *This would be confusing at best and would cripple our ability to teach, learn, and use taxonomic names in the field or in publications*. We would be confronted by a mass of polynomial names, tied together only by a tree graphic, with no agreed name (except a uninomial, conveying no hierarchy) to use for any particular species. (Nixon et al. 2003, p. 112, emphasis ours)

The Linnean hierarchy works superbly because it can pack vast amounts of information into single names by using a system of coordinate, mutually exclusive taxa that are recognizable, as such, simply by the standardized endings of their names. Thus, one needs to know only that an organism is a member of the spider family Oonopidae to know also that it is not a member of any other family of animals (or plants, for that matter), and that no member of any other family is also a member of the Oonopidae. Moreover, one can make all those inferences without having available a complete classification of the species included, either in the Oonopidae or in any other family. (Platnick 2012a, p. 360; see also Platnick 2009)

Finally,

The point, though, is that formal classifications are not the primary way in which taxon names are actually used, by anyone; names do not normally come with built-in recognizable indentations or with parenthetical lists of all their included taxa. In the real world, names are always presented in a context involving at most an incomplete listing of all the other known taxa on the planet, and they are usually presented in a context that does not involve even a complete listing of their own members. (Platnick 2012a, p. 360)

Types As Specimens

Ultimately, all definitions should refer to deposited specimens. Thus, the role of the specifiers approaches those of types in the Linnaean system.

(Pleijel & Rouse 2003, p. 167)

The *PhyloCode* includes *diagnosis* in its glossary: 'A brief statement of the features of a taxon that collectively distinguish it from other taxa with which it might be confused'. It also includes *definition*: 'A statement specifying the meaning of a name (i.e., the taxon to which it refers)'. Definitions are of this form:

Aves could be defined as 'the crown clade originating in the most recent common ancestor of *Struthio camelus* Linnaeus 1758 (*Palaeognathae*) and *Vultur gryphus* Linnaeus 1758 (*Neognathae*)'.

One might argue, instead, that the *definition* of any taxon is its *diagnosis* and the meaning of definition in the *PhyloCode* merely an explanation for the defined taxon. Having any statement of a taxon's characters require specimens, actual things to examine. In this sense, the type method seems to be perfect for that job[14]:

> ... the PhyloCode advocates have been forced to admit for species, a system of typification with single types is one of the best (we believe THE best) way to document names. (Nixon et al. 2003, p. 112)

PhyloCode: Revival or Dead on Arrival?

My conclusion is simple. Why would anyone be interested in forcing themselves into the straightjacket of the Phylocode, which offers no advantages, and could stifle their research if their tree is rejected in favour of another study, when the Linnaean codes have proved adaptable and reasonable adjuncts to systematic work?

(Benton 2007, p. 654)

In reality, the *PhyloCode* is, if not dead, moribund. At the time of writing (2019) the most up-to-date (and publically available) version of the *PhyloCode* is dated 2010 (Version 4c – Version 5 beta, with no date, is complete and 'differs substantially from the version currently posted ...'. This latest version was approved by the Committee on Phylogenetic Nomenclature in January 2014. A beta version is available on request, but see footnote 14). There is an International Society for Phylogenetic Nomenclature (ISPN) that, among other things, hosts the Committee on Phylogenetic Nomenclature, and there are supposedly two printed books on the subject: one is to be the printed version of the code, being called simply the *PhyloCode*, the other comprises a set of taxon names that have apparently followed the principles outlined in that Code (version unknown) and called *Phylonyms: a Companion to the PhyloCode*[15]. Quite who will use these phylonyms is an open question – if, that is, they ever appear. Still, the ISPN helpfully provides a guide to the literature on the *PhyloCode* but, like everything else connected with this organisation, is hopelessly out of date – the most recent addition to the literature is 2011 (http://phylonames.org/literature/).

Rather than dwell too much on this 'revival' of nomenclature, the most appropriate response to de Queiroz (1988), and all subsequent pro-*PhyloCode* commentary, can be found in the words of Quentin Wheeler:

---

[14] For an informed account of the type method see Witteveen (2014, 2016).

[15] With respect to *Phylonyms*, 'As reported at the Halifax meeting, progress has been slower than anticipated. However, two steps taken before and during that meeting have helped to expedite completion of the book: A 10-member editorial board was appointed to help review manuscripts, and a deadline was established for receipt of contributions (July 1, 2009)' (*PhyloCode*, ver. 4c, p. 11, www.ohio.edu/phylocode/PhyloCode4c.pdf). Also: 'This code will take effect on the publication of *Phylonyms: a Companion to the PhyloCode*, and it is not retroactive' (*PhyloCode*, ver. 4c, p. 46; ver. 5c, 2). Since writing this, both *PhyloCode* (de Queiroz & Cantino, 2020) and *Phylonyms: A Companion to the PhyloCode* (de Queiroz et al. 2020) have been published.

> Linnaean nomenclature [LN] is stable enough to say what we know, flexible enough to accommodate what we learn; independent of specific theory, yet reflective of known empirical data; compatible with phylogenetic theory, but not a slave to it; particular enough for precise communication, general enough to reflect refuted hypotheses. LN is an effective international, inter-generational, and trans-theoretical system of classification that was forged and tested by those describing the earth's biota, not touting political slogans. It has weathered more worthy adversaries than the Phylocode and will be in wide use long after the latter is a curious footnote to the history of taxonomy. (Wheeler 2001, p. 15)

By acknowledging the historical and theoretical separation of systematics and phylogenetics we avoid aphyletic taxa, homoplasy and moribund phylogenetic classifications. Systematics, that is biological classification, comes first, followed by phylogenetics when reading the cladograms as phylogenies, and so on. Many have come to recognise this separation (e.g., Naef, Nelson and Platnick), while others (Hennig, Avise, de Queiroz), have been eager to merge them in order to appeal to a wider audience, beyond that of taxonomists (e.g., evolutionary biology). The latest attempt at merging systematics and phylogenetics at a nomenclatural level has essentially failed, showing once again that existing taxonomy and nomenclature has worked since pre-evolutionary times, works now and looks set to work well into the future (Box 14.2).

Box 14.2 A Note on the *PhyloCode* and Philosophy

Much has been made by the *PhyloCode* enthusiasts of the view that Aristotelian definitions of taxa, versions of some kind of essentialism – taxa *defined* by characters – are the root cause of so much that was wrong in taxonomy prior to Darwin's evolutionary views. An early summary of this position was stated thus:

> Although other forms of definition have been suggested, traditionally, taxon names have been treated *as if they are defined by characters possessed by those organisms considered to be members of the taxa*. Under such character-based definitions, taxa have been treated as if they are classes of organisms ... in that their characters have been viewed as intensions, that is, those properties connoted by the taxon names. Consequently, traditional definitions of taxon names have been considered intensional definitions .... Such definitions have their roots in the Aristotelian form of definition .... Most modern taxonomists have abandoned the notion of essences ... and many have rejected the notion that taxa are classes of organisms. Nevertheless, *elements of the Aristotelian form of definition have persisted in modern biological taxonomy in that the names of taxa continue to be treated as if they are defined by lists of organismal characters* ... (de Queiroz & Gauthier 1990, p. 308, ellipses in place of references to protect the guilty; emphasis ours)

Box 14.2 *(cont.)*

To some (maybe most) taxonomists, these statements must seem odd, idiosyncratic and bizarre for such a decidedly empirical science as taxonomy (some supporters of the *PhyloCode* think so too, e.g., Pleijel & Härlin 2004). A contrast, with a more accurate version of that history, is provided by Winsor, an historian who has studied these issues:

> The current picture of the history of taxonomy incorporates A. J. Cain's claim that Linnaeus strove to apply the logical method of definition taught by medieval followers of Aristotle. Cain's argument does not stand up to critical examination. Contrary to some published statements, there is no evidence that Linnaeus ever studied logic. His use of the words 'genus' and 'species' ruined the meaning they had in logic, and 'essential' meant to him merely 'taxonomically useful.' The essentialism story, a narrative that has most pre-Darwinian biologists steeped in the world view of Plato and Aristotle, is ill-founded and improbable. (Winsor 2006, p. 2)

We have dealt with the 'essentialism story' (the name given by Winsor to this misrepresentation of the history of taxonomy) in more detail in Chapter 4 but here a few more quotations might help fix our viewpoint with respect to the *PhyloCode* philosophy in general. Winsor notes of Linnaeus:

> Unfortunately, the essentialism story has tended to dampen interest in the rich and complex story of exactly how taxonomists, both before and after Darwin, coped with the real-world challenges of comparing and identifying organisms. To loosen its grip, *we could begin by recognizing that maligning Linnaeus distorts our understanding of the entire history of systematics.* (Winsor 2006, p. 6)

And to close this brief comment, the views of a palaeontologist:

> Phylogenetic nomenclature stems from philosophical, not practical considerations. The *PhyloCode* team, and other reformers, are most welcome to continue the philosophical debate, and the rest of the world will observe their byzantine disquisitions at a distance. Mediaeval scholars debated endlessly how many angels might fit on the end of a pin, a debate that was incapable of resolution if angels were infinite and infinitesimal in dimensions. Phylogenetic nomenclature similarly offers many opportunities for interesting philosophical debate, but it is patently an absurd proposition as a practical system. (Benton 2000, p. 647)

References

Benton, MJ. 2000. Stems, nodes, crown clades, and rank-free lists: is Linnaeus dead? *Biological Reviews* 75: 633–648.

de Queiroz, K. & Gauthier, J. 1990. Phylogeny as a central principle in taxonomy: phylogenetic definitions of taxon names. *Systematic Zoology* 39: 307–322.

Pleijel, F. & Härlin, M. 2004. Phylogenetic nomenclature is compatible with diverse philosophical perspectives. *Zoologica Scripta* 33: 587–591.

Winsor, M. 2006. Linnaeus's biology was not essentialist. *Annals of the Missouri Botanical Garden* 93(1): 2–7.

References

Avise, JC. & Liu, J-X. 2011. On the temporal inconsistencies of Linnaean taxonomic ranks. *Biological Journal of the Linnean Society* 102: 707–714.

Benton, MJ. 2007. The Phylocode: beating a dead horse? *Acta Palaeontologica Polonica* 52: 651-655.

Béthoux, O. 2007a. Propositions for a character-state-based biological taxonomy. *Zoologica Scripta* 36: 409–416.

Béthoux, O. 2007b. Cladotypic taxonomy revisited. *Arthropod Systematics & Phylogeny* 65: 127–133

Béthoux, O. 2007c. Cladotypic taxonomy applied: titanopterans are orthopterans. *Arthropod Systematics & Phylogeny* 65: 135–156.

Béthoux, O. 2010a. Optimality of phylogenetic nomenclatural procedures. *Organisms Diversity & Evolution* 10: 173–191.

Béthoux, O. 2010b. Alternative nomenclatural procedures as a potential benefit to natural history collections. A reply to Dubois in *Org Divers Evol* (2010) 10: 81–90. *Organisms Diversity & Evolution* 10: 341–432.

Cantino, PD. & de Queiroz, K. 2010. *PhyloCode: A Phylogenetic Code of Biological Nomenclature*. www.ohio .edu/phylocode/PhyloCode4c.pdf

Cantino, PD., Bryant, HN., de Queiroz, K., Donoghue, MJ., Eriksson, T., Hillis, DM. & Lee, MSY. 1999. Species names in phylogenetic nomenclature. *Systematic Biology* 48: 790–807.

Carter, JG., Altaba, CR., Anderson, LC., Campbell, CD., Fang, Z., Harries, PJ. & Skelton, PW. 2015. The paracladistic approach to phylogenetic taxonomy. *Paleontological Contributions* 21(12): 1–9.

Craske, AJ. & Jefferies, RPS. 1989. A new mitrate from the Upper Ordovician of Norway, and a new approach to subdividing a plesion. *Palaeontology* **32**: 69–99.

Dayrat, B., Schander, C. & Angielczyk, KD. 2004. Suggestions for a new species nomenclature. *Taxon* 53: 485–491.

de Queiroz, K. 1988. Systematics and the Darwinian revolution. *Philosophy of Science* 55: 238–259.

de Queiroz, K. 2006. The PhyloCode and the distinction between taxonomy and nomenclature. *Systematic Biology* 55: 160–162.

de Queiroz, K. & Cantino, PD. 2020. *PhyloCode: A Phylogenetic Code of Biological Nomenclature*. CRC Press, Boca Raton, FL.

de Queiroz, K., Cantino, PD. & Gauthier, JA. 2020. *Phylonyms: A Companion to the PhyloCode*. CRC Press, Boca Raton, FL.

de Queiroz, K. & Gauthier, J. 1990. Phylogeny as a central principle in taxonomy: phylogenetic definitions of taxon names. *Systematic Zoology* 39: 307–322.

Donoghue, PCJ. 2005. Saving the stem group: a contradiction in terms? *Paleobiology* 31: 553–558.

Dubois, A. 2007. Phylogeny, taxonomy and nomenclature: the problem of taxonomic categories and of nomenclatural ranks. *Zootaxa* 1519: 27–68.

Dubois, A. 2010. Nomenclatural rules in zoology as a potential threat against natural history museums. *Organisms Diversity & Evolution* 10: 81–90.

Ebach, MC. & McNamara, KJ. 2002. A systematic revision of the family Harpetidae (Trilobita). *Records of the Western Australian Museum* 21: 135–167.

Gascoigne, RM. 1991. Julian Huxley and biological progress. *Journal of the History of Biology* 24: 433–455.

Gegenbaur, K. 1870. *Grundzüge der vergleichenden Anatomie*, 2nd ed. Wilhelm Engelmann, Leipzig.

Greuter, W. 2004. Recent developments in International Biological Nomenclature. *Turkish Journal of Botany* 28: 17–26.

Haeckel, EHPA. 1866. *Generelle Morphologie der Organismen: allgemeine Grundzüge der organischen Formen-Wissenschaft, mechanisch begründet durch die von Charles Darwin reformirte Descendenz-Theorie.* G. Reimer, Berlin.

Haeckel, EHPA. 1894–1896. *Systematische Phylogenie: Entwurf eines natürlichen Systems der Organismen auf Grund ihrer Stammesgeschichte.* G. Reimer, Berlin.

Hennig, W. 1950. *Grundzüge einer Theorie der Phylogenetischen Systematik.* Deutscher zentralverlag, Berlin,

Hennig, W. 1966. *Phylogenetic Systematics.* University of Illinois Press, Urbana.

Hennig, W. 1969. *Die Stammesgeschichte der Insekten.* Waldemar Kramer & Co., Frankfurt am Main.

Hennig, W. 1981. *Insect Phylogeny* [translated and edited by Adrian C. Pont; revisionary notes by Dieter Schlee]. John Wiley, Chichester.

Hoßfeld, U., Olsson, L. & Breidbach, O. (eds) 2003. Carl Gegenbaur and evolutionary morphology. *Theory in Biosciences* 122(2-3).

Huxley, JS. 1957. The three types of evolutionary process. *Nature* 180: 454–455.

Huxley, JS. 1958. Evolutionary processes and taxonomy with special reference to grades. *Uppsala Universitets Årsskrift* 6: 21–39.

Jefferies, RPS. 1979. The origin of chordates – a methodological essay. In: House, MR. (ed.), *The Origin of Major Invertebrate Groups.* Academic Press, London, New York, pp. 443–447.

Joyce, WG., Parham, JF. & Gauthier, JA. 2004. Developing a protocol for the conversion of rank-based taxon names to phylogenetically defined clade names, as exemplified by turtles. *Journal of Paleontology* 78: 989–1013.

Kluge, AG. 2005. Taxonomy in theory and practice, with arguments for a new phylogenetic system of taxonomy. In: Donnelly, MA., Crother, BI., Guyer, C., Wake, MH. & White, ME. (eds), *Ecology and Evolution in the Tropics: A Herpetological Perspective.* The University of Chicago Press, Chicago, pp. 7–47.

Kuhn, JH., Wolf, YI., Krupovic, M., Zhang, Y-Z., Maes, P., Dolja, VV. & Koonin, EV. 2019. Classify viruses – the gain is worth the pain. *Nature* 566: 318–320.

Lanham, U. 1965. Uninominal nomenclature. *Systematic Biology* 14: 144.

Lee, MCY. & Skinner, A. 2007. Stability, ranks, and the *PhyloCode*. *Acta Palaeontologica Polonica* 52(3): 643–650.

McNeil, J. 1996. The BioCode: integrated biological nomenclature in the 21st century? In: Reveal, J.L. (ed.), *Proceedings of a Mini-symposium on Biological Nomenclature in the 21st Century.* University of Maryland, College Park, MD. www .plantsystematics.org/reveal/pbio/ nomcl/mcne.html

Medawar, P. 1968 [1996] *The Strange Case of the Spotted Mice: and Other Classic Essays on Science.* Oxford University Press, Oxford.

Minelli, A. 2013. Zoological nomenclature in the digital era. *Frontiers in Zoology* 10(1): 4.

Mishler, BD. & Wilkins, JS. 2018. The Hunting of the SNaRC: A Snarky solution to the species problem.

Philosophy, Theory, Practice in Biology 10: 1.

Naef, A. 1919. *Idealistische Morphologie und Phylogenetik.* Gustav Fischer, Jena.

Naef, A. 1921-23. Die Cephalopoden (Systematik). In: *Pubblicazioni della Stazione Zoologica di Napoli.* Fauna e Flora del Golfo di Napoli, Monograph 35 (I-1). R. Friedländer & Sohn, Berlin, pp. 1-863.

Naef, A. 1972. *Cephalopoda. Fauna and Flora of the Bay of Naples* [*Fauna und Flora des Golfes von Neapel und der Angrenzenden Meers-Abschitte*]. Monograph 35, Part I, [Vol. I], Fascicle I. Smithsonian Institution Libraries, Washington, UK.

Naomi, S-I. 2014. Proposal of an integrated framework of biological taxonomy: a phylogenetic taxonomy, with the method of using names with standard endings in clade nomenclature. *Bionomina* 7: 1-44.

Nelson, G. 1974. Darwin-Hennig classification: a reply to Ernst Mayr. *Systematic Zoology* 23: 452-458.

Nelson, GJ. 2016. What we all learned from Hennig. In: Williams, DM., Schmitt, M. & Wheeler, Q. (eds), *The Future of Phylogenetic Systematics: The Legacy of Willi Hennig.* Systematics Association Special Volume Series. Cambridge University Press, Cambridge, pp. 200-212.

Nelson, GJ. & Platnick, NI. 1981. *Systematics and Biogeography: Cladistics and Vicariance.* Columbia University Press, New York.

Nicolson, D. 1991. A history of botanical nomenclature. *Annals of the Missouri Botanical Garden* 78: 33-56.

Nicolson, N., Challis, K., Tucker, A. & Knapp, S. 2017. Impact of e-publication changes in the *International Code of Nomenclature for algae, fungi and plants* (Melbourne Code, 2012) - did we need to "run for our lives"? *BMC Evolutionary Biology* 17: 116. https://doi .org/10.1186/s12862-017-0961-8 (see erratum at *BMC Evolutionary Biology* 17:156).

Nixon, KC. & Carpenter, JM. 2000. On the other "phylogenetic systematics". *Cladistics* 16: 298-318.

Nixon, KC., Carpenter, JM. & Stevenson, DW. 2003. The PhyloCode is fatally flawed, and the Linnaean system can easily be fixed. *The Botanical Review* 69: 111-120.

Parker, CT., Tindall, BJ. & Garrity, GM. 2015. *International Code of Nomenclature of Prokaryotes.* 2008 Revision, Microbiology Society. https:// doi.org/10.1099/ijsem.0.000778

Patterson, C. & Rosen, DE. 1977. Review of ichthyodectiform and other Mesozoic teleost fishes and the theory and practice of classifying fossils. *Bulletin of the American Natural History Museum* 158(2): 85-172.

Platnick, NI. 2009. [Letter to Linnaeus]. In: Knapp, S. & Wheeler, QD. (eds), *Letters to Linnaeus.* The Linnean Society of London, London, pp. 171-184.

Platnick NI. 2012a. The poverty of the phylocode: a reply to de Queiroz and Donoghue. *Systematic Biology* 61: 360-361.

Platnick NI. 2012b. The information content of taxon names: a reply to de Queiroz and Donoghue. *Systematic Biology* 62: 175-176.

Pleijel, F. & Rouse, GW. 2003. Ceci n'est pas une pipe: names, clades and phylogenetic nomenclature. *Journal of Zoological Systematics and Evolutionary Research* 41: 162-174.

Rieppel, O., Williams, DM. & Ebach, MC. 2013. Adolf Naef (1883-1949): On foundational concepts and principles of systematic morphology. *Journal of the History of Biology* 46: 445-510.

Simpson, GG. 1945. The principles of classification and a classification of mammals. *Bulletin of the American Museum of Natural History* 85: 1–350.

Simpson, GG. 1961. *Principles of Animal Taxonomy*. Columbia University Press, New York.

Wheeler, QD. 2001. Clever Caroli: Lessons from Linnaeus (invited comments). *Plant Press (Washington)* 4(2): 14–15

Williams, DM., Schmitt, M. & Wheeler, Q. (eds) 2016. *The Future of Phylogenetic Systematics: The Legacy of Willi Hennig*. Cambridge University Press, Cambridge, UK.

Willmann, R. 1987. Phylogenetic systematics, classification and the plesion concept. *Verhandlungen des Naturwissenschaftlichen Vereins in Hamburg (NF)* 29: 221–233.

Willmann, R. 2003. From Haeckel to Hennig: the early development of phylogenetics in German-speaking Europe. *Cladistics* 19: 449–479.

Witteveen, J. 2014. Naming and contingency: the type method of biological taxonomy. *Biology and Philosophy* 30: 569–586.

Witteveen, J. 2016. Suppressing synonymy with a homonym: the emergence of the nomenclatural type concept in nineteenth century natural history. *Journal of the History of Biology* 49: 135–189.

Further Reading

General Matters

Ebach, MC. 2017. *Reinvention of Australasian Biogeography: Reform, Revolt, Rebellion*. CSIRO Publishing, Clayton.

Nyhart, LK. 1995. *Biology Takes Form: Animal Morphology and the German Universities, 1800-1900*. Chicago University Press, Chicago.

Rieppel, O. 2016. *Phylogenetic Systematics: Haeckel to Hennig*. CRC Press, Boca Raton, FL.

Stevens, PF. 1994. *The Development of Biological Systematics: Antoine-Laurent de Jussieu, Nature, and the Natural System*. Columbia University Press, New York.

Williams, DM., Schmitt, M. & Wheeler, Q. (eds) 2016. *The Future of Phylogenetic Systematics: The Legacy of Willi Hennig*. Cambridge University Press, Cambridge, UK.

Codes, Taxonomy and Nomenclature

Barkley, T., DePriest, P., Funk, V., Kiger, R., Kress, W. & Moore, G. 2004. Linnaean nomenclature in the 21st century: a report from a workshop on integrating traditional nomenclature and phylogenetic classification. *Taxon* 53: 153–158.

Benton, MJ. 2000. Stems, nodes, crown clades, and rank-free lists: is Linnaeus dead? *Biological Reviews* 75: 633–648.

Dayrat, B. 2010. Celebrating 250 dynamic years of nomenclatural debates. In: Polaszek, A. (ed.), *Systema Naturae 250: The Linnean Ark*. CRC Press, Boca Raton, FL, pp. 185–239.

Forey, PL. 2001. The PhyloCode: description and commentary. *Bulletin of Zoological Nomenclature* 58: 81–96.

Hedberg, I. 2005. Species Plantarum 250 Years. Proceedings of the Species

Plantarum Symposium held in Uppsala August 22-24 2003. *Symbolae Botanicae Upsalienses* 33(3), 219 pp.

Heller, JI. 1964. The early history of binomial nomenclature. *Huntia* 1: 33–70.

Michel, E. 2016. (ed.) Anchoring biodiversity information: from Sherborn to the 21st century and beyond. *Zookeys* 550: 1–298.

Nixon, KC. & Carpenter, JM. 2000. On the other "phylogenetic systematics". *Cladistics* 16: 298–318.

Papavero, N., Llorente-Bousquets, J. & Abe, JM. 2001. Proposal of a new system of nomenclature for phylogenetic systematics. *Arquivos de Zoologia* 36: 1–145.

Pavlinov, IY. 2014a. Taxonomic nomenclature. Book 1. From Adam to Linnaeus [In Russian]. *Zoologicheskie Issledovania*, 12, 153 pp.

Pavlinov, IY. 2014b. Taxonomic nomenclature. Book 2. From Linnaeus to the first codes. [In Russian]. *Zoologicheskie Issledovania*, 15, 223 pp.

Pavlinov, IY. 2015a. Taxonomic nomenclature. Book 3. Contemporary Codes. [In Russian]. *Zoologicheskie Issledovania* 17, 59 pp.

Pavlinov, IY. 2015b. *Nomenclature in Systematics. History, Theory, Practice* [in Russian]. KMC, Moscow.

Polaszek, A. 2010. *Systema Naturae 250 – The Linnaean Ark*. CRC Press, Baton Rouge, FL.

Rieppel, O. 2006, The PhyloCode: a critical discussion of its theoretical foundation. *Cladistics* 22: 186–197.

Smith, A. 1994. *Systematics and the Fossil Record*. Blackwell Scientific Publications, Oxford.

Stevenson, DW. & Davis, JI. (eds) 2003. Approaches in examining the existing nomenclatural systems used in biology. *The Botanical Review* 69: 1–123.

Vaczy, C. 1971. Les Origines et les principes du developpement de la nomenclature binaire en botanique. *Taxon* 20: 573–590.

Watson, MF., Lyal, CHC. & Pendry, CA. 2015. *Descriptive Taxonomy: The Foundation of Biodiversity Research*. Cambridge University Press, Cambridge and New York.

Watt, JC. 1968. Grades, clades, phenetics, and phylogeny. *Systematic Zoology* 17: 350–353.

Wheeler, QD. (ed.) 2008. *The New Taxonomy*. Systematics Association Special Volume Series. CRC Press, Boca Raton, FL.

Witteveen, J. 2014. Naming and contingency: the type method of biological taxonomy. *Biology and Philosophy* 30: 569–586.

Witteveen, J. 2016. Suppressing synonymy with a homonym: the emergence of the nomenclatural type concept in nineteenth century natural history. *Journal of the History of Biology* 49: 135–189.

15

Further Myths and More Misunderstandings

What scientists *do* has never been the subject of a scientific, that is, ethological inquiry . . . It is no use looking to scientific 'papers', for they not merely conceal but actively misrepresent the reasoning that goes into the work they describe . . .

(Medawar [1967] 1968, p. 151, citing Popper 1963; our emphasis)

Would it not be better to win the confidence of others by frankly admitting 'I simply do not know', than to keep babbling on and cover oneself with embarrassment by endeavouring to find explanations for everything? [Ne faut-it pas mieux se concilier la confinace des autres par la sincérité d'un je n'en sais rien, que de balbutier des mots et se faire pitié à soi-même, en s'efforçant de tout expliquer].

(Diderot 1754, Thoughts on the Interpretation of Nature)

The science of classification is a frequently misunderstood field of research in which many myths abound (we discussed a few in Part II: Systematics: Exposing Myths). Reasons for this are numerous and we wish to close this book by briefly examining a few more.

For many, classification is shrouded in mystery and questions such as 'How do taxonomists find all those species?' have led philosophers of science to discuss species *concepts* rather than how taxonomists actually discover natural entities. The same is true for monophyletic taxa in general: much is made of *defining* monophyletic taxa rather than *discovering* them. Ask a room full of systematists to define monophyly and there will probably be at least five different definitions (see Vanderlaan et al. 2013). Yet, every single one of those individuals will most likely be able *to identify the same monophyletic taxon*. All that said, it seems what systematists *say they do* is often not *what they do* (*sensu* Medawar [1967] 1968, epigraph above; see also Winsor 2001), discovering monophyly being a case in point.

The problem is how semantics works in a scientific field that is mostly an activity (e.g., observation and thinking) followed by description (see Ebach 2005). Taxonomic activity, it seems, is obscured by the bony skulls of practitioners. Yet few non-taxonomists and non-systematists (e.g., philosophers of science and 'end users' *sensu* Godfray 2002, Godfray et al. 2007) have attempted to undertake taxonomic activity of their own in order to see what all the fuss is all about. Taxonomy, at the species level, is misunderstood by many. But what of taxonomy at higher taxon levels? Some taxonomists, well versed in describing and revising species, get completely flummoxed. The problem, again, is semantics. Cladistics is often confused with evolutionary biology, meaning that, for example, knowledge of evolutionary processes is *assumed* in order to find monophyletic taxa, or that evolutionary processes are *needed* in order to define monophyly. The *process before pattern* mantra is one myth that has infused generations of taxonomists with a 'cart before the horse' notion of science (see Brady 1981, Wilkins & Ebach 2014). It is well documented that patterns in nature have led to the assumption of natural evolutionary processes, not the other way around yet many (e.g., Crisp et al. 2011, Brummitt 2002, Fitzhugh 2008, *ad nauseam*) insist that it is the assumed processes that helps discover patterns, when clearly the opposite is true. Yet these myths, or popular interpretations, continue to prevail.

How is cladistics understood?

Misunderstandings I: Classification, Cladistics and Its Popular Interpretation

In the life sciences, taxonomy was once a hot area of inquiry. Every student of biology was expected to study it in some detail, and there were professionals who devoted their careers to the hierarchy of kingdom, phylum, class, order, family, genus and species. By the middle of the twentieth century, however, the subject was nearly played out; few professionals could be said to do taxonomy. All of that changed with the development of gene-sequencing technologies and molecular phylogenetics: today biological taxonomy is once again an exciting field.

New conferences, proceedings and learned journals such as *Cladistics* have breathed life into it.

(Jack Lynch, Great Vowels, Times Literary Supplement, *1 May, 2009)*

Medieval monks were canny at getting round the 'fish on Friday' thing. They made conies (rabbits) into honorary fish, because they lived in holes, like puffins. And puffins were obviously fish. Look, if angels can dance on the heads of pins, then puffins can be fish, okay?

(A.A. Gill, Ramadan, in: Table Talk, Sweet and Sour, Salt and Bitter, *2007)*

The numerical taxonomists classified creatures into species and higher categories by overall similarity, regardless of evolutionary history. The cladists argued that common ancestry – and therefore evolutionary history – is the only cogent basis for classifying. This was a bitter and arcane fight about which, trust me, you don't need the details.

(Quammen 2018, p. 249)

At any given moment, public opinion is a chaos of superstition, misinformation, and prejudice.

(Gore Vidal, 'Sex and the Law', Partisan Review, Summer, 1965)

If someone was to wade through the numerous commentaries on cladistics, both online and in print, they would probably emerge none the wiser – or at least somewhat confused. One might start with *Wikipedia* (https://en.wikipedia.org/wiki/Cladistics)[1] and gravitate to printed books. A relatively recent contribution to this literature is the popular book *Naming Nature: The Clash between Instinct and Science*, written by Carol Kaesuk Yoon (Yoon 2009[2]). The book was widely reviewed in newspapers, magazines and academic journals alike, providing a constellation of perspectives. One review, published in the *Times Literary Supplement* and so intended for a broader audience, summarised Yoon's book thus:

> Yoon's best chapters describe the discipline's long, painful divorce from observable reality - from living things classed intuitively along mostly morphological lines by smart amateurs, through numerical taxonomy (a largely forgotten art, in which the physical characters of each specimen got coded and fed into a mainframe computer, which then spat out calculations of the affinities among species), until finally, PhD-brandishing cladists were using DNA and every other molecule at their disposal to ascribe each organism its place in an endless evolutionary matrix; making a mockery of Linnaeus and intuition; and threatening to dispense with naming altogether in favour of some monstrosity called PhyloCode. (Jennie Erin Smith, Talk to the Animals, 14 May, 2010, Times Literary Supplement)

Written by Jennie Erin Smith, a reporter interested in 'science and natural history . . . zoos, museums, animals and conservation', the passage quoted above might appear to serve as a summary of apparent progress in systematics from the late 1990s to the present: from 'guesswork' based on external features, to numerical taxonomy, also based on external features, to DNA and 'PhD-brandishing cladists', who mock Linnaeus's efforts and those of the 'intuitive' persuasion (the

---

[1] It is important to note that many cladists, non-cladists and those critical of cladistics, have contributed to and/or edited the cladistics Wikipedia page with minimum editorial control. For example, one entry edited by mathematics student Josh Grosse on 23 September 2001 defined cladistics as '. . . a relatively recent method for uncovering the evolutionary relationships between living things, intended to provide explicit and testable hypotheses' (https://en.wikipedia.org/w/index.php?title=Cladistics&oldid=241731, accessed 5 June, 2018); on 15 November 2009, botanist Patrick Alexander admitted that 'The article at present seems to suggest that cladistics and the PhyloCode are inseparably linked . . . there are so many errors embedded throughout this article that I am not sure I can fix them without simply rewriting the whole thing!' The cladistics page appeared as Wikipedia's Main Page as Today's featured article on 4 May, 2004. (https://en.wikipedia.org/wiki/Talk:Cladistics, accessed 5 June, 2018).

[2] https://www.carolyoon.com/

guessers). Yet this trajectory may simply be a mistaken impression caused by other 'popular' contributions.

It is instructive to examine some of the comments made by a few scientists on Yoon's book:

> To *name* is to know is to be able to love, and that is biodiversity's last best hope: Such is the thesis of this compelling, quirky, beautifully written guide to our need to classify the world around us. Carol Kaesuk Yoon warns that as we lose our ability and interest to name the living world – as we abandon those tasks to scientists – we lose the living world itself. (David Takacs[3], author of *Philosophies of Paradise: The Idea of Biodiversity*)

> Carol Yoon charmingly reviews progress in taxonomy, the science of naming nature. Along the way, she reveals how personal familiarity with nature has been increasingly offloaded to experts. Yoon shows how the counterintuitive findings from science undercut people's confidence in their own observations, leading them to abdicate their ownership of nature to a cloistered intelligentsia. (Joan Roughgarden[4], author of *The Genial Gene: Deconstructing Darwinian Selfishness*)

> As we interact with nature, we use our distinctly human capacities to create names and categories for the creatures we encounter. Often this exercise becomes the fodder for cold scientific analysis and debate. In this lively account, Kaesuk-Yoon forces us to encounter our fundamental humanity as she explores the most basic, and most important, branches of science. (Neil Shubin[5], author of *Your Inner Fish: A Journey through the 3.5 billion Year History of the Human Body*)[6]

The key phrases are:

> . . . as we abandon those tasks to scientists – we lose the living world itself

> . . . the counterintuitive findings from science undercut people's confidence in their own observations, leading them to abdicate their ownership of nature to a cloistered intelligentsia

> . . . this exercise becomes the fodder for cold scientific analysis and debate

There's much of value in these comments, and we agree in an almost absolute sense that reducing comparative biology (taxonomy) to the products of analyses of 'DNA and every other molecule at their disposal to ascribe each organism its place in an endless evolutionary matrix' is, in the final analysis, misguided and doomed to failure. It is the connection to *cladistics* in these passages that disturbs. Where does it come from, this popular image of cladistics?

---

[3] An Associate Professor at the University of California, Hastings College of the Law.

[4] An ecologist and evolutionary biologist.　　[5] An evolutionary biologist.

[6] As far as we aware, none of these scientists have ever classified even a single organism.

Here we enter the world of the 'public intellectual', as opposed to the world of popular books, exemplified by Richard Dawkins[7], who offered a few direct comments on the subject of cladistics in his immensely popular *The Blind Watchmaker*[8]:

> My own interpretation is that they [cladists] enjoy an exaggerated idea of the importance of taxonomy in biology . . .' the way to truth lies in DNA, which avoids convergence, 'vague measurements', and 'meagre lists (of characters) provided by anatomy and embryology', and '. . . I don't mind much how people classify animals so long as they tell me clearly how they are doing it. (Dawkins 1986)

The key phrases are:

> . . . truth lies in DNA, which avoids convergence

> . . . 'vague measurements', and 'meagre lists (of characters) provided by anatomy and embryology'

These notions conflict with the summary given by Jennie Erin Smith. Dawkins comments are topped and tailed with:

> My own interpretation is that they [cladists] enjoy an exaggerated idea of the importance of taxonomy in biology

> I don't mind much how people classify animals so long as they tell me clearly how they are doing it.

These sentences betray the fact that Dawkins was probably never much interested in or took the time to understand classification. But where did Dawkins learn *his* cladistics, how did he arrive at these ideas? We can only speculate but one of Dawkins's students was Mark Ridley[9], author of *Evolution and Classification: The Reformation of Cladism* (Ridley 1986). Ridley's book was intended as a popular account of his critique of 'pattern cladistics'. Dawkins refers to another piece written by Ridley, a book review:

> As Mark Ridley more mildly said, in a review of the book in which Nelson and Platnick made the remark about Darwinism being false, Who would have guessed that all they really *meant* was that ancestral species are tricky to represent in cladistic classification? Of course, it is difficult to pin down the precise identity of ancestors, *and there is a good case for not even trying to do so.* But to make

---

[7] Dawkins probably needs no introduction but just in case: https://en.wikipedia.org/wiki/Richard_Dawkins

[8] Dawkins's book attracted the attention of those outside of professional biology. For example, Martin Amis, the English novelist, wrote 'I admire Richard Dawkins's *The Blind Watchmaker* – or Son of *Selfish Gene*: enough to say that it shares the grave wit and thrilling godliness of the earlier book' (Amis, Book of the Year, *Observer* newspaper).

[9] https://en.wikipedia.org/wiki/Mark_Ridley_(zoologist)

statements that encourage others to conclude that there never *were* any ancestors is to debauch language and betray truth. (Dawkins 1986, p. 286, our italics in the penultimate sentence)

Some of these sentences are puzzling: . . .debauch language and betray truth[10] . . . make statements that encourage others to conclude that there never *were* any ancestors . . .To modify Gore Vidal's comment above: At any given moment, popular *scientific* opinion is a chaos of superstition, misinformation, and prejudice.

Another review of Yoon's book, written by entomologist Wills Flowers (Flowers 2011), notes that:

Yoon focuses on the notorious Salmon-Lungfish-Cow incident, inflating it from a relatively minor exchange at a scientific meeting (Gee 2001, Williams & Ebach 2008) into a mythical crusade to crush traditional taxonomists.

The Salmon–Lungfish–Cow: '. . . a mythical crusade to crush traditional taxonomists'. This is Smith again, from the *TLS* review of Yoon's book:

And when the same 'regular folks', as Yoon is wont to call non-scientists, are presented with jarring truths drawn from largely invisible sources – the cladists' notorious pronouncements, for example, . . . that a lungfish may be closer, in evolutionary terms, to a cow than to a salmon – they stop caring about nature to a dangerous degree.

That final phrase jars a little: '. . . they stop caring about nature to a dangerous degree' – well, after all, the salmon does look a lot like a lungfish and not much like a cow. Cladistics might very well have been conceived of as an attack – but not on taxonomists, and somewhat ironically, especially not the 'traditional' ones[11]. It was largely an attack on the self-identifying *phylogeneticists*, at that time most *palaeontologists*. We can expand on this with reference to two diagrams, which might go some way to explaining why it seems virtually impossible to find a general article that explains cladistics[12] (Figure 15.1a and b).

The first rumblings concerning the Salmon–Lungfish–Cow debate arose with Beverley Halstead's (1978) report on the *26th Symposium of Vertebrate Palaeontology and Comparative Anatomy* held at the University of Reading, September

---

[10] Ridley wrote a shorter piece for the *New Scientist*, 'Can classification do without evolution' (Ridley 1983); responses came from Schoch (1984) and Kemp (1984), neither of those would have been considered pattern cladists *sensu* Ridley.

[11] By 'traditional' we mean those who work at the coal-face, those who sift through the hundreds of specimens that arrive at our natural history museums and botanical gardens, sift with a view to sorting and selecting. Those who, it might be said, work (almost) intuitively (Williams & Ebach 2017).

[12] Although those interested might consult Patterson (1985) and/or Nelson (2009) for straightforward popular articles.

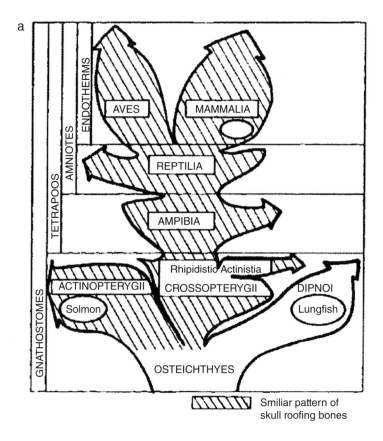

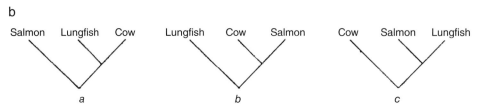

Figure 15.1 (a) Diagram attempting to convey what exactly happened with the salmon, lungfish and cow evolution, with reference to suspected ancestors and the origin of many higher taxa, including birds, mammals and reptiles (Halstead et al. 1979, with permission); (b) three simpler diagrams offering the three possible schemes of relationships between the Salmon–Lungfish–Cow (after Gardiner et al. 1979, with permission).

1978. Gardiner et al. (1979) offered a response to Halstead (1978), which inspired further comment (Halstead et al. 1979).

Figure 15.1a is from Halstead et al. (1979), a diagram that attempts to convey what exactly these authors thought happened with the salmon, lungfish and cow, complete with reference to suspected ancestors and the origin of many higher taxa, including birds, mammals and reptiles – for the uninitiated, the diagram is not that easy to understand.

Figure 15.1b from Gardiner et al. (1979) is their response to Halstead (1978). This diagram is simpler. It offers the three possible schemes of relationships between the Salmon–Lungfish–Cow. Accordingly, the scheme in the third diagram of Figure 15.1b (labelled as *c* in the original) was preferred by Halstead et al., while that in the first diagram (labelled as *a* in the original), was preferred by Gardiner et al. (1979). The arguments for or against any of the three schemes in Figure 15.1b may now not matter. What is important is that for each of the three schemes *evidence* might be found in support. We wish to make two points here.

First, with hindsight, a mistake made by the cladists was to draw the three contrasting schemes as trees, or tree-like diagrams. Once trees were used, it was easy for nearly everyone to mentally convert them into phylogenies of a simple kind. Once a tree is presented, then all sorts happen (see Chapter 3), leading to the often encountered statement that 'we need to get the phylogeny before we have the classification', when the information flows the other way: from classification to phylogeny (see Chapter 7). How could it be otherwise? So maybe it should have been made clearer: cladistics is about classification, *cladistics is classification*. If the three Salmon–Lungfish–Cow schemes had been set out as written classifications, then, who knows, subsequent history may have been less contentious. Still, the general point was 'Since we do not believe that evolutionary taxonomists disagree with us over the characters of lungfish, salmon and cattle, the disagreement is evidently over the meaning of the concept 'relationship'' (Gardiner et al. 1979).

Second, the Salmon–Lungfish–Cow debate was really about the relevance, or role, of the fossil record in discovering relationships, as well as the role of evidence in establishing relationships – that, in turn, was misunderstood by some as if it was a critique concerning the *actual fossils, the specimens* (the evidence), rather than in the stratigraphic sequence they appear (an interpretation).

We repeat: cladistics was not, and *never was*, an attack on taxonomists – but it was an attack on phylogeneticists. And at that time, many palaeontologists considered themselves to be *the* phylogeneticists. For the most part, today's phylogeneticists are not palaeontologists. It is those swamped in (phylo)genomic data who now occupy that role: those who reveal phylogenetic 'truth'-this time it is the tree-builders (e.g., Winther 2018, who anthologises one significant contributor).

What of Jennie Smith's DNA toting, PhD-brandishing cladists dispensing of names and mocking Linnaeus with 'some monstrosity called PhyloCode'? That can be modified a little. Instead, it is more accurate if rendered so: those DNA toting, PhD-brandishing *phylogenetists* dispensing of names and mocking Linnaeus with 'some monstrosity called PhyloCode'.

It would be more accurate to see cladists as the *defenders* of taxonomy and taxonomists, even, and including, the traditional ones[13]:

> For the last 50 years and more – even now continuing into the realm of nomen-clature – in the name of the modern and the new, Visionaries aim, as it were, to confine the past to a dustbin of history, and to bolt and lock the lid upon it. As if without it, we be in some way better, even born again more whole-some; as if Carl Linnaeus really were among the last of the Ancients, and not, rightly, the first of the moderns, and so related to us – of a group inclusive of us. (Gareth Nelson paraphrased in the *Annual Review of the Linnean Society, 2001*)

What should have been accepted is that in comparative biology (taxonomy) there is no privileged data source, no privileged method of analysis.

We noted two statements of Dawkins:

> My own interpretation is that they [cladists] enjoy an exaggerated idea of the importance of taxonomy in biology

Well, is it really exaggerated?

> I don't mind much how people classify animals so long as they tell me clearly how they are doing it.

Evidently it does matter – otherwise why would it have been the topic of such extended discussion – a discussion that began centuries ago. Our aim is to see that discussion continues into the twenty-first century. It really is that important.

By way of introduction to the following section, we might briefly return to Halstead's earlier critique (Halstead 1978). As far as we can establish, Halstead was the first to use the term 'cladistic revolution' (see the Introduction). He also closed his 1978 piece with the following:

> The cladists adhere to the tenets of Hennigism with religious fervour and are already entrenched in some of the major museums in the world.

This may very well be the first time words such as 'religious fervour' were used in the context of Hennig and cladistics – maybe Halstead is the original source of this kind of stereotyping.

Misunderstanding cladistics continues beyond that of cladists apparently attacking taxonomists. What if, then, a group of cladists were to take control of their own journal and state a preference for a particular method? If cladistics was understood as a particular method of classification, rather than a method of evolution, then no evolutionary biologist would ever have cared. Instead it created another flurry of accusations and abuse, this time out of the scientific journals and into social media – misunderstanding cladistics entered the twenty-first century.

---

[13] 'A common assertion that Hennig's methods are a rigorous version of "good old systematics" is not so far-fetched (see, e.g., Nelson & Platnick 1981)' (Eldredge & Novacek 1985, p. 67).

Misunderstandings II – #Parsimonygate: An Issue of Illumination

The cladists adhere to the tenets of Hennigism with religious fervour.

(Halstead 1978, p. 760)

At the beginning of 2016 *Cladistics*, the official journal of the Willi Hennig Society (hereafter WHS) published a short half-page editorial to explain its outlook and approach to manuscript content and submission to their journal (*Cladistics* 32, p.1, 2016). If nothing else, it was to the point:

> The epistemological paradigm of this journal is parsimony
>
> Phylogenetic data sets submitted to this journal should be analysed using parsimony
>
> Unless there is a pertinent reason to include multiple trees from alternative methods, a tree based on parsimony is sufficient as an intelligible, informative and repeatable hypothesis of relationships, and articles should not be cluttered with multiple, often redundant, trees produced from other methods
>
> All phylogenetic methods, including parsimony, may produce inconsistent or otherwise inaccurate results for a given data set. The absence of certain truth represents a philosophical limit of empirical science.

Reaction was swift, primarily via social media under the hashtag *Parsimonygate*. Most commentary could be cast aside as humorous abuse – as much as humorous abuse is tolerated on social media (the abuse largely took the view that 'cladistics', however defined or understood, is now really just a 'cult', 'sect' or 'religion', or some such thing, unconsciously following Halstead[14,15]). Some, notably

---

14 Here we draw attention to another gratuitous misunderstanding unconnected with the #parsimonygate debacle but related to the presentation of misinformation, from David Baum in his review of a book commemorating what would have been Willi Hennig's 100th birthday (Baum 2017). The numerous factual errors in his review aside, we note one passage: 'I would contend that the failure of cladistics to grow is not because its proponents have been bad proselytizers, in fact they have proven effective at fostering cult-like adherence in some circles'. As if the process of scientific enquiry is about conversion rather than comprehension. We noted that views suggesting Hennig's work attracts 'cult-like adherence'. Here's Yoon again: 'Taxonomists, one by one, realized its [phylogenetic systematics] power, each of them undergoing the equivalent of religious conversion. That the acceptance of Hennig's method was often described in near-religious, revelatory terms is not surprising. It required a kind of surrender, a giving of oneself over to Hennig, if you will' (Yoon 2009, pp. 250–251, and more of the same on pp. 262–263, with the suggestion that Felsenstein (2001) might have been the source for her words).

15 Cartmill (2018) comes to some interesting conclusions concerning the success of cladistics and his understanding of Kuhnian paradigm shifts: 'And just as Kuhn's account would have predicted, it was not forced on scientists by any new empirical findings during the 1960s and 1970s; it was brought about by something much more like a religious conversion experience ... The

evolutionary biologist Jonathan Eisen[16] (of phylogenomics fame, Eisen 1998, Eisen & Fraser 2003; he started the *Parsimonygate* hashtag), called for libraries to cancel their subscriptions to *Cladistics*, based on his claim that 'They said if you want to use another method, you have to show that it's *philosophically* better, not *scientifically* better ... That's why I said it seems like they're dropping science for dogma'[17]. Further discussion provided various accounts of how awful WHS meetings were – the main claim being that attendance at these affairs when presenting work that questioned any aspect of Wagner parsimony was received with some measure of abusive audience response. In sum total, none of the commentary painted a very warm picture of the WHS meetings – or of cladistics, for that matter[18].

It is not our intention to defend or attack the editorial in *Cladistics* – in our view, it is their right, as it is any other journal, to operate how they wish, and it is their right to state how that should occur in their editorials.[19]

---

disagreement [about how to classify] goes no deeper than a wrinkling of the nose. It is a matter of taste' (Cartmill 2018, pp. 678 and 685). No, not taste – it concerns evidence.

[16] https://en.wikipedia.org/wiki/Jonathan_Eisen

[17] www.wired.com/2016/02/twitter-nerd-fight-reveals-a-long-bizarre-scientific-feud/. The full account, like much else on the internet, has since disappeared with the exception of a few tweets and some excerpts captured in the article referenced above in the popular technology journal *Wired*. If by any chance anyone ventures to read the *Wired* article, they might like to glance at Conway-Morris (1994) and ponder the content (and style) of *that* review.

[18] It was interesting, nevertheless, that very few persons came to the society's defence.

[19] For what it's worth, we find considerably more disturbing notices in other journals, such as this from the *Journal of Phycology*, the official journal for the Phycological Society of America:

"Policy for Taxonomic Papers

. . .

3. For taxa above the rank of species a molecular phylogenetic analysis *must* be included demonstrating monophyly of the proposed new taxon and related sister taxa.

4. New species or subspecific taxa *must* include at least one diagnostic DNA sequence with the GenBank accession number listed (data may be new or derive from existing GenBank records, but must be from the type collection) and *must* be illustrated with photograph(s) or line drawing (s), including diagnostic features, if present.

. . .

Emphasis ours – and as an aside, '. . . demonstrating monophyly of the proposed new taxon and related sister taxa. . .' is just a little obscure.

Many societies and their journals also change definitions of well-known terms to suit the views of the period. For example, the journal *Systematic Biology*, the official publication of the Society of Systematic Biology, defines systematics as the 'the study of biological diversity and its origins. It focuses on understanding evolutionary relationships among organisms, species, higher taxa, or other biological entities such as genes, and the evolution of properties of taxa including intrinsic traits, ecological interactions, and geographic distributions'. Compare this to the definition offered by the Oxford English Dictionary:

But let us briefly return to what was Eisen's core statement, a paraphrase of a passage in the editorial and the statement that probably encouraged the online abuse: '... you have to show that [any other method is] *philosophically* better, not *scientifically* better'. The actual editorial statement was as follows: 'If alternative methods give different results and the author prefers an unparsimonious topology, he or she is welcome to present that result, but should be prepared to defend it on philosophical grounds'.

Still, that to one side, reflecting on Eisen's statement one might give pause to consider what it can possibly mean. How, exactly, would one go about finding '*scientifically* better' methodologies in the case of phylogeny reconstruction? A few things come to mind: a '*scientifically* better' methodology might be more accurate, providing more accurate results; it might reflect more accurately the real world that we are trying to document – we are sure there might be other factors, should we devote time to finding them but as Goloboff et al. noted: 'considering that methods can be justified only with statistical principles is itself a 'philosophical' position' (Goloboff et al. 2018, p. 431, see also Brower 2018).

In fact, what *appears* to be the issue is a *disagreement* on how to deal with data relevant to systematics. The disagreement might be thought of as between those who include models[20] (those who implement likelihood or Bayesian methods, etc.) and those who do not (those who implement Wagner Parsimony, or compatibility, perhaps). But this cannot really be the entire issue as Wagner parsimony and compatibility both include models, albeit rather simple ones (Chapter 8), so any distinction between methods that include models and those that do not is obviously false. The disagreement might be between those who invoke statistics at some point in their methods and those who do not. But this, too, is false. As, for example, Patterson (1982) and Nelson (1979), both considered to be cladists, invoked a statistical outlook, for congruence between characters in the former, for congruence between cladograms in the latter. So this is also false. We conclude, then, that there is no disagreement, once the methods (all of them) are understood for what they are, regardless of where that method came from or who created it. With respect to method, then, what seems at issue is some critical misunderstanding of what cladistics is, or what it became.

'Originally: = taxonomy n. 1. Now: the branch of biological science concerned with taxonomy and also the study of phylogenetic and biogeographic relationships on which such classification is based' (OED (2018), accessed 27 May, 2018).

In fact, the journals definition is closer to phylogenetics, namely:

'The branch of biology that deals with phylogeny, esp. with the deduction of the historical relationships between groups of organisms'.

It is not surprising, then, that *Systematic Biology* has not published any actual systematics articles for quite some time.

[20] Without going into too much detail, we define a model as 'opinions embedded in mathematics' (O'Neil 2016, p. 21), regardless of whether they are parametric or not.

One theme of this book, if not its major theme – which we hope to have emphasised – is so: to obtain 'accurate' results one needs a measure of some kind, and assuming that, in the case of phylogenetic relatedness, 'accurate' means a reflection of the real world, then this is simply not available to us nor will it ever be. We do not, and cannot, 'know' any phylogeny so we cannot judge how accurate our attempts at reconstructing it are, even if we build in some measure of uncertainty – as stated in the *Cladistics* editorial: 'The absence of certain truth represents a philosophical limit of empirical science'. Under these terms, we are forced to conclude that '*scientifically* better' *can only be accounted for as 'philosophically better'*. This is not all that helpful as philosophy simply helps us find *justification* for our approach – it cannot help us determine whether our approach is *valid* in any absolute sense. By that we mean in the sense of discovering something of the real world, *the nature of nature*, if you will.

With this in mind, we might again summarise our themes with respect to *our* view of cladistics: Wagner parsimony is *not* cladistics: it is just one method among many – the subject matter of cladistics is greater than one particular method. In our view, if cladistics is regarded as the study of *relationships* then this captures the *science of systematics* from Linnaeus to our present time, if not before Linnaeus, irrespective of method, irrespective of philosophy. As we have repeatedly noted in this book, every tree, every cladogram, however found, is a specific set of relationships, so in this sense every tree, every classification, however derived, is a cladogram. Relationships abound. How do these sets of relationships agree with one another? How much of the natural world can we actually discover from these snapshots of data?

If there is a disagreement, it is more fundamental. It is between those of us who eschew *all* methods and instead embrace explorations of the data rather than explorations of any model, be that the complex models of recent statistical approaches, to the relatively simple models of Wagner parsimony. Let us be clear: we reject all methods as producing, or possibly/potentially producing, artefacts, obscuring our understanding of nature as she is.

Finally, on a more personal note, we have ourselves both been subject to some direct and rather hostile attacks in the journal *Cladistics* (e.g., Farris 2014) and at one or two of the WHS meetings (whether we were present or not). It has left no visible marks on either of us, perhaps a few invisible ones, but nothing too damaging – but it has left us with the continued desire to be understood. And that is at the core of the #Parsimonygate reactions as we see it – a reaction fuelled on misunderstanding.

The desire to be understood by cladists and taxonomists is paramount as they do what we do. In this case, our misunderstandings occur at some methodological or theoretical level. This however fades when dealing with philosophers, namely non-biologists who prefer to read, or ask, what we do, but who rarely practise either taxonomy or systematics.

Misunderstandings III: Biologists and Philosophy, Philosophers and Biologists

Finally, I must acknowledge that I'm not sure myself under what circumstances, if any, scientists should take philosophers seriously.

(Beatty 1982, p. 33)

He might have added that what philosophers say should be handled even more gingerly, since at least until recently they seemed to be interested only in prescribing what scientists ought to be doing, and they showed indifference if not hostility to any truly empirical study of scientific activity.

(Rudwick 1988, p. 5, citing Medawar [1967] 1968)

During the week of 12 to 16 June, 1967 (Monday through Friday), the 4th Nobel Symposium was held in Stockholm, Sweden, its theme being *Current Problems of Lower Vertebrate Phylogeny*. This was also the title of the publication of the proceedings, produced a year later (Ørvig 1968). Its stated purpose was to celebrate the life and contributions of Erik Helge Osvald Stensiö (1891–1984) and to this end all but one of its contributions concerned the detailed anatomy of one or another group of 'lower vertebrates', fossil and Recent (see Patterson 1990 on Stensiö and Schultze 2005 on the symposium). The one exception was a contribution from Lars Brundin. Many years later, Schultze provided a retrospective on the symposium and of Brundin's contribution: 'Before the banquet and subsequently in the proceedings, L. Brundin, an entomologist of the Naturhistoriska Riksmuseet, Stockholm, gave a presentation on the application of Hennig's phylogenetic principles' (Schultze 2005: vi). As Schultze further commented, 'This had no impact on the other presentations subsequently published in the proceedings' (Schultze 2005: vi) – nevertheless, as is by now well known, Brundin's presentation had a marked effect on systematic biology, or more precisely, on palaeontology (see Introduction).

Just over 20 years later, the 70th Nobel Symposium was also held in Sweden, this time in Karlskoga, from August 29th to September 2nd 1988 (and also Monday through Friday). This time its stated theme was more general than the 4th, focusing on *The Hierarchy of Life*. Its proceedings were published under the same name a year later (Fernholm et al. 1989). Colin Patterson, a participant of both symposia, was asked to summarise the 1988 proceedings: 'My task is to take stock of the meeting and this volume' (Patterson 1989, p. 471). He began with an account of the changes in comparative biology between 1967, the 4th Nobel Symposium, and 1988, the 70th, noting three significant changes. The first was reflected in the title of each symposium and its resulting book, the former being focused on 'imprecisely formulated problems of phylogeny', the latter on 'concisely expressed ... dichotomous branching diagrams', the former concerned with phylogeny (process), the latter with hierarchy (pattern), or as we would put it: classification (Patterson 1989, p. 471). The third change was 'more politics and economics than science', reflecting on the participants, with the first symposium being dominated by Europeans, the second by North Americans (Patterson 1989, p. 472). But it was the second of Patterson's points that is pertinent here: the change in fields of interest of the participants. The first symposium was composed largely of palaeontologists (66%), alongside a few comparative anatomists and embryologists; the second had many fewer palaeontologists, about the same number of comparative anatomists, 'the remaining 60% of participants represent two new professions, philosophy of biology (David Hull) and molecular systematists ...', the latter replacing the palaeontologists (Patterson 1989, p. 471).

However accurate the claim that in the late 1980s philosophy of biology was regarded as a new profession[21], it was only David Hull who represented the field at the 70th Nobel meeting. It was much earlier, in the 1960s, that philosophy of biology began to intrude upon systematics – and that intrusion was spear-headed by Hull (1964) who eventually came to write *Science as a Process: An Evolutionary Account of the Social and Conceptual Development of Science*, an influential book on the recent history of systematics (Hull 1988). Hull's influence on systematics cannot be under-estimated[22]. Setting aside his book-length treatment, one of his early papers, 'The Effects of Essentialism on Taxonomy: Two Thousand Years of Stasis' (Hull 1965), had a marked, and detrimental, effect on the perception of systematics and taxonomists for many years after, the folly and inaccuracy of his position being exposed only relatively recently by Polly Winsor and others (Winsor 2003, p. 388, 'Although a number of authors over the years have expressed various reservations, corrections, and doubts about parts of the Cain-Mayr-Hull story ... it is still the established view'[23]; that established view is now slowly changing, see Chapter 4).

It is not our concern here to focus entirely on David Hull's perception of biological classification and related subjects, only to note that his final and, we have to conclude, mature view of the subject:

[21] Perhaps it is more accurate to think of the 1980s as a watershed for Darwin studies (e.g., Kohn 1985a, see the quotation from Kohn 1985b below and Loewenberg 1965 for an early account).

[22] His involvement with the Society of Systematic Zoology made him, perhaps, far too close to his subject to be a disinterested observer: Councillor, 1977–1979; President-elect, 1982–1983; President, 1984–1985; Past-President, 1986–1987. Hull never practised nor attempted to practise systematics – so close, yet so far.

[23] Remarkably, Hull's essay was reprinted on two separate occasions (in *Concepts of Species*, C. N. Slobodchikoff (ed.), Berkeley: University of California Press, 1976 and in *The Units of Selection*, Marc Ereshefsky (ed.), Cambridge: MIT Press, 1992. Hull eventually did, albeit rather meekly, respond to Winsor in Hull 2007).

> In biological classification, evolution supplies the theoretical orientation. The goal is to make the basic units of classification (taxonomic species) identical to the basic units of biological evolution (evolutionary species). The principle of order is supplied by phylogeny. Species splitting successively through time produce a phylogenetic tree. The primary goal of taxonomy since Darwin has been to reflect these successive splittings in a hierarchical classification made up of species, genera, families, and so on. (Hull 1998, www.rep.routledge.com/articles/thematic/taxonomy/v-1)

Our book has been devoted to showing that the above (and, as it happens, the entire contents of his online essay, if we were to be absolutely frank) is nothing but a rather naïve post-nineteenth-century version of biological classification – perhaps not just only simplistic, but *mistaken* in its every aspect (see our Section II, Chapters 3–6). Hull's influence has permeated all – or most – subsequent discussion in the philosophy of biology, at least where systematics is concerned.

Early in his career, Hull ended a lengthy essay on 'What Philosophy of Biology Is Not' with the following:

> In conclusion, there are many things that philosophy of biology might be. A philosopher might uncover, explicate, and possibly solve problems in biological theory and methodology. He might even go on to communicate these results to other philosophers, to scientists, and especially to biologists. He might show what consequences biological phenomena and theories have for other sciences and for philosophy or to show what consequences other sciences and even philosophy have for biology. These are some of the things which philosophers of biology might do. With rare exception, they have not. What philosophy of biology is not? It must be admitted that thus far it is not very relevant to biology, nor biology to it. (Hull 1969, p. 268)

We puzzle over its relevance too, which, from the vantage point of the beginning of the twenty-first century, seems to amount to the taxonomic community having to re-establish itself in the face of this philosophical intrusion (Part II: Chapters 3–6).

It would take some considerable time to review all the topics that appear currently to be of interest to philosophers (and historians) of biology today. Our own conceit to one side, we humbly suggest that it can be reduced to four topics: homology, species[24], natural selection and the *Modern Synthesis* – if not these four, then topics and persons associated with them (statistical methods in general;

---

[24] This would include the lengthy discussions that have been had on the individuality thesis, the understanding 'that species are individuals, and not classes, sets, or kinds' (Haber 2016, see also Rapini 2004), another topic bought to prominence by David Hull (1976), and the related topic of natural kinds (e.g., Rieppel 2013).

Fisher–Wright–Dobzhansky; population biology and population genetics; mathematical models of populations; mathematical models of natural selection; kinds of selection, and so on). Two of those subjects – species and natural selection – are intimately connected via the *Modern Synthesis*, its umbrella theory. One might reflect on those subjects – species, natural selection, the *Modern Synthesis*[25] – and find, yet again, the influence of Ernst Mayr[26] (see Chapter 5). One might see in these subjects Mayr's own preoccupation with species and their origin, as if that was the only subject that the study of evolution was concerned. One might also see in these topics Hull's concerns:

> The Linnaean hierarchy was developed for structural classifications … When the goal of systematics was changed to have classifications reflect genealogy, no one noticed the deep problems posed by any attempt to translate a historically connected tree into nested sets of taxa. The traditional Linnaean hierarchy is not up to the task. (Hull, 1998, www.rep.routledge.com/articles/thematic/taxonomy/v-1)

Again, there appears not to be a shred of truth in any of these statements, as we hope to have shown in this book.

One might link Hull (and, as it happens, a host of others) via Mayr to Darwin, the latter being the hapless hero of their melange of taxonomic nonsense. As one historian perceptively wrote:

> Darwin became a focus of detailed study only after the evolutionary synthesis [i.e., the Modern Synthesis], which enshrined Darwinian natural selection, was

---

[25] Even this is hard to define (see Stoltzfus 2017).

[26] 'As one whose first exposure to the history of biology, and to the history of science in general, was through the historical papers of Ernst Mayr (rather than the usual route through Kuhn, Koyré, Roger, or even Radl and Nordenskiold) …' (Sloan 1985, p. 145); 'As someone who had the privilege of being one of Ernst Mayr's students some twenty years ago and who has continued to benefit immensely from interacting with him ever since, I consider it a special honor to have the opportunity to participate in a collective evaluation of Mayr's multifold contributions to systematics, evolutionary theory, and the history and philosophy of biology' (Burkhardt 1994, p. 359). When Simpson, a palaeontologist, reviewed Mayr's *Growth of Biological Thought* (Mayr 1982), the title of his review is 'Autobiology', commenting that it was, for Mayr, 'an intellectual, psychological, and conceptual autobiography', not a history of biology (Simpson, 1982, p. 438). Autobiology appears to mean writing an account of one's own discoveries or contributions (autobiography) as if it is a legitimate account of the development of a period in biology. A later example – perhaps its author independently coined the same word – was given by Sturdy in relation to J.S. Haldane: 'Haldane's pursuit of a scientific career, and the choices he made about how to live life as a scientist, were thus inseparable from his embodied experience of his own biological life; in effect, autobiography was indistinguishable in Haldane's work from what we might call "autobiology"' (Sturdy 2011, p. 172).

consolidated and widely diffused. In other words, only after biologists legitimated
Darwin did historians rush to study him. (Kohn 1985b, p. 2)

Here, the word 'biologists' in the passage above could be easily replaced with
'Ernst Mayr' (or maybe even 'Richard Dawkins' as Mayr's *familiar spirit*). This is all
we have to, need to or want to say on these subjects – far be it from us to suggest
what philosophers *should* be focusing on. Here we simply note a few facts: like
countless generations of taxonomists[27], we understand 'species' to simply represent a particular *taxon*, like any other (genus, family, order, etc.), obviously, of
course, with no special meaning attached to the rank – including the rank of
species[28] (see Chapter 7); that natural selection is a process that governs the
dynamics of biological populations so, likewise, we have no need to discuss it or
comment further on its relevance to our subject; and finally, the *Modern Synthesis*
is a body of thought governed primarily (but not exclusively) by the process of
natural selection and Mayr's notion of species' origin that turns out to have little to
do with systematics or taxonomy as we understand that subject and as outlined in
this book (see also Wheeler 1995, 2004, 2008b[29]). This leaves *homology*.[30]

Most biologists and philosophers appear to agree that 'all useful comparisons in
biology depend on the relation of homology' (Patterson 1987, p. 18) and, as Ingo
Brigandt later noted, 'Homology is one of the most important concepts in
biology . . . . Unfortunately, philosophical discussion has not really focused on this
topic. This is astonishing because during the last decades there has been a
radiation of aspects and concepts of homology' (Brigandt 2002, p. 389). Further,
and five years later, Brigandt and Griffiths, when introducing their special volume
on homology in *Biology and Philosophy*, wrote:

Homology is one of the most important concepts in biology . . . . Having been
introduced in pre-Darwinian comparative biology, it continues to be fundamental
to taxonomy, phylogeny, and evolutionary biology. In recent times it has come to
play an important role in molecular and developmental biology. In addition to
figuring prominently in biological practice, the notion of homology is the subject of

---

[27] This 'modest number of men and women, a few of them clever or heroic but most of them quite
ordinary, and several of them displaying serious weaknesses of character or insight, how this odd
collection of people could have combined their efforts in such a way as to construct the
enormous body of knowledge about life's diversity that is our heritage today' (Winsor 2001,
pp. 242–243).

[28] It is more than we can bear, let alone inflict upon our readers, to reflect on the vast literature on
species.

[29] With respect to the workings of taxonomy, we include here contributions to the 'extended' or
'post-modern' synthesis (Koonin 2009; Pigliucci & Finkleman 2014; Huneman & Walsh 2017).

[30] And maybe the related Tree of Life conversation (e.g., LaPorte 2005, 2009; Velasco 2008, 2013,
2018).

extensive theoretical reflection among biologists …. Curiously though, homology has been discussed only sparsely by philosophers. (Brigandt & Griffiths 2007, p. 633)

As biologists, it is difficult to know where to begin when dealing with homology from a philosophical perspective – perhaps it is too difficult, so, with apologies, we begin with David Hull, in a paper entitled, conveniently enough, 'Ernst Mayr's Influence on the History and Philosophy of Biology: A Personal Memoir'[31]:

The contrast is marked in biology by the contrast between monophyletic and paraphyletic or polyphyletic taxa. At the level of characters this distinction is mirrored in the contrast between homologies and homoplasies. *'Monophyly' and 'homology' are wedded to the genealogical perspective.* (Hull 1994, p. 384, our italics)

Wedded to the genealogical perspective? Later, in the same journal, Bryant (a biologist), discussed homology in the context of the threefold parallelism:

Use of the terms 'character state' and 'character' follows Platnick (1979). Character state has no inherent phylogenetic connotations and is the appropriate term for attributes of taxa in linear hierarchies. From an evolutionary perspective, character states are individual features in a transformation series. *Character is a phylogenetic concept entailing homology due to derivation from a feature in a common ancestor*; a character may subsume a number of character states, and distributions of characters form nested hierarchies. (Bryant 1995, p. 198, our italics)

We noted above Brigandt's long essay on the subject of homology. In a lengthy footnote he writes:

Transformational approaches usually define homologues as characters that are obtained by a transformation series from a feature of the *common ancestor*, i.e., the change of structures in the course of evolution is taken explicitly into account. Taxic approaches determine homology by means of the *character distribution of a phylogenetic tree*. Two (apomorphic) characters are homologous if the least *common ancestor* had the same character (i.e., if the structure is a synapomorphy). If–on the other hand–two apomorphic characters are derived from a plesiomorphic character (in the least *common ancestor*), we are dealing with a homoplasy (due to convergent evolution). (Brigandt 2002, p. 389, footnote 1)

We draw attention here to two items (perhaps these are common to nearly all the quotations above): (1) the definitions he uses involve *common ancestors* and *common ancestry*[32]; (2) these definitions appear to take no account of what

---

[31] More autobiology?, Or perhaps, this time, autophilosophy?; see footnote 21.

[32] The footnote is not referenced, so we are unclear as to where these definitions arose or where taken from.

taxonomists actually do, what they undertake to discover taxa – this is the subject this present book addresses. In our view, all these philosophical discussions are made within the realm of *phylogenetic tree-builders* rather than *taxonomists* (e.g., MacLeod 2011; Currie 2012). Thus, it might seem much of this commentary, *with respect to taxonomy*, is somewhat redundant.

Misunderstandings in taxonomy and systematics occur both by historians and philosophers attempting to understand the biographical and theoretical nuances. Historians and philosophers are, in one sense, users of taxonomy in so far as how taxonomy is thought to be practised and how taxonomists interact. Another so-called end user, namely, those biologists who use taxonomic data (i.e., published names, cladograms, etc.) are intent on saving taxonomy without fully understanding what it does or how it works.

With historians and philosophers on one side and conservation biologists and ecologists on the other, taxonomic data is in high demand at the same time taxonomists are virtually unemployable. This, in part, has become known as the *taxonomic impediment.*

Misunderstandings IV: the Taxonomic Impediment, the End User and Counting Species

Some of the critics have proposed solutions to this 'taxonomic impediment' in the form of a triumvirate adjoining a unitary taxonomic cyberstructure + automated DNA barcoding + molecular phylogeny, which we consider to be nothing but a threefold myopia [sic]; one critic has even gone as far as to suggest that biologists who need systematists can circumvent this dependency by 'doing systematics themselves'.

(de Carvalho et al. 2008, p. 150; see also de Carvalho et al. 2007 and Dubois 2010)

The Threefold Myopia

The taxonomic impediment is real[33]. In reality, the lack of trained taxonomists and the lack of training *for* taxonomists have resulted in *several* impediments. In particular it is now generally recognised that there is a conflation of evolutionary processes ('process') with natural classification ('patterns'). That is, there is a confusion of the goals of evolutionary biology and conservation biology with the goals of taxonomy – there is the simple misunderstanding of nomenclature (the names) with classification (the relationships). The reasons for these impediments are given as a series of explanations that involve what are referred to as 'end users' (see below). What is more, the taxonomic impediment is rarely seen as a problem

---

[33] The literature on the taxonomic impediment is extensive. We suggest the following: Wheeler (2008a), de Carvalho et al. (2007) and Ebach et al. (2011). We have made some recommendations in the Further Reading section below.

in taxonomy but rather a problem for applied fields such as conservation biology or ecology that want to, or need to, use the names.

Counting numbers of species, rather than examining the plight of taxonomy itself, appears to be the biggest impediment, even though it directly affects fields outside of taxonomy (see below). For us, blaming taxonomy for 'impediments' that directly affect applied fields is simply a set of myths. A taxonomic impediment is something that affects taxonomy and not conservation biology or ecology, etc. Take, for example, Costello et al. (2013), who understand species descriptions and the resulting names purely as ways to benefit the disciplines of ecology and conservation biology:

> Once species are described, more detailed studies can look at populations and genetic and biochemical diversity. Species inventories draw attention to where taxonomic effort will discover most new species, including resources and ecosystems. Having a standard list of species names is essential for quality assurance in biological and ecosystem sciences and natural resource management. Another reason to discover species is to improve understanding of which and how many species will become extinct. (Costello et al. 2013, p. 413)

Taxonomy does not solely exist to service applied fields. To assume so yields the myth that somehow the sole *purpose* of taxonomy is to serve these various end users.

Myths like these arise because taxonomy has become less professionalised. That is, fewer people experienced in taxonomy undertake taxonomic practice. The lack of full-time professional taxonomists with years of experience is varied (see taxonomic impediment below). Instead, there are a growing number of biologists and palaeontologists who describe or revise taxa once, without any necessary experience or guidance – or, indeed, any training other than the ability to access handbooks or checklists. Thus Costello et al. (2013) were led to state[34]:

> Taxonomists are not in danger of extinction. They are increasing in numbers. (Costello et al. 2013, p. 416)

Indeed, *untrained* taxonomists are on the rise – but mostly as ecologists and conservationists who lack experience in describing and revising taxa and in turn actually de-professionalising the subject (de Carvalho et al. 2014). Describing a species from scratch, without understanding the organism's anatomy or ontogeny, without comparing it to all other known species of that genus, often results in poor descriptions and, eventually, invalid names (e.g., Crowther et al. 2014)[34].

The various protocols and procedures within taxonomy and nomenclature are necessary to ensure a stable classification, a rallying cry for many biologists

---

[34] Costello et al. (2013) yielded numerous commentaries. A selection is: Sluys (2103); Tancoigne & Dubois (2013); de Carvalho et al. (2014); Sangster & Luksenburg (2015).

frustrated with poor descriptions and taxonomies. The reason many biologists ignore taxonomic procedures and protocols may be due to the so-called *democratisation of science*, in which biologists are encouraged to practise all aspects of the field without any actual training. For instance, what would stop any palaeontologist from downloading vast numbers of sequences from GenBank, blindly aligning them and then processing them using some off-the-shelf phylogenetic software with an arbitrary number of settings (see Grant et al. 2003[35])? The resulting manuscript from such a study could be published anywhere, adding to the pile of molecular systematic literature that seemingly, for the moment, has no end. While such a publication would rightly (probably?) be ignored by the majority of molecular systematists, a newly described species using equally sloppy measures would not. But there is a catch – the name remains a valid entity until *someone* spots the poor diagnosis and either describes it anew, if indeed it is a new species, or synonymises it, if it is not. But who is that someone?

De-professionalisation is a problem throughout science and in taxonomy: that anyone can do taxonomy to a professional level and provide stable classifications is a myth. This will most likely increase poor classifications and destabilise nomenclatural names further. Moreover, non-taxonomists of this kind apply names to new species or genera with few ever doing taxonomic revisions, once again adding, rather than solving, the taxonomic impediment. It is a myth to assume that democratisation of taxonomy will somehow resolve all impediments for the so-called end user.

The End User Myth

Taxonomy is more than simply assigning names to specimens. The taxonomic process can be divided into its consistent parts:

1. Identifying taxa;
2. Discovering new characters;
3. Discovering new taxa; and
4. Revising existing taxa.

---

[35] 'To the pessimist, the fact that such a book could be published, distributed, sold, and recommended illustrates the extent to which point-and-click systematics has achieved predominance and suggests that the rising preeminence of phylogenetic systematics is now, more than ever, on the verge of self defeat. To the optimist, it underscores the broader scientific community's desire to incorporate phylogenetic systematics into their research programs and reveals a prime opportunity to educate colleagues and newcomers on the perils of point-and-click systematics' (Grant et al. 2003, p. 284).

The taxonomic process may be summarised as a system of classification that is self-checking and independent of any other science. Discovery of new characters leads to taxon discovery, which leads to taxon identification. These processes rely on observation alone. A self-checking process is taxon revision, which is made using cladistics as we understand it. The taxonomic process seems never-ending as new characters are used to describe and discover new taxa that in turn are revised in light of new evidence. The taxonomic process is iterative, cyclical and unfinished, as is any process of discovery.

The taxonomic process is also completely independent of any other method other than observation and cladistics. There is no need for theory, such as developmental (evo-devo)[36] or evolutionary theory, or for inferences, such as ancestor–descendant relationships or ancestors. As an independent scientific field, taxonomy as a whole has a single end user: the taxonomist. Compare this with other scientific fields such as astronomy, palaeontology, ecology or population genetics and they too lack end users other than the practitioners themselves. Where, then, does the idea originate that taxonomy has to be *for* ecologists, *for* conservationists, *for* other applied scientific disciplines?

> Taxonomy often pays insufficient attention to its 'end users', the ecologists, conservationists, pest managers and amateur naturalists who need or want to identify animals and plants. (Godfray 2002, p. 19)

The Merriam-Webster Dictionary defines an *end user* as 'the ultimate consumer of a finished product' (www.merriam-webster.com/dictionary/end%20user). In this case, 'end users' of taxonomy are those seeking the 'products' of taxonomy, namely, valid taxonomic names and classifications (*sensu* 'commodities') as well as taxonomists who can provide accurate identifications (*sensu* 'services'). The problem here is that classifications are for the most part hypotheses, many are still undergoing or need revision. In other words, these 'products' are classifications under construction, meaning that it is quite possible that there are often two or more valid classifications (i.e., hypotheses), which are either in conflict or in doubt. The taxonomic 'commodity' and 'service' is rarely up to the standard that end users would like, leading some to suggest:

> ... the governance of the taxonomy of complex organisms be brought under the purview of the International Union of Biological Sciences (IUBS). (Garnett & Christidis 2017, p. 25[37])

---

[36] And here, as well, philosophers enter, as well as the philosophically minded biologists (e.g., Brigandt 2007; MacLeod 2012; Baum 2013).

[37] There were numerous responses (e.g., Raposo et al. 2017).

Let us draw an analogy from astronomy. If the space and aeronautics industries developed the technology to send people to a distant star, they would no doubt consider themselves end users of astronomy. What if astronomers are unable to give precise details about habitable planets and their locations? Has astronomy failed to do its job? Obviously not, as astronomy simply has not made the necessary discoveries or revisions to those discoveries to divulge the information the end user requires. While astronomers may have discovered that the star Proxima Centauri has a planet, Proxima Centauri b, in its orbit, the discovery has been long coming – the star Proxima Centauri was first observed in 1915 while the discovery of its orbiting planet was only announced in 2016 (Anglada-Escudé et al. 2016). It may take some time to work out whether it is habitable or not. That discovery is part of an ongoing process in the same way taxa and classifications are being discovered. Discovery, at least in the historical sciences, is not an open and shut case, but a process of checks and balances. While this might irritate end users it should not alter the taxonomic process. Rather, it should be the end users who adapt to the changing fortunes of taxonomy.

The end user myth is the idea that taxonomy is a *service industry* that provides commodities and services to non-taxonomic end users. The resulting attitude is that

1. Taxonomy is ancillary to some larger more significant field (e.g., ecology, or evolutionary biology);
2. Taxonomic discoveries are not open to revision;
3. Taxonomic revisions are a sign that some classifications are unstable;
4. Taxonomy may be improved by automating the taxonomic process.

The first assumption completely undermines taxonomy as an independent scientific field. If end users are so dependent on taxonomies, then their fields are hardly independent. The second assumption treats taxonomy as an experimental science rather than one that is undergoing constant revision. The third ignores historical science altogether, and the fourth seeks to replace the taxonomic product with one that can be generated without the need for taxonomists[38].

---

[38] Janzen clearly articulates the issue of automatic identification. We do not question his motives just his interpretation of the science: 'Imagine a world where every child's backpack, every farmer's pocket, every doctor's office and every biologist's belt has a gadget the size of a cellphone. A free gadget. Pop off a leg, pluck a tuft of hair, pinch a piece of leaf, swat a mosquito, and stick it in on a tuft of toilet tissue. One minute later the screen says *Periplaneta americana*, *Canis familiaris*, *Quercus virginiana*, or West Nile virus in *Culex pipiens*' (Janzen 2004, p. 731). MacLeod wrote in his introduction to a book on the subject of automatic identification: 'Automated taxon identification is, in my opinion, a good servant, but should not be allowed to

The Taxonomic Toolbox Myth

If taxonomy is viewed as a process of discovery that involves our own senses, memory and experience, then we need tools in order to make this happen. For example, many trilobites are small, requiring a visual aid in the form of a stereo-microscope. Others are tucked away in clastic rocks, such as calcareous mud-stones, and need a good drill to break away the matrix. These tools are just that: tools. They simply aid in the examination of our specimens. We can even extend this to micrographs or photographs, which enable the taxonomist to see parts of the organism. What tools do not do is partake in the taxonomic process.

For example, we may use keys in order to help us locate the characteristics of taxa in order to identify them. The key does not do this for us; rather we use it, either correctly or incorrectly, to make a positive or false diagnosis. The taxonomist or person using the key makes the identification. A non-human tool cannot identify a new taxon. DNA barcoding was introduced in the early 2000s to do just this, identify new species. Taxonomists find new species by looking at specimens that are given a diagnosis, a lengthy comparison to other taxa along with a name. DNA barcoding identifies new species through the similarity of their DNA. Doing so makes DNA barcoding a replacement for, rather than a tool of, taxonomy (Ebach and Holdrege 2005). Yet, barcoders insist that barcoding is part of the taxonomist's toolbox and equivalent to a microscope, camera or drill. The claim made by barcoders is that barcoding identifies cryptic species that are indistinguishable using morphological characteristics. If this is the case, the identification was made by barcoding alone, meaning that the taxonomist had no part to play in the taxonomic process, thereby replacing him or her. For instance, Hebert et al. identified 10 new species from what was previously understood to be a single species of butterfly, *Astraptes fulgerator*, using barcoding techniques (Hebert et al. 2004) – but given that there are no diagnostic morphological characteristics, there is little for a taxonomist to further compare[39] (see Brower 2006 for a re-analysis of these data, where he found that '[a]t least three, but not more than seven mtDNA clades that may correspond to cryptic species are supported by the evidence'; Brower, 2010 later provided names for the 10 'species', and Brower and Warren,

become a bad master. Statistical procedures, mainly multivariate analyses, are a feature of some of the techniques. This is as it should be, but it is necessary that these techniques are properly understood and not misused. Computer programs can do no more than they are told to do and this is where problems can, and do, arise and where manual assessments cannot be done away with. Let us briefly look at some of these where unskilled practitioners can be at fault' (MacLeod 2008, p. x).

[39] 'Brower (2010) has provided names and diagnoses based on DNA barcodes for those ten species, but it is not clear which of them (if any) corresponds to the concept of *Astraptes fulgerator*' (Brower & Warren 2010).

2010, commented further on this 'flagship' example). Barcoding identifies 'new' taxa on the basis of degrees of similarity, something that, in the court of last resort, is purely arbitrary. The only justification for using such an indiscriminate number (often a percentage) is that DNA represents the building blocks of life – so are atoms, yet no one would be pedantic enough to make *that* argument.

DNA barcoding is a deliberate attempt to alter the taxonomic process by removing the taxonomist and making species identification somewhat arbitrary. If the assumption is that some percentage similarity is a valid taxonomic procedure, then comparative biology will ultimately fail as a rigorous and valid science – as Brower so eloquently put it:

> DNA barcoding is a tool, not a research programme. Stated plainly, if resources are cannibalized from systematics to support molecular parataxonomy, systematic training and research programmes will languish, the loss of systematic expertise will be accelerated, and the framework of natural history to which DNA barcodes are intended to link will be impoverished. (Brower 2006, p. 131)

Or lost forever.

The Counting Species Myth (de Carvalho et al. 2007)

Mr. Jones and me tell each other fairy tales.

(Counting Crows, Mr. Jones, 1993)

The biodiversity crisis, understanding how many species there are and how many are going extinct has, to all intents and purposes, fuelled the counting species myth. Like the end user myth, taxonomy is again regulated to the role of a service provider in producing a precise figure, in this case the number of living species. Unlike the end user myth, which was largely promoted by non-taxonomists, the counting species myth was devised by taxonomists, namely to justify taxonomy in an age of biodiversity loss:

> Our generation is the first to fully comprehend the threat of the biodiversity crisis and the last with the opportunity to explore and document the species diversity of our planet. (Wheeler et al. 2004, p. 285)

The problem here is that the taxonomic process is undermined as something that provides precise figures in short periods of time. Regardless of technology, taxon discoveries take time and are part of a process of checks and balances, no matter how fast they are catalogued, photographed and published. Taxonomy is there to discover nature, and not for reporting numbers. Yet, it is numbers that non-taxonomists are desperate for, for all sorts of biodiversity measurements and white papers to emphasise the need for more conservation. But is this really helping taxonomy?

Demand for taxonomic data and calls for it to be sped up, either through online databases or web-based services, shows a general lack of appreciation of what taxonomy actually does. Yes, taxonomists describe species, they also describe, genera and families and so on. But describing a new taxon is not the zenith of taxonomy. In fact it is the easiest part of it as you only need to discover something that has not been described before. Taxonomy gets harder when you need to revise a taxon that is known to be problematic, that is, artificial. With all the digital assistance in world, from online virtual libraries and electronic monographs, you need a trained taxonomist to do this over a lengthy period of time. The digital world can offer so much but it does not replace the experience of examining thousands of specimens. End users misunderstand that it is not the job of the taxonomist to supply a rolling update on species numbers. That so-called impediment is easily solved through various digital means. Training a taxonomist and keeping them employed is a lot harder – and that impediment is solved through investment by governments, universities, herbaria and museums. Yet, the money fails to arrive even with all the calls of a biodiversity crisis. We have wondered whether these misunderstandings that plague the history and philosophy of biology, evolutionary biology and even taxonomy itself, is to blame for the current situation? What institution would wish to hire a practitioner from a field that is hopelessly misunderstood? Even with an extinction approaching, many feel that investing in conservation and evolutionary biology will somehow provide answers and remedies to save taxa that may, just may, actually be real.

It might be considered an act of madness to attempt to write a book on the subject of cladistics, or at least a book explaining cladistics as its main aim, or even authoring a book with the word 'cladistics' in its title. We hope the preceding chapters have gone some way to explaining our approach – even so, some may continue to feel it is an exercise in madness. So be it. Perhaps, if we treat taxonomy and systematics as the *science of classification* rather than simply another aspect of some applied field of evolutionary biology, we will cement in the public's mind that taxonomy and systematics has a specific job to do: to classify.

References

Anglada-Escudé, G., Amado, PJ., Barnes, J., Berdiñas, ZM., Butler, RP., Coleman, GAL., de la Cueva, I., Dreizler, S., Endl, M., Giesers, B., Jeffers, SV., Jenkins, JS., Jones, HRA., Kiraga, M., Kürster, M., López-González, MJ., Marvin, CJ., Morales, N., Morin, J., Nelson, RP., Ortiz, JL., Ofir, A., Paardekooper, S-J., Reiners, A., Rodríguez, E., Rodríguez-López, C., Sarmiento, LF., Strachan, JP., Tsapras, Y., Tuomi, M. & Zechmeister, M. 2016. A terrestrial planet candidate in a temperature orbit around Proxima Centauri. *Nature* 536(7617): 437–440.

Baum, DA. 2013. Developmental causation and the problem of homology. *Philosophy, Theory, and Practice in Biology* 5: e403

Baum, DA. 2017. Does the future of systematics really rest on the legacy of one mid-20th-century German entomologist? *The Quarterly Review of Biology* 92: 450–453.

Beatty, J. 1982. Classes and cladists. *Systematic Zoology* 31: 25–34.

Brady, RH. 1981. Mind, models and Cartesian observers: a note on conceptual problems. *Journal of Social and Biological Systems* 4: 277–286.

Brigandt, I. 2002. Homology and the origin of correspondence. *Biology and Philosophy* 17: 389–407.

Brigandt, I. 2007. Typology now: homology and developmental constraints explain evolvability. *Biology and Philosophy* 22: 709–725.

Brigandt, I. & Griffiths, PE. 2007. The importance of homology for biology and philosophy. *Biology and Philosophy* 22: 633–641.

Brower, AVZ. 2006. Problems with DNA barcodes for species delimitation: 'ten species' of *Astraptes fulgerator* reassessed (Lepidoptera: Hesperiidae). *Systematics and Biodiversity* 4: 127–132.

Brower, AVZ. 2010. Alleviating the taxonomic impediment of DNA barcoding and setting a bad precedent: names for ten species of '*Astraptes fulgerator*' (Lepidoptera: Hesperiidae: Eudaminae) with DNA-based diagnoses. *Systematics and Biodiversity* 8: 485–491.

Brower, AV. 2018. Statistical consistency and phylogenetic inference: a brief review. *Cladistics* 34: 562–567.

Brower, AVZ. & Warren, A. 2010. *Astraptes fulgerator* (Walch 1775). Version 30, December 2010. http://tolweb.org/Astraptes_fulgerator/96653/2010.12.30 *in* The Tree of Life Web Project, http://tolweb.org/

Brummitt, RK. 2002. How to chop up a tree. *Taxon* 51: 31–41.

Bryant, HM. 1995. The threefold parallelism of Agassiz and Haeckel, and polarity determination in phylogenetic systematics. *Biology and Philosophy* 10: 197–217.

Burkhardt, RW. 1994. Ernst Mayr: Biologist-Historian. *Biology and Philosophy* 9: 359–371.

Cartmill, M. 2018. A sort of revolution: systematics and physical anthropology in the 20th century. *American Journal of Physical Anthropology* 165: 677–687.

Conway-Morris, S. 1994. Wonderfully, gloriously wrong. *Trends in Evolution and Ecology* 9(10): 407–408.

Costello, MJ., Wilson, S. & Houlding, B. 2013. More taxonomists describing significantly fewer species per unit effort may indicate that most species have been discovered. *Systematic Biology* 62: 616–624.

Crisp, MD., Trewick, SA. & Cook, LG. 2011. Hypothesis testing in biogeography. *Trends in Ecology and Evolution* 26: 66–72.

Crowther, MS., Fillios, M., Colman, N. & Letnic, M. 2014. An updated description of the Australian dingo (*Canis dingo* Meyer, 1793). *Journal of Zoology* 293: 192–203.

Currie, A. 2012. Convergence as evidence. *The British Journal for the Philosophy of Science* 64: 763–786.

Dawkins, R. 1986. *The Blind Watchmaker.* W. W. Norton & Company, Inc., New York [2nd edition 1996; numerous other editions exist, including a 25th anniversary edition and it has its own Wikipedia page, https://en.wikipedia .org/wiki/The_Blind_ Watchmaker#Reception].

de Carvalho, MR., Bockmann, FA., Amorim, DS., Brandao, CRF., de Vivo, M., de Figueiredo, JL., Britski, HA., de Pinna, MCC., Menezes, NA., Marques, FPL., Papavero, N., Cancello, EM., Crisci, JV., McEachran, JD., Schelly, RC., Lundberg, JG., Gill, AC., Britz, R., Wheeler, QD., Stiassny, MLJ., Parenti, LR., Page, LM., Wheeler, WC., Faivovich, J., Vari, RP., Grande, L., Humphries, CJ., DeSalle, R., Ebach, MC. & Nelson, GJ. 2007. Taxonomic impediment or impediment to taxonomy? A commentary on systematics and the cybertaxonomic-automation paradigm. *Evolutionary Biology* 34: 140–143.

de Carvalho, MR., Bockmann, FA., Amorim, DS. & Branda, CRF. 2008. Systematics must embrace comparative biology and evolution, not speed and automation. *Evolutionary Biology* 35: 150–157.

de Carvalho, MR., Ebach, MC., Williams, DM., Nihei, SS., Trefaut Rodrigues, M., Grant, T., Silveira, LF., Zaher, H., Gill, AC., Schelly, RC., Sparks, JS., Bockmann, FA., Séret, B., Ho, H-C., Grande, L., Rieppel, O., Dubois, A., Ohler, A., Faivovich, J., Assis, LCS., Wheeler, QD., Goldstein, PZ., de Almeida, EAB., Valdecasas, AG. & Nelson, G. 2014. Does counting species count as taxonomy? On misrepresenting systematics, yet again. *Cladistics* 30: 322–329.

Dubois, A. 2010. Zoological nomenclature in the century of extinctions: priority vs. 'usage'. *Organisms, Diversity & Evolution* 10: 259–274.

Ebach, M, 2005. Anschauung and the Archetype: the role of Goethe's Delicate Empiricism in Comparative Biology. *Janus Head* 8: 254–270.

Ebach, MC. & Holdrege, C. 2005. More taxonomy, not DNA barcoding. *Bioscience* 55: 822–823.

Ebach, MC., Valdecasas, AG. & Wheeler, QD. 2011. Impediments to taxonomy and users of taxonomy: accessibility and impact evaluation. *Cladistics* 27: 550-557.

Eisen, JA. 1998. Phylogenomics: improving functional predictions for uncharacterized genes by evolutionary analysis. *Genome Research* 8: 163–167.

Eisen, JA. & Fraser, CM. 2003. Phylogenomics: intersection of evolution and genomics. *Science* 300 (5626), 1706–1707.

Eldredge, N. & Novacek, MJ. 1985. Systematics and paleobiology. *Paleobiology* 11: 65–74.

Farris, JS. 2014. "Pattern cladistics" really means paraphyly. *Cladistics* 30: 236–239.

Felsenstein, J. 2001. The troubled growth of statistical phylogenetics. *Systematic Biology* 50: 465–467.

Fernholm, B., Bremer, K. & Jornvall, H. (eds) 1989. *The Hierarchy of Life.* Excerpta Medica, Amsterdam.

Fitzhugh, K. 2008. Abductive inference: implications for "Linnean" and "Phylogenetic" approaches for representing biological systematization. *Evolutionary Biology* 35: 52–82.

Flowers, RW. 2011. Review of *Naming Nature: The Clash between Instinct and Science. American Entomologist* 57: 115–116.

Gardiner, B., Janvier, P., Patterson, C., Forey, PL., Greenwood, PH., Miles, RS. & Jefferies, RPS. 1979. The salmon, the lungfish, the cow: a reply. *Nature* 277: 175–176.

Garnett, ST. & Christidis, L. 2017. Taxonomy anarchy hampers conservation. *Nature* 546: 25–27.

Gee, HE. 2001. *Deep Time: Cladistics, the Revolution in Evolution*. Fourth Estate, London. [There are various versions of this book, for example: Gee, H. E. (1999). *In Search of Deep Time: Beyond the Fossil Record to a New History of Life*. Comstock Publishing, Sacramento.]

Godfray, HCJ. 2002. Challenges for taxonomy. *Nature* 417: 17–19.

Godfray, HCJ., Clark, BR., Kitching, IJ., Mayo, SJ. & Scoble, MJ. 2007. The web and the structure of taxonomy. *Systematic Biology* 56: 943–955.

Goloboff, PA., Torres, A. & Arias, JS. 2018. Weighted parsimony outperforms other methods of phylogenetic inference under models appropriate for morphology. *Cladistics* 34: 407–437.

Grant, T., Faivovich, J. & Pol, D. 2003. The perils of 'point-and-click' systematics. *Cladistics* 19: 276–285.

Haber, MN. 2016. The individuality thesis (3 ways). *Biology and Philosophy* 31: 913–930.

Halstead, LB. 1978. Cladistic revolution – can it make the grade? *Nature* 276: 759–760.

Halstead, LB., White, EI. & MacIntire, GT. 1979. L. B. Halstead and colleagues reply. *Nature* 277: 176.

Hebert, PDN., Penton, EH., Burns, JM., Janzen, DH. & Hallwachs, W. 2004. Ten species in one: DNA barcoding reveals cryptic species in the neotropical skipper butterfly *Astraptes fulgerator*. *Proceedings of the National Academy of Sciences USA* 101: 14812–14817.

Hull, DL. 1964. Consistency and monophyly. *Systematic Zoology* 13: 1–11.

Hull, DL. 1965. The effects of essentialism on taxonomy: two thousand years of stasis. *The British Journal for the Philosophy of Science* 15: 314–326 & 16: 1–18 [reprinted in *Concepts of Species*, C. N. Slobodchikoff (ed.), University of California Press, Berkeley, 1976; *The Units of Selection*, Marc Ereshefsky (ed.), MIT Press, Cambridge, 1992].

Hull, DL. 1969. What philosophy of biology is not. *Journal of the History of Biology* 2: 241–268 [and *Synthese* 20: 157–184].

Hull, DL. 1976. Are species really individuals? *Systematic Zoology* 25: 174–191.

Hull, DL. 1988. *Science As a Process: An Evolutionary Account of the Social and Conceptual Development of Science*. University of Chicago Press, Chicago.

Hull, DL. 1994. Ernst Mayr's influence on the history and philosophy of biology: a personal memoir. *Biology and Philosophy* 9: 375–386.

Hull, DL. 1998. Taxonomy. In: *The Routledge Encyclopedia of Philosophy*. Taylor and Francis. doi:10.4324/9780415249126-Q102-1 (accessed 24 May 2018).

Hull, DL. 2006 [2007]. Essentialism in taxonomy: four decades later. *Annals of the History and Philosophy of Biology* 11: 47–58.

Huneman, P. & Walsh, DM. 2017. *Challenging the Modern Synthesis: Adaptation, Development and Inheritance*. Oxford University Press, Oxford.

Janzen, DH. 2004. Now is the time. *Philosophical Transactions of the Royal Society of London B* 359: 731–732.

Kemp, RM. 1984. Evolution and cladism. *New Scientist*, 5 January 1984, p. 42.

Kohn, D. 1985a. (ed.) *The Darwinian Heritage*. Princeton University Press, Princeton, NJ.

Kohn, D. 1985b. Introduction: a high regard for Darwin. In: Kohn, D. (ed.), *The Darwinian Heritage*. Princeton University Press, Princeton, NJ, pp. 1–5.

Koonin, EV. 2009. The Origin at 150: is a new evolutionary synthesis in sight? *Trends in Genetics* 25: 473–475.

LaPorte, J. 2005. Is there a single objective, evolutionary tree of life? *The Journal of Philosophy* 102(7): 357–374.

LaPorte, J. 2009. On systematists' single objective tree of ancestors and descendants. *Biological Theory: Integrating Development, Evolution, and Cognition* 4: 260–266.

Loewenberg, BJ. 1965. Darwin and Darwin studies 1959–1965. *History of Science* 4: 15–54.

MacLeod, M. 2011. How to compare homology concepts: Class reasoning about evolution and morphology in phylogenetics and developmental biology. *Biological Theory* 6: 141—153.

MacLeod, N. 2008. *Automated Taxon Identification in Systematics: Theory, Approaches and Applications*. CRC Press, Boca Raton, FL.

Mayr, E. 1982. *The Growth of Biological Thought: Diversity, Evolution, and Inheritance*. Harvard University Press, Cambridge, MA.

Medawar, PB. 1967. *The Art of the Soluble*. Metheun & Co. Ltd., London.

Nelson, GJ. 1979. Cladistic analysis and synthesis: principles and definitions, with a historical note on Adanson's *Familles des Plantes. Systematic Zoology* 28: 1–21.

Nelson, GJ. 2009. Cladistics. In: Milner, R. (ed.), *Darwin's Universe: Evolution from A to Z*. University of California Press, Berkeley, pp. 80–82.

OED [Oxford English Dictionary] Online. *systematics*, n. March 2018. Oxford University Press. www.oed.com/view/Entry/378323?rskey=BBS1hY&result=2&isAdvanced=false (accessed 27 May 2018).

OED [Oxford English Dictionary] Online. *phylogenetics*, n. March 2018. Oxford University Press. www.oed.com/view/Entry/143086?redirectedFrom=phylogenetics (accessed 27 May 2018).

O'Neil, C. 2016. *Weapons of Math Destruction*. Crown, New York.

Ørvig, T. (ed.) 1968. *Current Problems of Lower Vertebrate Phylogeny*. Fourth Nobel Symposium, June 1967, Stockholm. Interscience Publishers, John Wiley and Sons, Inc., New York, London, Sydney; Almqvist and Wiksell, Stockholm.

Patterson, C. 1982. Morphological characters and homology. In: Joysey, KA. & Friday, AE. (eds.), *Problems of Phylogenetic Reconstruction*. Academic Press, London, pp. 21–74.

Patterson, C. 1985. Cladistics. In: Campbell, B. & Lack, E. (eds), *A Dictionary of Birds*. A. & D. Poyser, London, p. 88.

Patterson, C. 1987. Introduction. In: Patterson, C. (ed.), *Molecules and Morphology in Evolution: Conflict or Compromise?* Cambridge University Press, Cambridge, UK, pp. 1–22.

Patterson, C. 1989. Phylogenetic relations of major groups: conclusions and prospects. In: Fernholm, B., Bremer, K. & Jôrnvaîl, H. (eds.), *The Hierarchy of Life*. Nobel Symp. 70. Excerpta Medica, Amsterdam, pp. 471–488.

Patterson, C. 1990. Erik Helge Osvald Stensiö. *Biographical Memoirs of Fellows of the Royal Society* 35: 363–380.

Pigliucci, M. & Finkleman, L. 2014. The Extended (Evolutionary) Synthesis debate: Where science meets philosophy. *Bioscience* 64: 511–516.

Platnick, NI. 1979. Philosophy and the transformation of cladistics. *Systematic Zoology* 28: 537–546.

Popper, K. 1963. *Conjectures and Refutations: The Growth of Scientific Knowledge.* Routledge & K. Paul, Oxford.

Quammen, D. 2018. *The Tangled Tree.* William Collins, London.

Rapini, A. 2004. Classes or individuals? The paradox of systematics revisited. *Studies in History and Philosophy of Science Part C: Studies in History and Philosophy of Biological and Biomedical Sciences* 35: 675–695.

Raposo, M., Stopiglia, R., Brito, GRR, Bockmann, FA., Kirwani, GM., Gayon, J. & Dubois, A. 2017. What really hampers taxonomy and conservation? A riposte to Garnett and Christidis (2017). *Zootaxa* 4317(1): 179–184.

Ridley, M. 1983. Can classification do without evolution? *New Scientist*, 1 December, p. 647.

Ridley, M. 1986. *Evolution and Classification: The Reformation of Cladism.* Longman, London.

Rieppel, O. 2013. Biological individuals and natural kinds. *Biological Theory* 7: 162–169.

Rudwick, MJS. 1988. *The Great Devonian Controversy: The Shaping of Scientific Knowledge among Gentlemanly Specialists.* University of Chicago Press, Chicago.

Sangster, G. & Luksenburg, JA. 2015. Declining rates of species described per taxonomist: slowdown of progress or a side-effect of improved quality in taxonomy? *Systematic Biology* 64: 144–151.

Schoch, RM. 1984. Cladism defended. *New Scientist* (12 January), p. 47.

Schultze, H-P. 2005. The first ten symposia on early/lower vertebrates. *Revista Brasileira de Paleontologia* 8: v–xviii.

Simpson, GG. 1982. Autobiology. *Quarterly Review of Biology* 57: 437–444.

Sloan, PR. 1985. Essay Review: Ernst Mayr on the History of Biology. *Journal of the History of Biology* 18: 145–153.

Sluys, R. 2013. The unappreciated, fundamentally analytical nature of taxonomy and the implications for the inventory of biodiversity. *Biodiversity & Conservation* 22: 1095–1105.

Stoltzfus, A. 2017. Why we don't want another "Synthesis". *Biology Direct* 12: 23. https://doi.org/10.1186/s13062–017–0194–1

Sturdy, S. 2011. The meanings of 'life': Biology and biography in the work of J. S. Haldane (1860-1936). *Transactions of the Royal Historical Society* 21: 171–191.

Tancoigne, E. & Dubois, A. 2013. Taxonomy: no decline, but inertia. *Cladistics* 29: 567–570.

Vanderlaan, TA., Ebach, MC. & Williams, DM. 2013. Defining and redefining monophyly: Haeckel, Hennig, Ashlock, Nelson and the proliferation of definitions. *Australian Systematic Botany* 26: 347–355.

Velasco, J. 2008. Philosophy and the Tree of Life: the metaphysics and epistemology of phylogenetic systematics. Ph.D. dissertation, University of Wisconsin.

Velasco, J. 2013. Phylogeny as population history. *Philosophy and Theory in Biology* 5:e402

Velasco, J. 2018. Universal common ancestry, LUCA, and the Tree of Life: three distinct hypotheses about the evolution of life. *Biology and Philosophy* 33: 31.

Wheeler, QD. 1995. The 'old systematics': classification and phylogeny. In: Pakaluk, J. & Slipinski, S. A. (eds), *Biology, Phylogeny, and Classification of Coleoptera: Papers Celebrating the 80th Birthday of Roy A. Crowson*. Muzeum I Instytut Zoologii PAN, Warszawa, pp. 31–62.

Wheeler, QD. 2004. Taxonomic triage and the poverty of phylogeny. *Philosophical Transactions of the Royal Society of London B Biological Sciences* 359: 571–583.

Wheeler, QD. 2008a. Taxonomic shock and awe. In: Wheeler, QD. (ed.), *The New Taxonomy*. CRC Press, Boca Raton, FL, pp. 211–226.

Wheeler, QD. 2008b. Undisciplined thinking: morphology and Hennig's unfinished revolution. *Systematic Entomology* 33: 2–7.

Wheeler QD., Raven, PH. & Wilson, EO. 2004. Taxonomy: impediment or expedient? *Science* 303: 285.

Wilkins, JS. & Ebach, MC. 2014. *The Nature of Classification: Relationships and Kinds in the Natural Sciences.* Palgrave Macmillan, New York.

Williams, DM. & Ebach, MC. 2008. *The Foundations of Systematics & Biogeography.* Springer-Verlag New York Inc., New York.

Williams, DM. & Ebach, MC. 2017. What is intuitive taxonomic practice? *Systematic Biology* 66: 637–643.

Winsor, MP. 2001. The practitioner of science: everyone her own historian. *Journal of the History of Biology* 34: 229–245.

Winsor, MP. 2003. Non-essentialist methods in pre-Darwinian taxonomy. *Biology and Philosophy* 18: 387–400.

Winther, R. 2018 (ed.), *Phylogenetic Inference, Selection Theory, and History of Science: Selected Papers of A. W. F. Edwards with Commentaries.* Cambridge University Press, Cambridge, UK.

Yoon, CK. 2009. *Naming Nature: The Clash Between Instinct and Science.* W.W. Norton & Company, New York, London.

Further Reading

Brower, AV. 2019. Background knowledge: the assumptions of pattern cladistics. *Cladistics* 35: 717–731 https://doi.org/10.1111/cla.12379

de Carvalho, MR., Bockmann, FA., Amorim, DS., Brandao, CRF., de Vivo, M., de Figueiredo, JL., Britski, HA., de Pinna, MCC., Menezes, NA., Marques, FPL., Papavero, N., Cancello, EM., Crisci, JV., McEachran, JD., Schelly, RC., Lundberg, JG., Gill, AC., Britz, R., Wheeler, QD., Stiassny, MLJ., Parenti, LR., Page, LM., Wheeler, WC., Faivovich, J., Vari, RP., Grande, L.,

Humphries, CJ., DeSalle, R., Ebach, MC. & Nelson, GJ. 2007. Taxonomic impediment or impediment to taxonomy? A commentary on systematics and the cybertaxonomic-automation paradigm. *Evolutionary Biology* 34: 140–143.

Inspiration for this paper was largely as a response to the contributions of Charles Godfray, and while they are not without some merit, his position reflects that of a 'user' rather than practitioner ('Godfray is a user of taxonomic end-products who has frequently been critical of the slowness with which modern taxonomy is furnishing these —especially species names—to ecologists, conservationists, 'biodiversity scientists', etc. . . . Godfray's criticism, echoed in other circles . . . is cast in what he has termed the 'second bioinformatics crisis', viz. that the alleged lethargy of modern taxonomy is mostly due to the lack of an adequate cyberstructure to disseminate its much needed products', p. 141). Having noted that, it is worth dipping into some of Godfray's papers to understand some of the technical possibilities for *data storage* (e.g., Godfray, HCJ. 2002. Challenges for taxonomy. *Nature* 417: 17–19; Godfray, HCJ. 2007. Linnaeus in the information age. *Nature* 446: 259–260; Godfray, HCJ., Mayo, SJ. & Scoble, MJ. Pragmatism and rigour can coexist in taxonomy. *Evolutionary Biology* 34: 309–311).

Evenhuis, NL. 2007. Helping solve the "other" taxonomic impediment: Completing the Eight Steps to Total Enlightenment and Taxonomic Nirvana. *Zootaxa* 1407: 3–12.

In relation to this paper, see: Flowers, RW. 2007. Taxonomy's unexamined impediment. *The Systematist* 28: 3–7; Flowers, RW. 2007. Comments on 'Helping Solve the "Other" Taxonomic Impediment: '*Completing the Eight Steps to Total Enlightenment and Taxonomic Nirvana*' by Evenhuis (2007). *Zootaxa* 1494: 67–68 ('Many taxonomists in my age cohort are now "molecular systematists", and I know that some of them became so because they saw that funding would be impossible otherwise. If funding became linked to describing new species, you can bet that many would switch back, some reluctantly but others gladly', p. 68).

Richards, RA. 2016. *Biological Classification. A Philosophical Introduction.* Cambridge University Press, Cambridge, UK.

A philosopher's viewpoint. Perhaps if less time was spent agonising over species, progress would be had.

Stevens, PF. 2006. An end to all things? – plants and their names. *Australian Systematic Botany* 19: 115–133.

'Thinking of naming systems as conventions may help clarify what we should be doing, if we are not to squander both the time and the reputation of systematics. Time is in short supply and our reputation not what it might be; solving the less cosmic issues may involve a self-discipline that also seems in short supply in the systematic community'.

Wheeler QD., Raven, PH. & Wilson, EO. 2004. Taxonomy: impediment or expedient? *Science* 303: 285.

This editorial was followed by a series of comments, all worth reading in the context of the future of taxonomy (*Science* 305: 1104–1107, includes five contributions).

Afterword

Taxonomy, the ability to classify the natural world and communicate what we observe, is an intuitive built-in process. Everyone does it to some degree, and with training and practice, may specialise in one or more taxon. The fact that taxonomy is an intuitive built-in process leads to a large number of non-taxonomists practising it at a specialised level leading to further taxonomic problems. Yet, simply naming a taxon does not make a taxonomist and studying taxonomy of any particular group does not necessarily mean natural taxa are the result. Yet, many taxonomists, with years of training and experience, insist that their taxa are natural when often character analysis has continually shown that their favourite genus or family is aphyletic (Chapter 6). At the same time, non-taxonomists with little or no training still describe and revise taxa. Why, then, does this happen – why in taxonomy rather than, say, astronomy or biochemistry? The answer is that classification is an intuitive built-in process and the more qualified someone is the more likely they believe they can describe a species or revise a genus. Moreover, since taxonomy is taught as a physical exercise (we observe and compare) little attention is given to what the theoretical foundations are, particularly the process of discovering natural taxa. A highly experienced taxonomist may know little of cladistics in the same way a medieval troubadour knows little about counterpoint. The same goes for making observations about taxonomy: philosophers of biology think taxonomists use species concepts when they describe their organisms, when clearly they do not. But philosophers of biology do not practise taxonomy and do not understand the private nature of describing a taxon, and that much of what we say we do is in fact what we think we do. To our knowledge no philosopher of biology has taken it upon themselves to practise biology and discover that much of what biologists say they do is fiction. With all this in mind, it is difficult to write a guide to taxonomy without picking apart the myths first. Because we are able to classify from birth, we pick up many ideas about classification along the way. Once we get to a level where we are trained, we still pick up 'ideas' from practitioners in other fields, such as evolutionary biology, who over time have nurtured myths about how and why we classify. From day one at taxonomy training school we are filled with more misinformation about biological classification than we are with actual taxonomic facts. We both teach biological classification at a tertiary level in the United Kingdom and Australia. Once students arrive they already assume cladograms are actual evolutionary trees, that nodes on those branching diagrams represent ancestors, and that those ancestors lived in ancestral areas. Unfortunately these ideas get in the way of distinguishing what is discoverable and what is interpreted, meaning that before we can train a taxonomist there is a certain amount of de-cluttering required. We hope that this book will go part way to achieving that. If students had no concept of classification, in the same way they may have no idea about metaphoric petrology, then our book would only be a fraction of its size. More importantly, taxonomy is taught primarily through training, through observation (e.g., dissection, measuring, microscopy, etc.) and not through books. Our book is merely a guide to de-cluttering the mind of myths and misconceptions. Get rid of these then biological classification and the discovery of natural taxa becomes easy – as well as a pleasure to learn.

Index

Systematics Association Special Volumes

---

[a] Published by Clarendon Press for the Systematics Association

[*] Published by Academic Press for the Systematics Association

[‡] Published by Oxford University Press for the Systematics Association

[**] Published by Chapman & Hall for the Systematics Association

[‡‡] Published by CRC Press for the Systematics Association